BIOTECHNOLOGY IN AGRICULTURE, INDUSTRY AND MEDICINE

GENETICALLY ENGINEERED CROPS: BIOTECHNOLOGY, BIOSAFETY AND BENEFITS

BIOTECHNOLOGY IN AGRICULTURE, INDUSTRY AND MEDICINE

Additional books in this series can be found on Nova's website under the Series tab.

Additional E-books in this series can be found on Nova's website under the E-books tab.

AGRICULTURE ISSUES AND POLICIES

Additional books in this series can be found on Nova's website under the Series tab.

Additional E-books in this series can be found on Nova's website under the E-books tab.

BIOTECHNOLOGY IN AGRICULTURE, INDUSTRY AND MEDICINE

GENETICALLY ENGINEERED CROPS: BIOTECHNOLOGY, BIOSAFETY AND BENEFITS

LINDSAY M. GROVER
EDITOR

Nova Science Publishers, Inc.
New York

For permission to use material from this book please contact us:
Telephone 631-231-7269; Fax 631-231-8175
Web Site: http://www.novapublishers.com

LIBRARY OF CONGRESS CATALOGING-IN-PUBLICATION DATA

Genetically engineered crops : biotechnology, biosafety, and benefits /
editor, Lindsay M. Grover.
p. ; cm.
Includes bibliographical references and index.
ISBN 978-1-61122-526-6 (hardcover)
1. Crops--Genetic engineering. I. Grover, Lindsay M.
[DNLM: 1. Crops, Agricultural--genetics. 2. Plants, Genetically
Modified. 3. Genetic Engineering. SB 123.57]
SB123.57.G47845 2010
631.5'233--dc22
2010042633

Published by Nova Science Publishers, Inc. ✝ New York

CONTENTS

PREFACE

Chapter 1 - Genetically engineered crops are those crop plants that are transformed with transgenes of desired traits into their nuclear genome using recombinant DNA technology. The common strategies used for plant transformation are by exploiting natural gene transfer by soil bacterium, *Agrobacterium tumefaciens* and by using gene gun where gold microcarriers coated with desiccated DNA are delivered using pneumatic pressure. Transgenic technology has expanded the scope of conventional breeding by breaking the restrictions inflicted by traditional cross-pollination and selection techniques. Additional advantages like reduction in time involved in developing a variety, avoiding linked undesirable genes and overcoming sexual incompatibility barriers are also attributed to transgenic breeding. Many genetically engineered crops are successfully created worldwide and are either cultivated in the field or waiting for approval. These include herbicide resistant soybean, rapeseed, sugar beet and corn, boll worm resistant cotton, European corn borer resistant corn, fruit borer resistant egg plant, ring spot virus resistant papaya, high sugar content sugarcane, nutrient enriched potato and rice and crop for the production of biopharmaceuticals. Genetically engineered crops offer a multitude of advantages over traditional crop varieties which include yield enhancement, suppression of pathogen, pest and weed infestations, reduction of the use of harmful pesticide and herbicide as well as reduction of the incidence of pesticide resistance in pests. Functional foods with high nutrient contents can inexpensively improve health of people especially in developing and underdeveloped countries. Transgenic crops also have been subjected for considerable arguments against their usage *viz.*, improved seeds and associated technologies make farmers to depend on multinational companies for their seed supply, biopharmaceuticals produced in food crops posses the risk of accidental ingestion, inadvertent introduction of allergens in food, pollen from herbicide resistant crops creating super weeds, reduction of biodiversity and disruption of ecosystem.

Chapter 2 - By enabling the development of new crop varieties with increased nutritional content, disease resistance, and stress tolerance, plant biotechnology plays an important role in addressing the food needs of the world's population. According to ISAAA Brief 39-2008, the area of land used globally for cultivating transgenic crops grew from 114.3 million hectares in 2007 to 125 million hectares in 2008. In 2007, the direct global farm income benefit from biotech crops was $10.1 billion (Brooke, 2009). So far, the majority of transgenic crops have been produced either by *Agrobacterium*-mediated transformation or by biolistic transformation. The gene controlling a trait of interest is usually delivered along with a selectable marker gene (such as an antibiotic-resistance gene) for later selection of the transgenic plants from the bulk of non-transformed plants. After serving this purpose, the selectable marker gene is no longer needed. However, this approach for producing transgenic crops has raised concerns from both the regulatory and public sectors over the presence of selectable marker genes in the food supply or their possible escape into the environment through the outcrossing of transgenic crops to related species. Therefore, research in plant transgenesis continues to focus on ways to generate transgenic crops free of selectable marker genes. Several approaches for producing selectable-marker-free transgenic crops have been reported. Among these approaches, one of the most powerful is the application of microbial site-specific recombination systems to the transformation technology. This article will discuss the promise of transgenic breeding for future agriculture, the public concerns over genetically modified plants, and the utilization of site-specific recombination systems (including the novel ParA system recently described by the authors' laboratory) for the production of marker-free transgenic plants. The use of site-specific recombination systems to tailor controlled transgene expression in plants and to genetically engineer plants for breeding programs will be discussed, as well.

Chapter 3 - Sugarcane borer, *Diatraea saccharalis*, is a major maize stalk-boring pest in many areas of the mid-south region of the United States. In Louisiana, transgenic maize expressing *Bacillus thuringiensis* (Bt) Cry1Ab protein was first commercially planted in 1999 and now is the primary tool for managing stalk boring pests. Resistance development in target pest populations is an important issue in the sustainable use of Bt crops. Since 2004, a coordinated research program has been established to assess the risk of *D. saccharalis* resistance to Bt maize. This paper summarizes the major accomplishments and presents up-to-date results of this research program. Field surveys showed that a complex of *D. saccharalis*; southwestern corn borer, *Diatraea grandiosella*; and European corn borer, *Ostrinia nubilalis* could occur in Louisiana maize fields. *D. saccharalis* is the dominant species across the state, accounting for 88% of the total maize borers sampled during 2004-2009. Both *D. saccharalis* and *D. grandiosella* were able to overwinter in the major maize growing areas of the state. During 2004-2009, a total of 2,061 feral individuals of *D. saccharalis* were collected from four geographical locations in Louisiana and examined for Cry1Ab resistance alleles using F_1/F_2 screening methods. Cry1Ab resistance allele frequency was low (0.0011) for the populations collected during 2004-2008, but increased in 2009 (0.0136). A highly Bt resistant strain (Cry1Ab-RR) of *D. saccharalis* was isolated from a field population by using the F_2 screen. The Cry1Ab-RR strain was able to survive and complete larval development on commercial Cry1Ab maize plants. Relative to a susceptible strain of *D. saccharalis*, the resistant strain showed an approximately 100-fold resistance to Cry1Ab and also exhibited a significant resistance to Cry1Aa and Cry1Ac. Larvae of the resistant heterozygotes demonstrated relatively high survivorship rates on Cry1Ab maize hybrids which were commonly planted in Louisiana, suggesting most Cry1Ab maize hybrids planted in the state did not express a "high dose" against *D. saccharalis*. Cry1Ab resistance in *D. saccharalis* was likely controlled by one (or a few linked) autosomal gene. However, effective dominance level of resistance varied depending on bioassay methods and maize hybrids, ranging from incompletely recessive to incompletely dominant. In addition, the Cry1Ab resistance in *D. saccharalis* was not associated with any fitness costs and the resistance was stable after 24 generations without selection. The results of the 6-year study suggest that there appears to be a high risk for development of Cry1Ab resistance in *D. saccharalis* if Cry1Ab maize continues to be widely used in the U.S. mid-south region.

Chapter 4 - Agricultural production of the world is severely affected by various stresses such as biotic and abiotic stress, contributes majorly by decreasing the average yields over 50% and biotic stress reduces the crop yield by 20 to 30%. Plants innate immunity is based on a surprisingly complex response that is highly flexible in its capacity to recognize and counteract different invaders. To combat invasion by microbial pathogens and herbivorous insects effectively, plants make use of pre-existing physical and chemical barriers as well as inducible defense mechanism that become activated upon attack. The plant hormone salicylic acid (SA), Jasmonic acid (JA) and ethylene are major players in the regulation of signaling networks that are involved in induced defense responses against pathogens, insects and abiotic stress as well. During fast few years, significant progress has been made in understanding the function of Nonexpressor of pathogenesis related genes 1 (NPR1), a key regulator of systemic acquired resistance (SAR), that is essential for transducing the SA signals to activate pathogenesis related (PR) gene expression. Therefore it has been a well known fact that NPR1 induce PR genes and thereby SAR. In the recent years, the involvement of NPR1 in abiotic stress is also being elucidated and it found to be induced by salinity, drought and low temperature by different regulatory pathways. Overall the signaling molecules mediated protein-protein (NPR1-PR) interaction induces the enhanced tolerance to biotic and abiotic stress in crop plants. This review would emphasize on understanding the multifunctional role of NPR1 in plants.

Chapter 5 - The rapid development of biotechnology and molecular biology has made the genetically modified (GM) crops became a part of agricultural production. From 1996 to 2010, a total of 144 GM crops events have been approved worldwide. Following the success in global market, safety issues of GM crops have become a public concern in the last decade. Therefore, GM crop safety assessment protocol has been developed following guidelines proposed by WHO/FAO and Codex. The current methodology of GM crops/foods safety assessment is relatively empirical and conservative. Thus, accurate hazard identification and risk assessment will continue to represent important part of safety assessment research. Today, many efforts have made in bioinformatics allergenicity prediction, animal models and the analysis of unintended effects. However, the knowledge gap between unintended effects and GM crop safety assessment will be the major challenge in the future.

Chapter 6 - Competition from weed plants causes great losses in agriculture production. It is considered that the yield loss due to competition from weeds far exceeds that of pests and diseases. It is important to remove competition by weeds at an early stage of crop growth. In this context, herbicides (post and pre emergent) play an important role in weed control. Different herbicides are known belonging to different chemical groups according to their mode of action. Glyphosate is one of the most used broad spectrum herbicides which act as a competitive inhibitor of the enzyme 5-Enol Pyruvyl Shikimate 3-Phosphate Synthase (EPSPS) hampering aromatic amino acid biosynthesis in plants. However, the effect of herbicides can be lethal to crop plants as well. To avoid this, herbicide tolerant crops would be an option. Both conventional breeding and transgenic technology can be used to produce herbicide tolerant crops. The herbicide tolerance therefore offers the opportunity to spray crops at the most effective time to kill weed species without damaging the crop plants. Various strategies conferring glyphosate tolerance in particular will be discussed in this chapter. A large number of glyphosate tolerant crops have been generated and commercialized. This chapter would provide an updated account of the glyphosate tolerant transgenic crops. The potential risks and benefits of the herbicide tolerant crops will also be emphasized. Further, the importance of the herbicide tolerant genes as selectable markers for stringent screening and evaluation of the transgenic crops will also be addressed.

Chapter 7 - Impacts of genetically modified crops (GMCs) on non-target organisms and ecosystems are a key issue of risk assessment of GMCs introduction since they released. Introduction and use of genetically modified crops brought a list of changes in GMCs-soil ecosystems. The main changes include three aspects, i.e. physiological alterations of transgenic crops, rhizosphere characteristics variations of transgenic crops, and changes in management and agronomic measures, which result from the insertion of exotic genes and their biological functions. Compared to their parental line, above changes of transgenic crops would lead to a great deal of subsequently changes in soil ecosystem and the entire field ecosystem, e.g. nutrient cycling, soil microorganisms (e.g. arbuscular mycorrhizal fungi, AMF), input of agro-chemicals and so on. In this review, the changes brought by introducing GMCs in plant-soil system under field and the resultant impacts on soil ecosystem were summarized, highlighting the impact mechanisms on different scales, and ecological implications of interactions between Bt transgenic crops (BTCs) and AMF.

Chapter 8 - Plants are endowed with diverse mechanisms including both passive (preformed) and active (inducible) defense responses that protect them from microbial attack. Inducible plant defense responses include phytoalexins, lytic enzymes, oxidizing agents, cell wall lignification, PR proteins etc. Phytoalexins are a chemically heterogeneous group of low-molecular-weight secondary metabolites with antimicrobial properties that play a key role in defense system. The concept of phytoalexins in plants was first developed in 1940. It has taken more than 70 years to gain a basic perceptive of how a small part of nature's plant antimicrobial compound is manufactured. However, role of these compounds in defense has been revealed through several studies *viz.,* rate of phytoalexins synthesis in relation to cessation of pathogen development, quantification of phytoalexins at the infection site and relationship of pathogen virulence to the phytoalexin tolerance. The application of molecular and recombinant DNA technologies has led to significant advances in host-pathogen interactions and disease resistance mechanisms. In addition, latest successes in genetic engineering of plants afford new strategies for directly manipulating these interactions and enhancing disease resistance. In this review the authors discuss engineering novel phytoalexins in heterologous plants that provide a unique strategy to develop plant innate defense systems.

Chapter 9 - RNA interference (RNAi) has become a powerful genetic approach to systemically dissect gene function on a genome wide scale. Double stranded RNA (dsRNA) mediated gene silencing has been proven as a potential approach to control several insect pests including fruit fly, nematode, cotton bollworm, and corn rootworm. RNA interference (RNAi) induced by hairpin RNAs are new line of defence against coleopteran and lepidopteron pests. It was shown that insect larvae fed with plant material expressing dsRNA for the insect genes triggered the RNAi pathway in insect larvae and blocked the expression of target genes. The resulting reduction in larval growth and subsequent death has opened up new options for insect control involving the insect genes perse. Hence, protein targets for insect control strategies can be minimized or completely ignored. Prospecting for novel resistance mechanisms depend on the plant generating an effective form of double-stranded RNA in the absence of an endogenous target gene without detrimental effect to itself.

Although the amazing popularity of RNAi as a biotechnology tool is certainly justified, the underlying biology is still being worked out and the relative advantages and disadvantages of the approach are being efficiently addressed using emerging Bioinformatic tools and many genome sequences. In this review the authors would like to highlight the attempts of pest control using gene silencing or RNAi technology, few strategies, target genes for RNAi in insects, target site of action, and probable likely targets for RNAi in insects and plants for potential durable pest resistance mechanisms.

In: Genetically Engineered Crops
Editor: Lindsay M. Grover
ISBN: 978-1-61122-526-6

Chapter 1

GENETICALLY ENGINEERED CROPS AND TIIEIR IMPLICATIONS

V. Balasubramanian and N. Sakthivel*
Department of Biotechnology, Pondicherry University,
Kalapet, Puducherry, India

ABSTRACT

Genetically engineered crops are those crop plants that are transformed with transgenes of desired traits into their nuclear genome using recombinant DNA technology. The common strategies used for plant transformation are by exploiting natural gene transfer by soil bacterium, *Agrobacterium tumefaciens* and by using gene gun where gold microcarriers coated with desiccated DNA are delivered using pneumatic pressure. Transgenic technology has expanded the scope of conventional breeding by breaking the restrictions inflicted by traditional cross-pollination and selection techniques. Additional advantages like reduction in time involved in developing a variety, avoiding linked undesirable genes and overcoming sexual incompatibility barriers are also attributed to transgenic breeding. Many genetically engineered crops are successfully created worldwide and are either cultivated in the field or waiting for approval. These include herbicide resistant soybean, rapeseed, sugar beet and corn, boll worm resistant cotton, European corn borer resistant corn, fruit borer resistant egg plant, ring spot virus resistant papaya, high sugar content sugarcane, nutrient enriched potato and rice and crop for the production of biopharmaceuticals. Genetically engineered crops offer a multitude of advantages over traditional crop varieties which include yield enhancement, suppression of pathogen, pest and weed infestations, reduction of the use of harmful pesticide and herbicide as well as reduction of the incidence of pesticide resistance in pests. Functional foods with high nutrient contents can inexpensively improve health of people especially in developing and underdeveloped countries. Transgenic crops also have been subjected for considerable arguments against their usage *viz.*, improved seeds and associated technologies make farmers to depend on multinational companies for their seed supply, biopharmaceuticals produced in food

* Corresponding author: Dr. N. Sakthivel, Professor and Head, Department of Biotechnology, Pondicherry University, Kalapet, Puucherrry 605014, India. Tel. :+91 413 2654430; fax: +91 413 2655255, E-mail address: puns2005@gmail.com

crops posses the risk of accidental ingestion, inadvertent introduction of allergens in food, pollen from herbicide resistant crops creating super weeds, reduction of biodiversity and disruption of ecosystem.

Key words: Transgenic crops, *Agrobacterium tumefaciens*, Gene gun, Bt cotton, Golden rice, Super weed.

1. INTRODUCTION

Genetic engineering is the direct manipulation of an organism's genomic DNA or organelle DNA to make it useful for either science or technology. The history of genetic engineering is time immemorial where the events of animal and plant breeders using various methods to alter and change genes to their advantage had been recorded in various instances. But genetic engineering in modern sense is of recent origin and is based on laws of genetics propounded by Gregor Johann Mendel whose work was rediscovered in 1900 (LeVine, 2006). One event which gave a breakthrough in modern genetic engineering is the successful cloning of genetically engineered molecules in foreign cells. In 1973, Stanley Cohen and Herbert Boyer, the founding fathers of genetic engineering first demonstrated the technique of DNA cloning by ligating DNA fragment harboring frog ribosomal RNA in the plasmid vector pSC101 and successfully transferring it into the bacterium *Escherichia coli* and expressing it (Cohen et al., 1973). In fact the genesis of the term "genetic engineering" can be traced to 1951 when it was first used in "Dragons Island" the science fiction novel authored by Jack Williamson. With a phenomenal progress in research in the past four decades the art and science of genetic engineering is simplified and standardized to suit the objectives of the work. Major steps involved in genetic engineering are as follow,

Identification and isolation of DNA fragment harboring gene of interest: The prime most activity in genetic engineering is obtaining the copy of the DNA fragment harboring gene of interest which is to be ultimately incorporated into the genome of the organism to be modified. Isolation can be done by simple amplification by polymerase chain reaction (PCR) provided the required sequence information is at disposal. Other means is to construct genomic or cDNA libraries and cull out the clone of interest from them. In those cases where the gene is to be lifted from domain bacteria to domain eukaryote or *vice versa* additional modifications involving removal of introns or optimization of codons or ligating to suitable promoters are to be carried out appropriately.

Ligation of an isolated donor DNA fragment into an appropriate vector genome: This step involves ligation of isolated DNA fragment with an appropriate vector. In majority of cases ligation is achieved by T4 DNA ligase enzyme. The vectors employed can be a plasmid or a modified phage DNA or a cosmid or a phagemid or a phasmid or modern vectors like artificial chromosomes. The choice of vector is determined mainly by the size of the DNA fragment to be ligated and downstream use of the clone. Tailor made vectors are also available to suite the need of researcher's objective. The ligated product is appropriately called as recombinant DNA.

Transfer of recombinant DNA into appropriate host: The recombinant DNA is to be introduced into target cell using appropriate technology. The technique depends on the vector and the target cell. In case of bacterial cells the commonly used techniques are calcium chloride mediated transformation and electroporation whereas plant cells are efficiently transformed using *Agrobacterium tumefaciens* or gene gun.

Selection of transformed/transfected cells: This step involves identification of cells that have received the desired insert at the appropriate orientation from those which have failed. Modern cloning vectors have built in selectable markers to accomplish selection (Kirubakaran and Sakthivel, 2008). The selectable markers are more frequently any appropriate antibiotic resistant genes and in some cases herbicide resistance genes. In addition to this many vector system do include blue/white screening to identify transformants with inserts. This technique involves alpha complementation concept in a medium supplemented with IPTG (Isopropyl-beta thio galactopyranoside) and X-Gal (5-Bromo-4-Chloro-3-Indoly-β-D-Galacto-pyranoside). In spite of this, the transformants have to be further confirmed with high end techniques like amplification by polymerase chain reaction (PCR), restriction analysis, Southern hybridization and DNA sequencing analysis. When the host cell transformed with recombinant vector divides, copies of it are apportioned into progeny which further increase in number by replications. Thus on the emergence of a colony, each constituent cell contains hundreds of recombinant DNA accomplishing the task of gene cloning (Primrose and Twyman, 2006: Dale and Schantz, 2007).

2. GENETIC ENGINEERING *VS.* CONVENTIONAL PLANT BREEDING

Conventional plant breeding is purposeful manipulation of desired plant species with an objective of creating a new variety so that they become more suited for human needs. It all started at least 10,000 years ago with the domestication of wild plants possessing desired characters. Many of the modern crop plants are a result of domestication done during ancient times (Abbo, et al., 2001). The rediscovery of Mendel's laws of genetics in the beginning of twentieth century provided the scientific basis for modern plant breeding. It is also well established that the genes controlling various traits of plants are located on chromosomes and conventional plant breeding can be considered as the manipulation of the combination of chromosomes thereby manipulating gene combination (Gepts, 2002; Singh, 2008).

In general, there are three main procedures adapted in conventional plant breeding to manipulate the plant genome towards the desired direction. The first procedure called pure line-selection which is the ancient and basic procedure in plant breeding applied mainly for self-pollinated crop plants. This procedure involves initial selection of large number of phenotypically superior plants from a genetically variable population. This is followed by evaluating the selected plants by growing them in individual progeny rows. After appropriate eliminations, the selected progenies are evaluated for performance under different environmental conditions for several years. Finally, the selected progenies are evaluated along with well established varieties for yield potential and other agronomic characters and successful progenies are released as variety (Rommens, 2007; Singh, 2008). The second and most frequently employed procedure is called hybridization. It is done by deliberate crossbreeding of different plant lines to bring the desired traits into one plant line. This procedure initially involves generation of homozygous inbred lines by repeated self-pollination. The selected inbreds are later crossed and resulting progenies are selected for desired combination of traits. Hybridization uses "heterosis", the phenomenon of increased vigor by F1 (first filial generation) progeny (Chopra, 2000; Singh, 2008). The third procedure uses the phenomenon called polyploidy, the condition in which the cells contain an increased numbers of chromosomes or chromosome set. Polyploidy is a naturally occurring condition in many crop plants but can be artificially induced by using the chemical called colchicine which leads to a doubling of the chromosome number by inhibiting chromosome segregation (Adams and Wendel, 2005; Lin et al., 2010). Generally, the main effect of polyploidy is increase in size and genetic variability which is desired in many horticultural crops. On the other hand, polyploid plants often exhibit reduced fertility and growth rate.

Although initially considered as artificial, mutation breeding is also acceptable as standard method in crop improvement. Mutations induced by chemical of physical mutagens can create new and novel genetic variations. The mutants obtained are tested and selected for desired traits. This method relieves scientists who depend only on naturally existing genetic variability. Some of the disadvantages of mutation breeding are that the site of the mutation cannot be controlled when chemicals or radiation are used as agents of mutagenesis and percentage of mutants obtained with desirable trait is very negligible (Waugh et al., 2006; Shu and Lagoda, 2007).

Conventional plant breeding primarily utilizes selection and hybridization as principle tools to alter genetic set up which is already present within a species. It harnesses the natural processes like pollination and asexual reproduction to achieve the breeding objectives. It handles those characters which are present within the genetic potential of the species for millennia.

The new tool in the hands of scientists to improve crop varieties in desired direction is 'genetic engineering' or 'transgenic technology'. This modern procedure involved in developing transgenic crop starts with the extraction of genomic DNA of organism found to have the gene of interest. This is followed by construction of genomic library, culling out the clone which harbours the DNA fragment possessing gene of interest and modification of the gene with appropriate promoter and terminator sequences. In some cases even codons in the coding sequence of gene is modified to avoid codon usage bias between the source and target species. Appropriately modified gene is now packed in suitable vector and transformed into a bacterial host and allowed to multiply to make thousands of copies. The multiplied recombinant vector is then introduced into the target plant cells using suitable gene delivery method. The transformed cell undergoes tissue culture protocols for regeneration of the transgenic plant. The final step in the process is to backcross newly developed transgenic plant with a well established local variety and to select progenies that express the inserted gene in a desired manner. The duration involved in taking a transgenic plant from lab to land depends upon the gene involved in the work, crop species and variety handled, available resources at the disposal of scientist, and statutory regulatory measures prevailing in the country concerned.

Genetic engineering is not only used to introduce novel genes but also to deliberately silence targeted plant genes by RNA based method popularly referred to as RNA interference or RNAi or cisgenics (Matzke et al., 2001). Genetic engineering together with genomics also strengthens molecular marker assisted selection (MAS), which is valuable for reducing the time taken in developing a variety by classical breeding (Collard and Mackill, 2008). The option of genetic engineering of crop plant is opted only in circumstances where the trait to be introduced is not present in the germplasm of the crop or not available to breeder by any other reasons or the trait has proved very difficult to improve by conventional breeding methods or when it will take a very long time to introduce and/or improve such trait in the crop plant by conventional breeding methods (Singh, 2008). It can also be deciphered that genetic engineering cannot displace conventional plant breeding but only reinforce it for achieving useful practical outcomes. As genetic engineering of crops crosses all natural species barriers for combining desired genes, its products are popularly called "transgenic crops" or "genetically modified crops".

3. Gene Transfer Techniques

Plant transformation involves delivering and integrating a fragment of DNA harboring a gene controlling a specific trait of interest into the genome of crop plant. On successful integration and stable inheritance, the foreign gene is termed as "transgene". The transgene can be sourced either from another related plant or from a completely different species from the taxonomic spectrum. Since the early reports of successful plant transformation with tobacco in 1984 (De Block et al., 1984; Horsch et al., 1984; Paszkowski et al., 1984), more than 144 plant species traversing almost all major phylogenetic lineages of the plant kingdom have been reported to be successfully engineered genetically (Birch, 1997). The list includes cereals, millets, pulses, vegetables, fruits, ornamental and medicinal, trees and pasture crops. The objectives of plant transformation experiments are of two, as research tool to understand basic plant science and as technology for crop improvement to serve the need of mankind. Since early 1980s, various novel plant genetic engineering techniques have been reported with success. These include systems using *Agrobacterium* mediated gene transfer (Alimohammadi and Bagherieh-Najjar, 2009; Opabode, 2006), Biolistic procedure involving high velocity bombardment of DNA coated microprojectiles using gene gun (Sanford, 2000; Taylor and Fauquet, 2002), direct gene transfer into protoplasts by osmotic or electric shock (Karesch et al., 1991), acupuncture (Stokes, 2001), electrofusion and electroporation (Van Wert and Saunders, 1992), electrophoresis of embryos (Giesbach, 1994), microinjection (Neuhaus and Spangenberg, 1990), pollen-tube pathway (Song et al., 2007), silicon carbide- and liposome-mediated transformation (Rao et al., 2009). Of these systems of plant transformation, only *Agrobacterium*-mediated transformation, delivery of DNA into protoplasts and the biolistic procedure have been used commonly for crop improvement.

The success of developing and deploying transgenic crops is decided by various parameters. These include an amenable target plant tissue, an effective DNA delivery technique, appropriate methods to select and confirm transgenic tissues, regenerability of the target tissue into a complete plant in a higher frequency, a simple, efficient and reproducible genotype independent method to avoid somaclonal variation and sterility in ensuing generations (Hansen and Wright, 1999; Khan and Liu, 2009).

3.1. Biological Method of Plant Transformation

Members of the genus *Agrobacterium* are ubiquitous bacterial component of soil microflora. They are saprophytic in nutrition, surviving primarily on decaying organic matter in soil. *Agrobacterium* is more studied as causal agent of neoplastic diseases in plants. The genus *Agrobacterium* belongs to order Rhizobiales and family Rhizobiaceae which include a number of species. Based on the genome analyses, chromosome structure, metabolic activities and plants infected they are broadly categorized in three biovars. *A. tumefaciens* causing crown gall and *A. rubi* causing cane gall belong to biovar I. *A. rhizogens*, the causal agent of hairy root disease in dicotyledonous plants belong to biovar II and *A. vitis* that causes galls in grapevines belongs to biovar III. Strains belonging to biovar I and biovar III harbour a Ti (Tumour inducing) plasmid, whilst strains of biovar II harbour a Ri (Root inducing) plasmid (Young et al., 2001; Farrand et al., 2003; Anand and Mysore, 2006). Of these species, *A. tumefaciens* plays a significant role in understanding molecular biology and remains a method of choice for transforming plant cells despite the development of sophisticated alternative gene transfer methods.

A. tumefaciens is Gram negative, motile and rod shaped bacterium causing crown gall disease in wide range of dicotyledons plants and gymnosperms by eliciting neoplastic growth at the site of infection. (Bouzar and Moore, 1987). The disease is characterized by a tumour-like growth or gall on the infected plant, often at the junction between the root and the shoot. But for the crown region, the bacterium is also capable of causing galls at the sites of injury on the stems and roots of numerous dicotyledonous plants, although necrosis of infected root is not reported from any crown gall disease caused by this bacterium (Smith et al., 1911). Crown gall is an agriculturally important disease worldwide causing significant economic loss in fruit and nut trees, grapevines, cane berries, chrysanthemum, rose, and other nursery crops (Lopatin, 1939; DeCleene and DeLey, 1976; Burr and Katz, 1984). The disease is very severe in nurseries but generally not a menace in established orchards unless infected nursery stock is planted. Characteristic symptom of crown gall disease is development of tumorous outgrowth (gall) of roots, crowns, and occasionally on trunks and scaffolds. Diversion of water and nutrients from developing tissue due to the rapidly developing unproductive gall is the main reason behind reduced yield (Klein and Link, 1955). Initially the galls are soft and smooth but as they mature and enlarge they turn hard, rough, woody and dark. Mature galls on cracking, serve as sites of secondary infection by phytopathogens (e.g. *Pseudomonas syringae* and *Armillarea mellea*) and pests (e.g. stem borers) and increases the host plants susceptibility to abiotic stresses. Crown gall also decreases the grafting ability of host plant as the tumor tissue developing at graft union prevents fusion of stock and scion tissues (Scroth et al., 1988).

Though the disease came under limelight in 1850s as pathogen of grape wine (Lioret, 1956) but the mechanism behind pathogenesis was fully deciphered only in the first decade of 1900 (Smith and Townsend, 1907). Once the root of a dicotyledonous plant gets wounded by grafting or transplanting or by subterranean insects, the root exudates containing amino acids, organic acids, and sugars are released into rhizosphere soil. These chemicals serve as chemoattractants and lure *A. tumefaciens* towards wound site, the process termed as chemotaxis. On reaching the wound site the bacterium initially adheres weakly through the synthesis of acetylated polysaccharides and later reinforces the attachment through the extrusion of cellulose fibrils.

The molecular events following the adhesion are very unique as it involves interkingdom transfer of DNA from *A. tumefaciens* into plant's nucleus, the process defined as 'genetic colonization' (Chilton et al., 1977). The pathogenesis of *A. tumefaciens* involves transfer, stable integration and expression of a set of genes from a large tumor-inducing (Ti) plasmid of *A. tumefaciens* into the genome of a plant (Nester et al., 1984; Binns and Thomashow,

1988). The size of the various Ti plasmids characterized range in the order of 200-800 kbp. The transferred DNA (T-DNA) originally a part of T-region when present in Ti plasmid is of 10-30 kb in size. The T-DNA contains two types of genes: the oncogenic genes responsible for production of auxins and ctokinins which are responsible for gall formation and the genes responsible for production of opines which are synthesized from amino acids and sugars by condensation. Opines are secreted from crown gall cells and can be utilized only by *A. tumefaciens* as carbon and nitrogen source. Ti plasmid also harbors *trans* acting virulcnce (*vir*) genes which are responsible for reception of external signal from wounded site, processing, packaging, transporting and integrating of T-DNA in plant genome. Other genes present in Ti plasmid are genes for opine catabolism and bacterium-plant plasmid conjugative transfer genes (Zupan and Zambrysky, 1995). A successful infection establishes a monopolized ecological niche in the rhizosphere of infected plant which can be inhabited only by *A. tumefaciens* (Wei et al., 2000; Escobar and Dandekar, 2003).

Three aspects of pathogenesis by *A. tumefaciens* have been harnessed for plant genetic engineering. Firstly, crown gall disease is a natural genetic engineering process involving successful transfer, stable integration of T-DNA in plant genome and expression of T-DNA genes. Secondly, the T-DNA genes express only in plant cell and have no role in transfer process. Third and important aspect is any foreign DNA irrespective of its origin when appropriately placed between the T-DNA borders will be transferred to plant cells as like native genes of *A. tumefaciens*. These facts allowed scientists to conceive and construct *A. tumefaciens* based plasmid vectors and strains for plant transformation.

Decades of research on these aspects have resulted in establishment of several *Agrobacterium* strains, plasmid vectors and protocols catering the transformation of various plant species (Tzfira and Citovsky, 2006; Khan and Liu, 2009). *Agrobacterium*-mediated transformation has several merits over other plant transformation methods and it is the method of choice to transform plants. It accounts for about 80% transgenic plants produced so far. The merits include (a) Many plants which were recalcitrant for transformation are now been tamed with suitable techniques (Cao et al., 2009), (b) In majority of transformation events, only a single or low copy number of the T-DNA is integrated in the transformed plant genome with less DNA rearrangement (Hamilton et al., 1996) (c) The choice of vector systems with T-DNA borders containing various reporter and selectable marker genes is multitude, allowing scientists to choose the most appropriate combination to insert heterologous genes (d) The method has been refined even to transfer large fragments of DNA, including bacterial artificial chromosomes (BAC) (Hamilton et al., 1996) (e) *In planta* transformation techniques have been developed (Trieu et al., 2000).

3.2. Physical Method of Plant Transformation

A score of physical DNA delivery methods have been developed so far to circumvent the host-range restrictions of *Agrobacterium*-mediated transformation. Many techniques such as gene gun bombardment, microinjection, electroporation and PEG-mediated transformation have been reported successful in transforming plants. Of these techniques, gene gun bombardment is the most widely used method for genotype-independent plant transformation (Veluthambi et al., 2003). The gene gun method is referred in many ways as the particle gun method, the microprojectile method, the particle acceleration method and the bio-blaster method. This methodology was developed by Sanford and co-workers and was successfully demonstrated in 1987. The inventors of the process have coined the term 'biolistic' (biological ballistics) to describe both the process and any associated apparatus used to shoot biological materials into living targets (Sanford, 1988).

In this method the DNA of choice is precipitated onto microscopic particles like gold or tungsten, which are then accelerated into target plant cells using either electric discharge or pneumatic pressure. The velocity used for acceleration is sufficient to penetrate the cell wall but below that which will also cause lethal damage. The target DNA is transported into cell's interior where it detaches from the microprojectile and integrates into the nuclear or chloroplast genome. The target of gene delivery can be suspension cultures, tissues in culture, whole plant parts, etc (Newell, 2000). There are two variants of gene gun method used successfully in delivering DNA into plant genome. The first is electrical discharge particle acceleration method called ACCELL technology. This technology uses high voltage electrical discharge to vaporize a droplet of water. DNA-coated microprojectiles are spread on a mylar sheet and positioned within a vacuum chamber. Shock waves generated by electrical discharge into the water droplet are channeled to accelerate the macro- and microcarrier complex into a stopping screen and project the microprojectiles into the target cells (McCabe and Christou, 1993). The second is PDS-1000/He, Biolistic particle delivery system. The device is powered by a burst of helium gas that accelerates the macrocarrier upon which DNA-coated gold particles (microprojectiles) are uniformly smeared. These gold particles bombard over target cells with fixed velocity and integrate the smeared DNA in target tissue (Kikkert, 1993). The advantages of particle bombardment system are: (a) It is species independent and plants which are recalcitrant to *Agrobacterium*-mediated transformation can be transformed as it avoids the complex interaction between bacterium and plant tissue. (b) Protocols are developed for plants which lack good regeneration systems. (c) Hardship of handling specialized vectors is avoided. (d) Co-transformation with multiple DNA fragments/plasmids can be achieved by co-bombardment, thus eliminating the necessity of constructing a single large plasmid containing multiple transforming sequences. (e) False positives resulting from reporter gene expression in *Agrobacterium* are avoided. (f) Organelle transformation can be achieved only by particle bombardment. (Taylor and Fauquet, 2002; Veluthambi et al., 2003).

4. Genetic Engineering of Crops

4.1. Pathogen Resistant Crops

4.1.1. Fungal Resistant Crops

Till date, more effort has been devoted to create transgenic plants that are resistant to fungi than any other plant pathogen. Various strategies have been tried successfully to develop fungal-resistant transgenic crops which are briefly explained below with few relevant examples.

Heterologous expression of genes whose products are directly toxic to pathogens or that hamper their growth inside the host plant: This strategy includes using genes of pathogenesis related (PR) proteins such as hydrolytic enzymes (chitinases, glucanases), antifungal proteins (osmotin- and thaumatin-like), antimicrobial peptides (thionins, defensins), ribosome inactivating proteins (RIP), and phytoalexins (Melchers and Stuiver 2000; Rommens and Kishore 2000). PR proteins are several classes of novel proteins that are coded by the host plant induced by various types of pathogens such as viruses, viroids, bacteria, and fungi and those induced by the application of chemicals that mimic the effect of pathogen infection or induce similar stresses (Bol et al., 1990; Stintzi, 1993). PR-proteins were first discovered in 1970 from leaf extracts of tobacco reacting hypersensitively to Tobacco mosaic virus (TMV) (Van Loon and Van Kammen, 1970). It is now well recognized that diverse group of monocotyledonous and dicotyledonous plants possess proteins with immunological and amino acid sequence similarities to tobacco PR-proteins. Based on the sequence similarities, serologic or immunologic relationships, and enzymatic properties all recognized PR proteins are classified in 17 families, regardless of the original plant species (Muthukrishnan et al., 2001; Liu et al., 2005).

PR-2 family β-1,3-endoglucanases and the PR- 3, PR -4, PR -8, and PR -11 endochitinases, act against fungi by hydrolyzing glucans and chitins respectively which are the principle components of fungal cell wall. There are quite a number of reports on successful genetic engineering of various crop plants with chitinases of plant and non-plant origin showing enhanced resistance against fungal pathogens. Indica rice cultivar was transformed with a class I endochitinase gene, *McCHIT1* from bitter melon (*Momordica charantia*). Transgenic rice lines over expressing the gene showed an improved resistance to the fungal pathogens *Magnaporthe grisea* causing blast and *Rhizoctonia solani* causing sheath blight (Li et al., 2009). Endochitinase (CHIT42) coding gene *chit*1 was isolated from the entomopathogenic fungus *Metarhizium anisopliae*. Transgenic tobacco transformed with the coding sequence of this gene driven by CaMV 35S promoter was found to be resistant against *R. solani*, a soil borne pathogen that causes damping off, seedling blight, and root rot (Kern et al., 2010). Transgenic Indian mustard (*Brassica juncea*) lines expressing class I basic glucanase sourced from tomato under the control of CaMV 35S promoter were created. These transgenic lines were found very much resistant to Alternaria leaf spot caused by *Alternaria brassicae* (Mondal et al., 2007).

There are also reports of co-expression of chitinases and glucanases which further strengthens the transgenic plant's resistance against multitude of fungal pathogens. Binary vector harbouring the rice chitinase (*chi11*) gene under maize ubiquitin promoter and the tobacco β-1,3-glucanase gene under CaMV 35S promoter in the same T-DNA was constructed and used for transformation. Rice plants transformed with this construct showed constitutive and increased levels of expression of these two genes. These transgenic lines showed an enhanced and persistent resistance to *R. solani* which causes sheath blight (Sridevi, 2008).

PR-5 family includes antifungal thaumatin-like proteins, osmotins, zeamatins, perm-eatins, similar to alpha-amylase/trypsin inhibitors. The PR-5 proteins act by inducing fungal cell leakiness, through interaction with the plasma membrane resulting in the formation of transmembrane pores (Kitajima and Sato, 1999). Oil seed rape (*Brassica napus*) was transformed using *Agrobacterium* with chimeric gene containing barley cDNA for the thaumatin-like protein Hv-TLP8 under the control of CaMV 35S promoter and terminator sequences. Stable transgenic lines were obtained which showed improved resistance to the fungus*Plasmodiophora brassicae* which causes club root disease (Reiss, 2009).

Both PR-12 (Defensins) (Lay and Anderson, 2005) and PR-13 (Thionins) (Bohlmann, 1994) have broad antibacterial and antifungal activities. Plant defensins are small (45–54 amino acids) highly basic polypeptides containing eight conserved cysteine residues linked to disulphide bridges. Plant defensins have globular structure composed of a cystein stabilized αβ (CSαβ) motif which consist of an α-helix and a triple stranded β-sheet, organized in a βαββ architecture and are stabilized by four disulfide bridges (Liu et al., 2007). The structure-activity relationships and modes of action for most of the plant defensins remains to be understood thoroughly. Transgenic tomato plants expressing chimeric chili defensins gene driven by CaMV 35S promoter exhibited an enhanced resistance to the fungal pathogens *Fusarium* sp. and *Phytophthora infestans* than the control plants (Zainal et al., 2009).

Thionins are small, hydrophobic, basic plant proteins (45 to 50 amino acids) which include three or four conserved disulphide linkages. They are able to inhibit a broad range of plant-pathogenic fungi and bacteria by electrostatic interaction with membrane phospholipids, leading to pore formation (Molina et al., 1993; Stec, 2006). The potential use of thionins in plant protection was well demonstrated with transgenic tomato. Transgenic tomato plants transformed with the *Arabidopsis thaliana* thionin (*Thi2.1*) gene showed a consistent multiple disease resistance against Furarium wilt caused by the fungus *Fusarium oxysporum* f. sp. *lycopersici* and bacterial wilt caused by *Ralstonia solanacearum*, the two very important soil-borne vascular diseases difficult to control by conventional approaches (Chan et al., 2005).

Ribosome-inactivating proteins (RIPs) are cytoxic N-glycosidases that specifically depurinate the conserved α-sarcopin loop of larger rRNAs, thus inactivationg the ribosome ultimately stalling the process of translation. They are reported to be present in a wide variety of plant species and within a variety of different tissues. The reported RIPs are grouped in three types: type I has a single polypeptide chain while type II is a heterodimeric protein composed of 'A chain' and 'B chain'. 'A chain' is functionally equivalent to type I chain and 'B chain' is a galactose binding lectin. Type III is a single chain containing an extended carboxyl-terminal domain. Type III RIPs are synthesized as inactive precursors (proRIPs) that are post translationally processed for the formation of active site. Although the discovery of RIPs is more than 100 years old, their actual biological function(s) in a plant still remains open to speculation. Some RIPs are also reported to exhibit other enzymatic activities such as RNase, DNase, phospholipase, and superoxide dismutase (SOD). RIPs play an important role in plant defense as they hamper the multiplication and growth of several pathogenic virus, fungi, and bacteria either alone or along with other defense-related proteins. RIPs inhibit the fungal pathogens by directly inhibiting their growth rather than by depurinating ribosomes. For instance, RIPs may regulate protein expression by targeting mRNA instead of ribosomes (Nielsen and Boston, 2001; Park et al., 2004). Transgenic tobacco lines expressing antifungal ribosome-inactivating protein (PhRIP I) from *Phytolacca heterotepala* showed an enhanced resistance to fungal pathogens *Alternaria alternata* and *Botrytis cinerea* causing brown spot or leaf spot disease and grey mold diseases respectively (Corrado et al., 2005).

Other weapons in the plant's arsenal against fungal pathogen which are used successfully to engineer plants to defend against fugal pathogen are phytoalexins. These are low molecular weight antimicrobial secondary metabolites produced in a broad range of plant species on induction by pathogen infection, under abiotic stresses and elicitors. They are chemically diverse groups each characteristic of any given plant species. They fall into different classes including terpenoids, glycosteroids and alkaloids (Grayer and Kokubun, 2001). Genes coding for stilbene synthase involved in synthesizing stilbene phytoalexin trans-resveratrol were sourced from the grapevine (*Vitis vinifera*). These genes were transferred to tomato by means of *A. tumifaciens*. Considerable accumulation of the phytoalexin trans-resveratrol, the product of stilbene synthase, was detected on fungal inoculation in transgenic tomato resulting in enhanced resistance to the fungus *P. infestans* (Thomzik et al., 1997). Chimeric genes coding for resveratrol synthase *vst1* and *vst2* from grapevine (*V. vinifera*) were constructed under the control of stress-inducible grapevine promoters. These two genes were transformed into bread wheat and stable expression was obtained. The transgenic wheat lines showed a significant decline in disease incidence on inoculation with the fungal pathogens *Puccinia recondita* f. sp. *tritici* and *Septoria nodorum* which causes stem rust and the glume blotch diseases of wheat respectively (Serazetdinova et al., 2005).

Heterologous expression of gene products that destroy or neutralize a component of the pathogen's arsenal: Interaction between a fungal pathogen and a host plant is a complex process in which a variety of molecules of fungal pathogen and host plant origin are involved. The interaction between them determines whether the outcome is a disease (fungus succeeding in colonizing the host plant tissues) or resistance (the plant succeeding in defending the fungal invasion) (Yoder and Turgeon, 1996). Pathogenesis by successful plant pathogenic fungi involves production of a gamut of extracellular hydrolytic enzymes and toxins that aid in invading the host tissue. These enzymes are collectively called cell wall-degrading enzymes (CWDE). These enzymes and toxins involve in overcoming external physical barriers of plant cell including wax, cuticle and cell walls thus aiding tissue invasion and pathogen dissemination. Additionally these compounds also serve as extracellular signals in eliciting an array of host cell defense responses (Kikot et al., 2009).

Several strategies are designed to inactivate these virulence factors to confer resistance to phytopathogenic fungi in transgenic plants. Polygalacturonase-inhibiting protein (PGIP) is an extracellular, leucin-rich repeat glycoprotein present in the cell wall of many plants and involved in primary plant defense by interacting with fungal endopolygalacturonases. The inhibition of fungal endopolygalacturonases is an initial and crucial event during plant defense against fungi (Manfredini et al., 2005). Chimeric gene consisting of coding sequence of the pear fruit PGIP (*pPGIP*) with the CaMV 35S promoter and the U1 W translational enhancer of TMV was constructed for regulated *pPGIP* expression. Tomato plants were transformed with this construct and stable expression was confirmed in cell wall protein fraction. Compared to nontransgenic plants these transgenic tomato plants showed considerable reduction in expansion of disease lesions and the associated tissue maceration in ripe fruits. This confirms the inhibition of necrotrophic fungus *B. cinerea* on ripe tomato fruit (Powell et al., 2000).

Phytotoxic metabolites like mycotoxins and oxalic acid are also produced by fungal pathogens to facilitate the process of infection. One another strategy to confer plants with fungal pathogen resistance is to engineer them with ability to degrade these metabolites. Trichothecenes are the mycotoxins produced by a host of fungal species including *Fusarium*, *Myrothecium*, *Trichoderma*, *Trichothecium*, *Stachybotrys* etc. They are also produced on a variety of grains like wheat, oats and maize by various *Fusarium* species such as *F. graminearum*, *F. sporotrichioides*, *F. poae* and *F. equiseti*. Trichothecenes belong to sesquiterpene compounds and acts by inhibiting protein synthesis in eukaryotes by reacting with ribosomes and stimulate the development of plant diseases as a phytotoxin by interfering with the expression of defense related genes (Mitterbauer et al., 2004; Desjardins, 2009).

One of the serious diseases of cereal crops is Fusarium head blight caused by the necrotrophic fungi *F. graminearum* and *F. culmorum*. These fungi produce the trichothecene mycotoxin deoxynivalenol (DON) and its derivatives during the course of pathogenesis. These pathogens also produce trichothecene 3-*O*- acetyltransferase coded by the gene *Tri101* for self protection. Expression of trichothecene resistance gene *Tri101* in transgenic rice significantly reduced the phytooxic action of DON on and rice plants (Ohsato et al., 2007). Transgenic tobacco expression trichothecene-degrading enzymes from *F. sporotrichioides* showed decreased tissue damage and enhanced seedling emergence in presence of trichotecene (Muhitch et al., 2000).

Other toxic product that many pathogenic fungi release inside the plant cell is oxalic acid. The proposed mechanisms of oxalic acid's actions in infection are acidification to facilitate cell wall-degrading enzyme activity, pH mediated tissue damage, and sequestration of Ca^{2+} ions (Dutton and Evans, 1996). Oxalate oxidases are glycoproteins that act on oxalic acid and breakdown it into CO_2 and H_2O_2. H_2O_2 serve as signal molecule in eliciting plant defense responses and also it improves the tolerance of the host plant against the fungal toxin (Thompson et al., 1995). Expression of barley oxalate oxidase in transgenic peanut (*Arachis hypogaea*) resulted in enhanced tolerance to the fungus *Sclerotinia minor*, the causal agent of Sclerotinia blight (Livingstone et al., 2005).

Heterologous expression of gene products that can reinforce the structural defenses in the plant: Lignification of plant cells around sites of infection of fungal pathogen or lesions can considerably slow down the spread of the invading pathogen into the surrounding cells. This strategy can yield transgenic plants showing resistance towards a broader spectrum of pathogens including fungi rather than for few selective pathogens as in case of other strategies (Nicholson and Hammerschmidt, 1992). It involves exploiting the enzyme peroxidase which catalyses the final stages of polymerization of phenolic derivatives into lignin and also suspected to play a role in suberization or wound healing. The fact is proven in two ways, one with fact that transgenic potato expressing tryptophan decarboxylase was observed to be more susceptible to *P. infestans*. This is due to the reason that the aromatic amino acid was redirected in transgenic plants resulting in decreased levels of lignin (Yao et al., 1995) and second, a higher level of lignin was observed in transgenic potato following expression of the H_2O_2-generating enzyme glucose oxidase. This transgenic plant was also reported to show enhanced tolerance to several fungal pathogens (Sitbon et al., 1999).

Expression of resistance gene (R) products involved in the hypersensitive response (HR) and in interactions with avirulence (Avr) factors: Plants have evolved with sophisticated recognition systems to detect proteins produced during infection by specific races of pathogens. These proteins, referred as effectors or elicitors are encoded by the avirulence (*avr*) genes of pathogens are recognized by plant disease resistance (*R*) genes. This recognition triggers the array of plant defenses responses including the hypersensitive reaction (Flor, 1971). This understanding of molecular basis of host pathogen interaction resulted in the cloning of both the avirulence (*avr*) and the corresponding resistance (*R*) genes which have been successfully applied to protect crop plants from pathogens. More than 30 resistance genes have been cloned conferring resistance to a breadth of pathogens including bacteria, fungi, viruses and even namatodes and aphids (Hammond-Kosack and Jones, 1997; Williamson, 1999; Takken and Joosten, 2000; Houlb, 2001, Hulbert, 2001). Gene cassette consisting of coding sequence of the elicitor cryptogein (*Crypt*) from the pathogen

Phytophthora cryptogea and a pathogen-inducible promoter from tobacco 'hsr203J' gene was transferred and expressed in tobacco. The transgenic tobacco showed a controlled HR response on infection to *Peronospora parasitica* var. *nicotianae* and also non specific response to *Thielaviopsis basicola*, *Erysiphe cichoracearum* and *B. cinerea* (Keller et al., 1999).

Expression of genes whose products release signals that regulate plant defenses: Once the elicitor molecule is recognized by the receptor, a cell signaling cascade is triggered by activation of various activator molecules present in host cell. These activator molecules then interact with signal molecules such as salicylic acid (SA), H_2O_2, ethylene and jasmonic acid (JA), which in turn trigger defense responses in plants. The defense responses in plants include production of an array of PR proteins, phytoalexins and hypersensitive response (HR) (Heath 2000; McDowell and Dangl, 2000). Over-expression of many of these signaling molecules has succeeded in making transgenic plants resistant to many fungal pathogens. Expression of a gene for H_2O_2-generating enzyme glucose oxidase from *Aspergillus niger* in potato is found to improve fungal resistance in various degrees on infection by *P. infestans*, *Alternaria solani* and *Verticillium dahlia*. As H_2O_2 is phytotoxic in increased level, it needs to be handled with caution (Wu et al., 1997). The over-expression of SA in transgenic tobacco plants by bacterial transgenes has shown to enhance PR-protein production and provide resistance to fungal pathogens (Verberne et al., 2000). There are many reports on the role of ethylene and jasmonic acid as signaling molecules in response of plants to necrotrophic pathogen. They work either independently or show antagonism to SA-mediated responses (Thomma et al., 1999). They are also suspected to be involved in mediating defense response signaling in the gene-for gene interaction (Dong, 1998). Mutant or transformed plants that are non responsive to either jasmonate or ethylene were found to be more susceptible to infection by root- and foliar infecting fungi confirming a role for these signal molecules in certain host–pathogen interactions (Thomma et al., 1999). However transgenic studies giving unambiguous answer to disease resistance by these two molecules are yet to be reported.

4.1.2. Bacterial Resistant Crops

Cloning and expression of antibacterial proteins of non-plant origin: Many genes of non-plant origin encoding antibacterial proteins have been cloned and expressed in plants with success to confer resistance to bacterial diseases. It includes genes for lytic peptides which are small proteins with an amphipathic α-helical structure which form pores in bacterial membranes (Broekaert et al., 1997). Transgenic tobacco plant expressing gene for lytic peptide cercopins and their synthetic analogues showed enhanced resistance to *R. solanacearum* which causes bacterial wilt (Jaynes et al., 1993) and *P. syringae* pv. *tabaci* which causes tobacco wild fire (Huang, 1997). Transgenic apple plant expressing another class of lytic peptide attacin showed a reduced susceptibility to *Erwinia amylovora*, the causal agent of fire blight (Norelli et al., 1994). Transgenic popular trees expressing the gene coding for D4E1, a synthetic antimicrobial peptide consisting of just 17 amino acid residues showed reduction in symptoms caused by *A. tumefaciens*, *Xanthomonas populi* pv. *populi* (Mentag et al., 2003). Other antibacterial proteins used with success include genes for lysozyme and lactoferrin. Lysozymes are a universally distributed family of enzymes that are found in many tissues and secretions of humans, animals, plants and as well as in bacteria and bacteriophages. The lysozyme chews the murein layer of bacterial peptidoglycan ultimately resulting in the lysis of cell wall of both Gram-negative and Gram-positive bacteria. Different lysozyme genes which were cloned and transferred to plants to confer bacterial resistance include Hen egg-white lysozyme (HEWL), T4 lysozyme (T4L), T7 lysozyme and human and bovine lysozyme genes (Nakajima et al., 1994, Mourgues et al., 1998; Tripathi et al., 2004). Human Lactoferrin is an iron-binding glycoprotein known to have antibacterial properties. It was originally isolated from milk where it is found in high concentrations. It is normally detected in other exocrine secretions exposed to normal flora. Expression of full-length human lactoferrin (HLF), as well as the N-lobe, the N-terminal half molecule (HLFN), in rice plants conferred enhanced resistance against bacterial seedling blight disease caused by *Burkholderia plantarii* (Takase et al., 2005).

Exploiting mechanism by the plant inhibiting bacterial pathogenicity or virulence factors: The compounds which are currently harnessed are toxins, pectic enzymes, exopolysaccharides and hormones. Using toxin resistance genes has been successfully employed to build resistance against *P. syringae* pv. *tabaci*. This bacterium causes wildfire of tobacco by producing pathogenic toxin called, tabtoxin. Transgenic tobacco plant expressing a tabtoxin resistance gene (*ttr*) coding for an acetyltransferase isolated from *P. syringae* pv. *tabaci* under the control of CaMV 35S promoter showed no chlorotic symptoms caused by tabtoxin treatment or with infection by *P. syringae* pv. *tabaci* (Anzai et al., 1989). The seed borne bacterial pathogen *P. syringae* pv. *phaseolicola* also causes the halo blight disease of common beans (*Phaseolus vulgaris*). This pathogen causes chlorosis in the infected plant by the production of a phytotoxin called phaseolotoxin. Octicide, the active component of phaseolotoxin binds strongly to the enzyme ornithine carbamoyltransferase (OCT), and inhibit it irreversibly. OCT plays a critical role in urea cycle by converting ornithine and carbamoyl phosphate to citrulline. The event of OCT inhibition leads to a buildup of ornithine and a shortage of arginine, leading to chlorosis and eventual death of the tissue. Virulent strains of *P. syringae* pv. *phaseolicola* posses two sets of OCT genes, *argF* and *argK* genes coding for phaseolotoxin-resistant (ROCT) ornithine carbamoyltransferase and phaseolotoxin-sensitive carbamoyltransferase (SOCT) (Mosqueda et al., 1990). ROCT is orthologous to SOCT, but it has changes in key conserved amino acids around the carbamoyl phosphate and the ornithine binding sites. These amino acid changes decrease the binding affinity of octicidine to ROCT, resulting in a phytotoxin-resistant enzyme (Templeton et al., 2005). Transgenic plants transformed with *argK* gene coding for ROCT showed 100% resistance to phaseotoxin (Hatziloukas and Panopoulos, 1992).

Enhancing natural plant defenses: Crop plant's defense against bacterial pathogen has been successfully reinforced by encouraging natural plant defenses. The molecular steps which are exploited for the purpose are production of elicitors, expression of resistance (*R*) genes, enhanced production of reactive oxygen species (ROS) and expression of plant defense genes. The gene for plant defense elicitor cryptogein (*Crypt*) was obtained from the pathogen *P. cryptogea* by PCR amplificaton and was transformed into tobacco under the control of promoter of the rice phenylalanine ammonia-lyase (PAL) gene. Transgenic tobacco lines showed significantly enhanced resistance against wild fire bacteria, *P. syringae* pv. *tabaci* (Donghua, et el., 2004). The most cited and successful example of engineering *R* genes in crop plants for bacterial resistance is definitely the *Xa21* gene from rice (Song et al., 1995), which confers resistance to all races of *Xanthomonas oryzae* pv. *oryzae*. The gene *Xa21* isolated from a resistant line has been introduced into a wide range of susceptible rice varieties and was repeatedly reported to provide resistance under laboratory conditions as well as in the field (Tu et al., 2000; Zhai et al., 2000; Zhai et al., 2001). The resistance (*R*) gene *Bs2* of pepper is a member of the nucleotide binding site–leucine-rich repeat (NBS-LRR) class of *R* genes. It exclusively recognizes and confers resistance to strains of the bacteria *Xanthomonas campestris* pv. *vesicatoria* that harbour the corresponding bacterial avirulence gene, *avrBs2*. This bacterium causes bacterial spot disease in tomato. Transgenic tomato plants expressing the pepper *Bs2* gene are found to efficiently suppress the growth of *X. campestris* pv. *vesicatoria* (Tai et al., 1999).

An important early response associated with plant defense mechanism is a rapid, transient but massive generation of different reactive oxygen species (ROS), which includes the superoxide anion radical (O_2^-), hydroxyl radical (OH^-) and hydrogen peroxide (H_2O_2). The accumulation of hydrogen peroxide and related ROS is involved in triggering local hypersensitive cell death, exerting direct antimicrobial activity against pathogens and participating in the reinforcement of the plant cell wall. Moreover, H_20_2 also act as a diffusible intercellular signal for the induction of defense related genes in surrounding tissues. Potato plants were transformed with *A. niger* gene for glucose oxidase (GO) which catalyses the oxidation of glucose to gluconic acid and hydrogen peroxide. This transgenic potato showed increased level of hydrogen peroxide which also resulted in a simultaneously increased resistance to *Erwinia carotovora* ssp. *carotovora* (Wu et al., 1995). The same gene in transgenic cabbage conferred resistance to *X. campestris* pv. *Campestris* (Lee, et al., 2002). Even though the production of glucose oxidase was constitutive and extracellular, an outburst in the hydrogen-peroxide level was detected only after bacterial infection. Pectinolytic enzymes of bacterial origin would have discharged the intracellular glucose, which, in turn, served as a substrate for extracellular glucose oxidase. PR proteins have been well studied as a major defense response in several plants, both in R gene-mediated resistance and in systemic acquired resistance (SAR). The over expression of heterologous plant defense genes coding for PR proteins or phytoalexins or plant antimicrobial peptides are used commonly to enhance bacterial disease resistance in transgenic plants. Transgenic tobacco expressing alpha-thionin gene from barley was found to have enhanced resistance to two pathovars of *P. syringae* (Carmona et al., 1993). Another promising approach is to control the activation of SAR, leading to plant immunity (Hunt and Ryals, 1996). Researches are now focusing on harnessing this phenomenon for developing bacterial pathogen resistant transgenic plants.

Another strategy used successfully for making plants bacterial resistant is to induce artificially the programmed cell death at the site of pathogen infection commonly referred as hypersensitive reaction (HR) (Dangl et al., 1996). HR lesions act as physical barriers that limit the spread of the pathogen within the infected tissue (Goodman and Novacky, 1994). Following artificial system of rapid-cell-death induction have been developed and used successfully for resistance against bacteria. A number of natural and induced mutants showing rapid accumulation of PR proteins, salicylic acid, and formation of HR lesions in the absence of an actual pathogen are reported and some of these mutants are termed as 'disease lesion mimic' mutants. These mutants were also found to be resistant to pathogen attack (Greenberg et al., 1994). Transgenic tobacco and tomato plants were created with chimeric bacterio-opsin (*bO*) gene under the control of a wound-inducible promoter. *bO* transgene encodes a proton pump and confers a lesion mimic phenotype to the transformed plants. Inducible expression of *bO* was reported to enhance the resistance of tobacco plants to *P. syringae* pv. *tabaci*, and *R. solani* and tomato plants to *P. syringae* pv. *tomato* (Rizhsky and Mittler, 2001). These transgenic plants also accumulated high levels of salicylic acid, which is a key chemical signal of a pathogen-induced SAR.

4.1.3. Viral Resistant Crops

Deployment of virus-resistant transgenic plants in an important strategy to implement effective and sustainable control against major virus diseases as other options of disease control is very limited in this case. Viral resistance is commonly achieved through genes and DNA/RNA sequences of viral origin (Sanford and Johnston, 1985). Recent strategies include using antiviral genes from other sources too.

Coat-protein mediated resistance: Coat proteins (CP) of a virus play important roles in various aspects in the life cycle of a virus which includes encapsidation, virus replication, dissemination, cell-to-cell and/or systemic movement (Callaway et al., 2001). It was in mid 1980s the breakthrough report appeared declaring success that the expression of a viral coat protein (CP) gene in a transgenic plant confers resistance to the donor virus (Powell-Abel et al., 1986). Since then a lot of similar reports have appeared in regular intervals on creation of virus-resistant transgenic plants (VRTPs) using CP genes targeting all major group of plant RNA viruses. One of the recent reports on using this strategy used the CP gene of Rice tungro spherical virus (RTSV). This CP gene was expressed in rice cultivar Pusa Basmati-1 which in turn showed a reduced titer of viral DNA and concomitant reduction in tungro symptoms (Ganesan et al., 2009). It was also reported that transgenic lines of *Nicotiana benthamiana* expressing CP of Cucumber mosaic virus (CMV) was generated by *Agrobacterium* mediated transformation. These transgenic lines showed persistent resistance and remained symptomless to CMV through out the life after virus challenge. These lines also showed resistance against two closely related strains of CMV (Srivastava and Raj, 2008). Despite extensive studies in past two decades, the molecular mechanisms behind CP-mediated resistance (CPMR) are not completly deciphered. Furthermore, mechanisms of CPMR are different in different viruses (Bendahmane et al., 2007).

Replication enzyme mediated resistance: This approach involves engineering virus resistance in plants by using genes encoding viral RNA dependent RNA-polymerases or replicase and has been shown to produce very effective and stable resistance (Canto and Palukaitis, 1999). Success with strategy was first reported for TMV (Golemboski et al., 1990) and later with replicases from other viruses such as CMV (Carr et al., 1994) and the Potato Virus X (Longstaff et al., 1993). Replicase gene mediated resistance is very strain-specific and operates only against very closely-related viruses, even more so than CPMR, but found to confer resistance at all stages of viral RNA life cycle (Carr and Zaitlin, 1991; Carr et al., 1992). Various cultivars of rice grown in Africa susceptible to rice yellow mottle virus (RYMV) were transformed with a transgene encoding the RNA-dependent RNA polymerase of RYMV. Transgenic rice plants showed stable resistance to RYMV for at least three generations (Pinto et al., 1999). Potato variety Desirée was transformed with complete sense Potato leafroll virus (PLRV) replicase gene under the control of CaMV 35S promoter. On infection with PLRV using infested aphids the signs of initial infection was observed but later the plants recovered showing undetectable virus levels even 40 days after infection (Ehrenfeld, 2004).

Replication-associated protein (Rep) mediated resistance to single-stranded DNA viruses: Single stranded DNA plant viruses do not code for DNA polymerase but needs an interaction between a viral replication-associated protein (Rep) and host plant's DNA polymerase. There are a number of reports on using Gemini virus Rep proteins to generate viral resistance in crops (Hanley-Bowdoin et al., 1999; Gutierrez, 2000). Transgenic tomato expressing 210 amino acids (Rep-210) from the N-terminal of Tomato yellow leaf curl

Sardinia virus (TYLCSV) Rep was generated. This transgenic tomato was found to strongly challenge the infection with TYLCSV (Lucioli et al., 2003)

Movement-protein (MP) mediated resistance: After initial entry through insect vectors the ability of plant viruses to move to adjacent cells is crucial in pathogenicity, virulence and host range. The movement of virus across the plant cells which have strong cell walls needs 'movement proteins' which are encoded by many if not all plant viruses. It is also observed that when the process of virus movement is impeded the plant acquires certain level of resistance to virus infection (Maule, 1991).

Expression of dysfunctional or mutant movement proteins have been reported to confer broader resistance to the wild-type (wt) virus by competition between wt virus-coded MP and the preformed modified MP. Transgenic tobacco plants expressing a mutant form of MP genes *BC1* of tomato mottle Gemini virus (TMoV) was generated. These were tested for virus resistance by exposure to viruliferous whiteflies carrying TMoV. These plants showed Gemini virus in inoculated leaves but was not readily detected in leaves beyond the inoculation sites in the highly resistant plants. This confirms resistance conferred by movement protein (Duan et al., 1997).

RNA interference (RNAi) technology: RNA interference (RNAi) is mechanism within living cells that is involved in regulation of gene expression. It includes both endogenously induced gene silencing effects of miRNA as well as silencing triggered by foreign dsRNA. It is also known by other names like 'post transcriptional gene silencing', 'transgene silencing' and 'quelling'. It involves formation of a double stranded ribonucleic acid (dsRNA) that particularly silences the expression of genes with sequences, which are complementary to it. The dsRNA formed inside the cell binds to an endonuclease enzyme complex called 'dicer' which cleaves it into short fragments of 20 to 25 base pairs with a few unpaired overhung bases at both ends. The product of dicer cleavage, the short dsRNA fragments is called small interfering RNAs (SiRNAs). Thus, miRNA and SiRNA share same cellular machinery as well as functional analogy. One of the two strands of each fragment (SiRNA or miRNA), is then incorporated into the RNA-induced silencing complex (RISC). The catalytic active component of the RISC complex is known as argunaute protein which acts as endonuclease and mediates SiRNA induced cleavage of the target mRNA strand. Consequently, one of the RNA strands (antiguide strand or passenger strand) is degraded while the other is selected as a guide strand which remains bound to RSIC complex. When a complementary mRNA is located by an RSIC bound guide strand it binds to it and is cleaved and degraded. The expression of the gene corresponding to the mRNA is silenced. This process is known to spread systemically throughout the organism despite an initial limited molar concentration of siRNA (Fire et al., 1998; Bagasra and Prilliman, 2004; Voinnet, 2005; Ding and Voinnet, 2007). As for as the biological significance of RNAi is concerned, it plays an important role in the immune response to viruses and other foreign genetic material and especially in plants it prevents self-propagation by transposons (Stram and Kuzntzova , 2006). *A. thaliana* is found to express multiple dicer homologs which can counter numerous types of viruses which it encounters during its life cycle (Blevins et al., 2006).

RNAi technology has been applied to develop virus-resistant plants through expression of virus derived double-stranded RNAs which in turn are processed into small interfering RNAs (siRNAs) by the host's RNA silencing machinery. Inverted repeat (IR) transgenes of Nia-protease (*Pro*) gene sequence of potato virus Y (PVY) was constructed which yield a long double-stranded (ds) RNA precursors of siRNAs. Utilization of such IR transgene constructs in transformation yielded upto 90% of transgenic tobacco plants resistant to the virus (Smith et al., 2000). One of the drawbacks of RNA-mediated resistance is that it has a long sequence and is ineffective against viruses whose sequence differs from that of the transgene by more than 10% (De Haan et al., 1992). But transgenic plants showing resistance simultaneously to four different tospoviruses (Tomato spotted wilt virus (TSWV), Groundnut ring spot virus (GRSV), Tomato chlorotic spot virus (TCSV) and Watermelon silver mottle virus (WSMoV)) was generated by fusing 150-nt fragments of viral sequences of these tospoviruses in a single small chimeric IR construct. This strategy resulted in a high frequency of multiple resistant plants (Bucher et al., 2006).

Plantibodies: The term "plantibodies" refers to be the products of transgenic plants expressing antibodies and antibody fragments *in planta*, which has many applications. In this technology, plants are used as antibody factories as they produce large amounts of clinically viable proteins using their endomembrane and secretary systems. These antibodies are purified later from plant tissue. Although the immune system linked to antibodies in animals is not naturally present in plants, affinities of selected antibodies can be high enough to disrupt essential functions of a viral protein in plants (De Jaeger et al., 2000). The first successful demonstration of plantibodies was in 1989 with transgenic tobacco expressing mouse antibody (Hiatt, 1989). Various developments in the recent past like generation of single-chain variable fragment (scFv) antibodies, the development of the phage display approach and the generation of synthetic scFv libraries have greatly improved the applicability of plantibody strategy (Ziegler and Torrance, 2002). The first successful use of plantibodies against plant virus was against Artichoke mottle crinkle virus described reduced susceptibility to Artichoke mottled crinkle virus (AMCV) by expressing scFv directed against the CP of the virus (Tavladoraki et al., 1993). Since then this approach is progressing in slow and steady pace because of the instability of many scFvs in plant cells. Recently transgenic approach to control Citrus tristeza virus (CTV) was tried successfully with plantibodies. Two different scFv constructs were developed from the monoclonal antibodies 3DF1 and 3CA5, specific against CTV p25 major coat protein. This mixture is able to detect all CTV isolates characterized so far. Mexican lime lines co-transformed with these constructs were able to show clear protective effect against CTV when aggressively challenged by graft inoculation (Cervera et al., 2010).

4.2. Insect Resistant Crops

Insect resistant transgenic crops are the most widely cultivated genetically modified crops next only to herbicide-resistant transgenic crops. Strategies that have been tried successfully to develop insect-resistant transgenic crops are briefly explained below with relevant examples.

Insecticidal proteins of Bacillus thuringiensis: *Bacillus thuringiensis,* commonly known as Bt, is a soil living, Gram-positive, aerobic, and sporulating bacterium. During the process of spore formation Bt forms crystal proteins or Cry proteins which are crystalline insecticidal δ-endotoxins. These crystalline proteins are highly selective in insecticidal action even in low concentration. (Bravo et al., 2007). When the susceptible insect larvae ingest the protein produced by Bt, the function of their digestive systems is disrupted, producing slow growth and, ultimately, death. It is not harmful to humans, other mammals, birds, fish, or beneficial insects. Due to this reason *Bt* strains and their insecticidal crystal proteins have been accepted worldwide as an eco-friendly biopesticide and have been used extensively in agriculture, horticulture, forestry, animal health and mosquito control for half a century (Schnepf, 1995).

Traditional way of using Bt as spray has lot of disadvantages like short life span of bacterium, easy and early denaturation of Cry protein by sunlight, physical removal of spray by means of wind and water and large scale escape of burrowing insects. Moreover as an ingredient of commercial sprays, Bt is relatively expensive and must be applied exactly where and when the target insects are feeding and they must consume it quickly before it disappears. These limitations founded the idea of transgenic expression of genes for appropriate Cry protein at sufficient level to kill the target larvae. Heterologous expression of Cry protein in transgenic crops offer several advantages like the level of toxin expression sufficient enough to kill the target pest, the toxin do not escape to environment thus harming only target insects and the expression of the protein can be modulated using appropriate genetic engineering techniques (Ranjekar et al., 2003).

First successful expression of *cry* genes with proven resistance to pest attack was reported with tobacco in 1987. The tobacco plants transformed with truncated *Cry1Aa* and *Cry1Ab* toxin genes were found to be resistant to the larvae of tobacco hornworm (*Manduca sexta*) but with expression level not sufficient enough to control the target pest (Vaeck et al., 1987; Jouanin et al., 1998). The lacuna was appropriately addressed by modification of cry genes with altering the codons of bacteria for usage in higher plants, removal of other aspects like putative polyA type signals and splice sites which destabilize *Bt* mRNAs in plant cells. Expression of these modified genes *cry1Ac* in cotton and *cry3Aa* in potato conferred an enhanced protection against lepidopteron and coleopteran pests respectively. This was accomplished by the researchers at Monsanto Company (USA) (Ranjekar et al., 2003). Since then many number of crops including rice, corn, cotton, groundnut, soybean, broccoli, brinjal, potato, canola, tomato, and cabbage were reported to be successfully transformed with various versions of modified *cry* genes (Shelton et al., 2002; Tabashnik et al., 2003; Romeis et al., 2006; Duan et al., 2010). In fact, Bt technology is one of the commercially successful technology which started with "Bollgard", the bollworm-resistant cotton in USA in the year 1996.

Proteinase inhibitors: Plant proteinase inhibitors (PIs) are one of the most abundant proteins that have been reported from tissues of various plants, animals and microorganisms. Especially in plants they are an important weapon against phytophagous insects and microorganisms (Koiwa et al., 1997). PIs act by inhibiting the proteases present in insect guts or secreted by microorganisms, causing a reduction in the availability of amino acids necessary for their growth and development (De Leo et al., 2002). Other physiological roles of PIs perform in plants include are, regulation of endogenous proteinase levels before and during germination of seeds which ultimately control protein turnover by proteolysis (Baumgartner and Chrispeels, 1976), protein stabilization during seed development and maturation (Domoney et al., 1995) and to protect the plant from microbial infection and insect attack following wounding (McGurl et al., 1995). Studies in soybean have revealed that they also play a role in programmed cell death (apoptosis) (Kosslak et al., 1997). Huge amount of report is available on the role of PIs as defensive compounds against pests by direct assay or by expression in transgenic crop plants. Many PIs had been studied and characterized and their genes have been successfully isolated. One of the early reports on plant genetic engineering using *PI* gene was by transferring trypsin inhibitor gene from *Vigna unguiculata*

(*CpTI*), to tobacco which conferred resistance to wide range of insect pests which include *Heliothis*, *Spodoptera* (lepidoptera), *Diabrotica*, *Anthonomnous* (coleoptera) and *Locusts* (orthoptera) (Hilder et al., 1987). PIs also have added advantage that they are non-toxic to mammals on consumption and in fact they reinforce the nutrition of transgenic plant by adding more cysteine and lysine (Ryan, 1989). Transgenic Indica cultivar of rice 'Swarna' was transformed with a potato chymotrypsin inhibitor (*pin2*) gene using *A. tumefaciens*. This transgenic rice showed enhanced levels of resistance to yellow stem borer (Ramana Rao et al., 2009).

Alpha amylase inhibitors: During storage of seeds, especially those of legumes the storage pests devour the stored products by producing α-amylolytic enzymes (α-1,4-glucanohydrolases) and transforming starch into mono- and disaccharides, which are used as sources of metabolic energy (Franco et al., 2000; Pelegrini et al., 2006). Plants counter the insect attack by producing antifeedent compounds especially α-amylase inhibitors. It prevents the growth of these insects by reducing carbohydrate digestion (Payan, 2004; Sivakumar et al., 2006; Ramalho et al., 2007). Based of their tertiary structure and substrate specificity α-amylases are divided into different families such as lectin-like, knottin-like, cereal-type, kunitz-like, γ-purothionin-like, and thaumatin-like (Franco et al., 2002). Targeting carbohydrate metabolism in insects through α-amylase inhibitors is another strategy tried successfully in many leguminous seed crops to control storage pests. Indian cowpea (*V. unguiculata*) cultivar, Pusa Komal was successfully transformed using *A. tumefaciens* with common bean (*P. vulgaris*) α-amylase inhibitor-1 (*αAI-1*) gene. The expression of the transgene was under bean phytohemagglutinin promoter control resulted in accumulation of αAI-1 in transgenic seeds. The transgenic protein showed an active inhibition of porcine α-amylase *in vitro*. These transgenic lines were also found to strongly inhibit the development of the storage pests *Callosobruchus maculatus* and *Callosobruchus chinensis* in insect bioassay. Although αAI-1 also inhibits human α-amylase, cooked peas will not have a negative impact on human energy metabolism (Solleti et al., 2008).

Lectins: Lectins are a class of carbohydrate-binding proteins found in all living organisms ranging from viruses and bacteria to plants and animals (Sharon and Lis, 1989). They exclusively recognize diverse sugar structures and mediate a variety of biological processes such as cell–cell and host–pathogen interactions, serum–glycoprotein turnover, and innate immune responses (Lis and Sharon, 1998; Vijayan and Chandra, 1999). They vary widely in size, structure, molecular organization and constitution of combining sites with respective receptors on gut epithelial cells of insects (Eisemann, 1994). Some of the lectins are toxic on ingestion to insects especially against insects belonging to coleoptera, diptera, lepidoptera and homoptera. They are toxic to insects on ingestion by virtue of binding to gut surface resulting in local leision (Carlini and Grossi-de-Sa, 2002). Chimeric gene consisting of coding sequence of garlic (*Allium sativum*) leaf lectin gene (*asal*) coding for mannose binding homodimeric protein (ASAL) with CaMV 35S promoter was constructed. The recombinant vector carrying the chimeric gene was introduced by *Agrobacterium*-mediated transformation into elite indica rice cultivars which have a proven record of susceptibility to various sap-sucking insects, *viz*., brown planthopper (BPH), green leafhopper (GLH) and whitebacked planthopper (WBPH). These transgenic plants showed stable integration and expression of the transgene and *in planta* assay showed an enhanced level of entomotoxic effects of ASAL on BPH, GLH and WBPH resulting in the diminished survival, development and fecundity of the insects (Yarasi et al., 2008).

4.3. Herbicide Tolerant Crops

Modern agriculture prefers managing weeds by using selective herbicides (herbicides that kill only a specific weed therefore allowing it to safely come into contact with a planted crop) on the crops which are naturally tolerant to that herbicide. For instance the herbicide triazine can be used on monocots to control dicotyledonous weeds as monocots are naturally resistant to monocots. Similarly the herbicide 2,4-dichlorophenoxyacetic acid (2,4-D) act only on dicot plants as they are broad-leaved and can be used as a selective herbicide in monocot crops like rice, wheat and maize (Bhat and Chopra, 2006). Developing herbicide resistant crops through traditional breeding is a cumbersome process where very little success has been reported. But with the advent of genetic engineering and increased knowledge on the mechanism of herbicide-plant interaction have opened various vistas to engineer crops to tolerate herbicides. The development of herbicide-resistant transgenic crops (HRTCs) is one of the first commercial applications of plant genetic engineering (Tsaftaris, 1996; Freyssinet and Cole, 1999). Once a transgenic crop expressing gene/s for herbicide is developed and approved for cultivation farmers can use any broad spectrum (non-selective) herbicides for weed management. These broad-spectrum herbicides will not harm the transgenic crop and only the susceptible weeds are destroyed. Commonly used broad-spectrum herbicides such as glufosinate and glyphosate are comparably biodegradable, display low levels of toxicity, and to date, weeds have shown minimal resistance to repeated application. Use of genes conferring herbicide resistance had been sourced from various sources especially bacteria and resistant lines of plants were successfully developed in crop varieties. Some of them are explained below,

Bromoxynil tolerance: Bromoxynil (3,5-dibromo-4-hydroxybenzonitrile) is a nitrile herbicide and works by inhibiting electron transport of photosystem II (PSII) of photosynthesis. It inhibits photosynthesis by binding to the QB-binding niche on the D1 protein of the photosystem II complex in chloroplast thylakoid membranes, thus blocking electron transport from QA to QB; this stalls the processes of CO_2 fixation and production of ATP and $NADPH_2$. It is selective in action and harms more dicotyledonous plants than monocotyledonous plants (Fedtke and Duke, 2004). A gene *bxn*, from the soil bacterium *Klebsiella ozaenae* was found to encode a nitrilase that converts bromoxynil to its primary metabolite 3,5-dibromo-4-hydroxybenzoic acid a non-phytotoxic product. Chimeric version of the gene was constructed with the coding region of the gene driven by light-regulated tissue-specific promoter, the ribulose bisphosphate carboxylase small subunit. Transgenic tobacco expressing this chimeric gene in leaves was found to be resistance to high levels of a commercial formulation of Bromoxynil (Stalker et al., 1988).

Glyphosate tolerance: Glyphosate (*N*-(phosphonomethyl) glycine) is a broad-spectrum, systemic, and non selective herbicide. It is absorbed through foliage and translocated to growing points and due to this it is effective only as post-emergence herbicide. It controls plants by inhibiting the enzyme 5-enolpyruvylshikimate-3-phosphate synthase (EPSPS) involved in the synthesis of the aromatic amino acids: tyrosine, tryptophan and phenylalanine. This results in reduced aromatic amino acids and deregulation of pathway which results in massive flow of carbon into the pathway, with high level of shikimic acid accumulation and its derivatives. The enzyme EPSPS, is found in plants, bacteria and fungi, but not in animals (Franz et al., 1997; Duke et al., 2003). A number of reports had been published on successful generation of glyphosate resistant transgenic crops using various mechanisms and associated genes. An *Agrobacterium* gene *CP4* was found to encode a highly efficient glyphosate-resistant form of EPSPS. Transgenic crops developed using this gene were found to be resistant to glyphosate because they continue to synthesize aromatic amino acids as glyphosate does not bind to bacterial EPSPS (Padgette et al., 1996). The bacterium *Ochrobactrum anthropi* encodes an enzyme glyphosate oxidoreductase (GOX). This enzyme degrades glyphosate to glyoxylate and aminomethylphosphonate (AMPA), two compounds that are non-toxic to plants (Pline-Srnic, 2006). This gene is also transferred to many crop plants either individually or along with other herbicide resistant genes with positive results (Green, 2009). As EPSPS is found in chloroplasts also, increasing the efficiency of the choloroplast transfer proteins (CTP) to transfer EPSPS into chloroplast has found to enhance the chloroplast concentration of glyphosate resistant EPSPS molecules. This permits the insensitive EPSPS molecule to out compete the plant's endogenous EPSPS molecules for access to glyphosate. One of the recent mechanisms exploited to detoxify glyphosate *in planta* is by acetylation. The enzyme, glyphosate acetyltransferase sourced from soil bacterium *Bacillus licheniformis* converts glyphosate into the non-toxic acetylglyphosate, before glyphosate can reach and inhibit the EPSPS enzyme. Scientists use gene shuffling technique, to optimize both the enzyme's acetylation efficiency and its specificity before incorporating the gene into plants (Castle et al., 2004).

Glufosinate tolerance: Glufosinate is a broad-spectrum herbicide which resembles glutamine, a neurotransmitter chemical used to transmit nerve impulses in the brains of animals and acts on plants by inhibiting the enzyme glutamine synthetase. During the normal processes of plant metabolism, nitrate (NO_3^-) is absorbed and converted to ammonium ions (NH_4^+). Later it reacts with glutamate to form glutamine. Glutamine is used as reactant in many other cellular reactions which are important for photosynthesis. Glufosinate disrupts the action of glutamine-synthetase which catalyses reaction between ammonium and glutamate, thereby preventing this reaction (Hoerlein, 1994). Thai pineapple (*Ananas comosus* `Phuket') was transformed by with the recombinant vector AHC25 harboring β-glucuronidase (*gus*) reporter gene and the bialaphos resistance (*bar*) gene for glufosinate tolerance. Transformed plants showed tolerance to the commercial herbicide Basta, containing glufosinate ammonium as the active component in both *in vitro* and glasshouse conditions. The transgenic plants also showed resistance in field conditions even when sprayed with Basta containing Glufosinate ammonium at a concentration upto 1400 mg^{-1} (Sripaoraya et al., 2001). Sugarcane cultivar NCo310 was transformed with the *pat* (phosphinothricin acetyltransferase) gene, which confers resistance to the herbicide Buster containing glufosinate ammonium. Various transgenic lines successfully expressing the genes were obtained and one of the line showed stable expression of *pat* gene even after three rounds of vegetative propagation with no reduction in agronomic traits (Snyman et al., 1998; Leibbrandt and Snyman, 2003).

Sulfonylurea/ imidazolinone tolerance: Sulfonylurea herbicides are a new class of broad-spectrum herbicides with the general structure, R'-SO_2NH-CONH-R. The R group attached to the sulfur atom of the sulfonyl moiety can be an aliphatic, aromatic, or heterocyclic group, whereas that attached to the terminal nitrogen atom of the urea moiety can be a substituted triazine or pyrimidine ring. They are known for good crop selectivity, and very low animal toxicity. They inhibit the acetolactate synthase (ALS) enzyme (also known as acetohydroxy acid synthase, or AHAS), thus blocking the biosynthesis of the branched-chain amino acids valine, leucine and isoleucine. This impedes the cell division and growth in plants (Brown,

1990; Hay, 1990). The gene encoding a mutant form of ALS, which is sulfonylurea herbicide-resistant was isolated. The mutant form is found to have a point mutation which changes the amino acid position 196 from proline to serine. This mutant gene of ALS under the control of CaMV 35S promoter was transformed in tobacco which showed stable expression of the gene and resistance to applied sulfonylurea herbicide chlorsulfuron (Odell et al., 1990).

4.4. Delayed Ripening of Fruits and Berries

4.4.1. FruitsFruits are edible seed vessels or receptacles developed from a mature ovary after fertilization. Fruits along with vegetables and nuts constitute a commercially important and nutritionally indispensable food commodity. Nutritionally, fruits are known for their high energy, roughage value, minerals, vitamins (B-complex, C and K in some instances), *β*-carotene (pro-vitamin A), dietary fiber and phenolics (antioxidants) (Wargovich, 2000). No other class of food can parallel fruits on the variety of aroma, taste, colors, succulence, flavor and texture offered by them. The process of fruit ripening lead to the development of a soft and edible ripe fruit with desirable quality attributes which are in fact aimed at appeasing the seed dispersing agents. It is a highly co-ordinated, genetically programmed, and an irreversible phenomenon involving a series of physiological, biochemical, and organoleptic changes that results in favorable transformations in the texture, levels of pigments (especially carotenoids and flavonoids), sugars, acids, and aromatic volatile compounds (Brady, 1987). Fruits are classified as climacteric and non-climacteric based on their respiratory pattern and ethylene biosynthesis during ripening. Climacteric fruits are capable of ripening even after they are detached from parent plant. They are usually harvested at full maturity. At the beginning of ripening process the respiration rate increase to a climacteric peak, after which it decreases. It is characterized by the production of ethylene and a significant increase in CO_2 production. On the other hand non-climacteric fruits cannot continue ripening once detached from the parent plant. During the process of ripening their respiration pattern and ethylene production is initially low and decline gradually (Gamage and Rehman, 1999). One of the challenges the producer face is the loss of their produce due to negative consequences of over-ripening: over softening and decay. Traditionally post-harvest loss of fruits is managed by early harvest of fruits, by controlling the postharvest storage atmosphere and by using appropriate varieties which show slow or late ripening of fruits. Though these measures can help to certain extend, they can't help in all cases and often result in poor quality products. This problem is addressed by suitably modifying fruits by genetic engineering to delay ripening. The strategies which are used to develop fruits with delayed ripening are regulation of ethylene production, control of ethylene perception and suppression of polygalacturonase activity.

Regulation of ethylene production: Ethylene, the only gaseous hormone in plant system plays an important role in ripening of fruits. The biosynthesis of the ethylene in plant cells begins with the conversion of sulphur containing amino acid amino acid methionine to S-adenosyl-L-methionine (SAM or Adomet). This step is catalyzed by the enzyme MetAdenosyltransferase. In the second stage the enzyme ACC synthase (ACS) converts SAM to 1-aminocyclopropane-1-carboxylic-acid (ACC). This is the rate limiting step of the ethylene biosynthesis process and regulation of this enzyme is the key for ethylene biosynthesis. In the third and final step ACC is broken to form ethylene in the presence of oxygen by the action of the enzyme ACC oxidase (ACO), formerly known as the Ethylene-forming enzyme (EFE) (Yang and Hoffman, 1984). Delayed ripening of fruits has been successfully achieved by down regulating ethylene biosynthesis. Strategies which are employed till date are, suppression of ACC synthase gene expression by antisense RNA technology (Eckes, 1992; Theologis et al., 1992), insertion of the ACC deaminase gene from the nonpathogenic soil bacterium *Pseudomonas chlororaphis*, which hydrolyses ACC to alpha-ketobutyrate and ammonia thus reducing the amount of ACC available for ethylene production (Klee et al., 1991), insertion of the SAM hydrolases from *E. coli* T3 bacteriophage which hydrolyse SAM to homoserine thus reducing the precursor metabolite for ethylene production (Good et al., 1994) and suppression of ACC oxidase gene expression by anti-sense RNA technology (Hamilton et al., 1990). ACO gene of papaya (*Carica papaya*) was isolated using PCR amplification and was ligated to CaMV 35S promoter to drive its expression in sense orientation. Papaya plants were transformed using gene gun and stable transgenic lines were obtained. A reduction in mRNA level for ACC oxidase in transgenic fruits was confirmed by Northern blot analysis. A sharp decline in ethylene and CO_2 production in transgenic plants was observed which ultimately resulted in delayed ripening rate (Lopez-Gomez et al., 2009).

Control of ethylene perception: Once ethylene is successfully produced it signals the onset of fruit ripening process. It also serves as signal molecule for many other physiological processes including seed germination, flower development, stem and root elongation, flower initiation, organ senescence and abscission, certain plant defense mechanisms, and a number of interactions with other plant hormones. The ETR1 gene coding for an ethylene receptor was isolated from *A. thaliana* (Bleecker et al., 1988). It was also observed that plants with modified ETR1 lack the ability to respond to ethylene (Schaller and Bleecker, 1995). Tomato ethylene receptor *LeETR4* is also found to play an important role in fruit ripening as its suppression resulted early ripening, with no significant change in fruit size, yield and flavour-related chemical composition (Kevany et al., 2008). Based on these observations it is suggested that delayed ripening of fruits can also be achieved by engineering the ethylene receptors of plants.

Suppression of polygalacturonase activity: Pectins are a major component of the plant cell wall and comprise one of the two major coextensive networks in which cellulose microfibrils are embedded (Carpita and Gibeaut, 1993). A wide range of enzymes are known to catalyze pectin modification and disassembly during various developmental processes particularly those that require cell separation. The best characterized enzymes are *exo*- and *endo*-polygalacturonases, pectate lyase, pectin methylesterase, and β-galactosidase (De Veau et al., 1993; Prasanna, et al., 2007). Fruit softening by breakdown of pectin is another vital event associated with fruit ripening (Fischer and Bennett, 1991). The enzyme polygalacturonase (PG) is found to play the pivotal role in the process.

The most cited example in using PG for delayed ripening is the 'Flavr Savr' transgenic tomatoes, which were developed by Calgene, Inc. and commercialized in U.S.A in 1994. Antisense expression of PG gene in the transgenic tomato caused a substantial delay in the fruit ripening. But this tomato was pulled out of market in 1997 due to public uproar against its transgenic origin.

4.4.2. Berries

Ripening of coffee beans in a coffee plant is not uniform where a few coffee beans ripen each day for many months. Depending on the variety concerned and climate prevailing, harvesting of coffee bean may even span round the year. Harvesting of coffee beans is done either by hand or machine Manual picking of coffee beans is very labor intensive as the best quality beans must be picked just after ripening. Mechanical harvesting involves stripping all berries from the branches regardless of ripeness. This should be followed by sorting of ripe and unripe berries. In any case it is obviously preferable that the beans get ripe at the same time. Scientists from University of Hawaii have successfully modified coffee strains so that they do not synthesize ethylene. These transgenic coffee plants reach the same not quite ripe stage and stop to wait for the triggering signal. They can be further ripened uniformly by spraying ethylene so that the berries can be harvested at once either mechanically or manually. Currently a Hawaii based company, Integrated Coffee Technologies Incorporated (ICTI), has patented the technology (US Patent No. 5,874,269, "Purified proteins, recombinant DNA sequences and processes for controlling the ripening of coffee plant, issued on February 23, 1999). They do not plan to begin marketing it in for few years. As of yet, nobody has tasted the GM coffee. Coffee species *Coffea arabica* and *Coffea canephora* were successfully transformed with *ACC oxidase* gene in antisense orientation and their expression was confirmed. But details on the effect of inhibition of these genes on fruit maturation and ethylene production in transgenic plants are yet to be deciphered (Ribas et al., 2006).

4.5. Nutritionally Enhanced Food Crops

A staple food is one that is eaten regularly and in such quantities as to constitute the dominant part of the diet and supply a major proportion of energy and nutrient needs. Staple food crops vary from place to place based on the topography of the land, type of soil and climate prevailing. They are usually cereal grains or starchy root vegetables that can be stored for a long time. For example, rice in the Indian sub continent and Far East, maize in Zambia and cooking bananas in Uganda. No single staple food can alone provide all the nutrients required to allow the human body to grow and be healthy. One of the viable options to tackle the problem of nutritionally poor staple crops is by enhancing their nutritional qualities through a strategy known as 'biofortification', which should translate into improved diets. Biofortification was traditionally done by mineral fertilization of field during crop cultivation and developing suitable crop varieties by conventional breeding/mutagenesis. After the advent of plant genetic engineering techniques many staple crops and many other food crops have been modified with enhanced nutrient content in quick succession (Christou and Twyman, 2004, Gomez-Galera et al., 2010).

4.5.1. Minerals

Genetic engineering of plants with traditional metabolic pathway cannot be applied to fortify the plants with organic nutrients as they are not synthesized in the plant but obtained from the immediate environment. Other approaches like enhancing the ability of plant to uptake and transport mineral nutrition to harvestable tissues and increasing the amount of bioavailable mineral accumulating in the plant have been successfully demonstrated in many cases. Majority of the effort on this track had been devoted to iron and zinc which are frequently deficient in human diets especially in developing and underdeveloped countries. Fe and Zn are essential micronutrients, with numerous cellular functions in plants and animals. (Lee et al., 2009). Ferritin is a ubiquitous iron-storage protein reported from a wide variety of animals, plants, and bacteria. It is a large protein (450 kDa) consisting of 24 homologous or heterologous subunits. It stores up to 4500 iron atoms in its central cavity. As storage protein it provide iron for the synthesis of iron proteins such as ferredoxin and cytochromes and at the same time prevent damage from free radicals produced by iron/dioxygen interactions (Theil, 1987; Andrews et al., 1992; Theil, 1990). The iron content of rice is improved using soybean ferritin gene. The entire coding region of soybean ferritin gene was ligated to rice seed storage protein glutelin promoter, *GluB-1* and the construct was transformed into *Oryza sativa* cultivar Kita-ake by *Agrobacterium*-mediated transformation. Stable accumulation of the ferritin protein in rice seed was confirmed by Western blotting analysis and its specific accumulation in the endosperm by immunologic tissue printing. The iron content of T1 seeds was found to increase threefold than the control plants (Goto et al., 1999). Nicotianamine (NA) is a chelator of heavy metals and plays a key role in metal homeostasis inside plant cell. Rice grains were also reported fortified with enhanced level of available iron using nicotianamine synthase (NAS) gene. Rice NAS gene, *OsNAS3* under the influence of CaMV 35S enhancer elements was transformed into rice plant. The transgenic plants expressing this NAS were found to accumulate more Fe and Zn. Anemic mouse fed with these plants recovered early to normal levels of hemoglobin and hematocrit early compared to controls. This confirms the bioavailability of iron in transgenic rice (Lee et al., 2009).

4.5.2. Oils

Fatty acids play many important roles in human body including cardiovascular health, inflammatory responses, and blood pressure regulation. Thus they are also another target of biofortification. The strategies pursued successfully in this context are to synthesize the very-long-chain polyunsaturated fatty acids (VLC-PUFAs) like arachidonic acid (ARA), eicosapentenoic acid (EPA) and docosahexenoic acid (DHA), which are usually obtained from fish oils (Zhu et al., 2007) and to increase the levels of the essential fatty acids like linoleic acid and α-linolenic acid. Transgenic *A. thaliana* was generated expressing three genes encoding Δ^9-elongase, Δ^8-desaturase and Δ^5-desaturase from *Isochrysis galbana*, *Euglena gracilis* and *Mortierella alpina* respectively. Expression of these transgenes reconstructed the ω3/6 Δ^8-desaturation biosynthetic pathways for the formation of C20 polyunsaturated fatty acids, ARA and EPA in vegetative tissues in an increased level with no deleterious effect on plant growth (Qi, et al., 2004). The fatty acid composition of *Brassica* seed oil has been suitably changed using genetic engineering to make them safer for human use. The transgenes which are used in *Brassica* transformation works are those coding for the enzymes which play crucial roles in the formation of the fatty acids and in the incorporation of the fatty acids into triacylglycerols (TAG). The list includes acetyl-CoA carboxylase (ACCase), β-ketoacyl-[ACP] synthase (KAS), acyl-ACP thioesterase, desaturase, elongase, and acyltransferase. Genetic engineering has succeeded in reducing the saturated fatty acid levels of *B. napus* using various genes including KAS, desaturases, and thioesterases. Expression of KAS II gene in *B. napus* (Bleibaum et al., 1993) and KAS IV gene in *A. thaliana* (Slabaugh et al., 1998) was reported with altered C16:0 levels in the respective transgenic plants. Recently, *B. napus* coexpressing *Cuphea pullcherima* KAS I and KAS IV sequences as well as a safflower D9-desaturase, was reported with reduced levels of saturated fatty acids to below 3.4%, compared with 6.0% in nontransformed *B. napus* plants (Scarth and Tang, 2006).

4.5.3. Proteins

Modification of protein in food crop is done by altering the amino acid composition of a single protein to enhance its nutritional value or by increasing the protein content in the edible part by expressing an entirely new protein (Dunwell, 1999). Altering the amino acid composition becomes mandatory because most food crops are deficient in certain essential amino acids that cannot be *de novo* synthesized by humans; for example, cereal grains such as rice and wheat are deficient in lysine, tryptophan and threonine, whereas legumes such as peas and beans are deficient in methionine and cysteine. The pea (*Pisum sativum*) storage protein legumin gene (*legA*) which is rich in lysine content is expressed successfully in rice (Sindhu Singh et al., 1997) and wheat (Stoger, et al., 2001) endosperms rectifying the amino acid deficiency in these cereals. Similarly, transgenic potato was generated by expressing *AmA1* gene from the Prince's feather (*Amaranthus hypochondriacus*), which encodes seed albumin. The transgenic potato expressing the transgene was shown to double the protein content and increase the levels of several essential amino acids (Chakraborty et al., 2000). This potato is popularly referred as 'protato' by media. There are also interesting reports of expression of various human proteins in food crops. Gene encoding the human milk protein β-casein was introduced into potato under the control of an auxin inducible promoter which showed detectable amount of the protein in leaves and tubers (Chong et al., 1997). Similarly, the human alpha-lactalbumin (Takase and Hagiwara, 1998) and lactoferrin (Salmon et al., 1998) proteins have also been expressed in transgenic tobacco.

4.5.4. Starch

Starch is by far the most important storage carbohydrate isolated from plants. Starch not only serves the need of energy in our diet but also for serving a wide range of industrial applications such as coating of textiles and paper, as a thickening or gelling agent in the food industry. The major source of starch for human use is the respective staple crop of the region but starch for industrial use is mainly obtained from corn, wheat, cassava and potato (Heyer et al., 1999). Biosynthesis of starch involves a host of enzymes which include four different starch synthases that are involved are in starch synthesis by adding glucose to the growing end and the branching enzymes that introduce α-(1,6) branches in the α-(1,4)-glucan that characterize the amylopectin portion of starch. Of the four starch synthases one of them is tightly bound to the growing starch granule, the others are soluble enzymes (Kossmann et al., 1999). Other enzymes that play essential role in starch metabolism are debranching enzymes, disproportioning enzymes, starch phosphorylases and the degradative amylases. Transgenic potato plants were generated by expressing antisense RNA so that the activity of granule bound starch synthase in stalled. This led to the production of amylose-free starch that has excellent co-polymerization capability. This modified starch can serve as excellent raw material in food industry in the production of soups and sauces (Visser et al., 1997). The antisense repression of R1-enzyme, which is involved in starch modification also results in altered starch but with reduced phosphate content and gelation capacity. This makes it a ready raw material for production of films (Lorberth et al., 1998).

4.5.5. Vitamins

40–50% of the world's population especially those hailing from developed and underdeveloped countries suffer from diseases caused by a lack of essential minerals and vitamins in their diet (Graham, 2001). As most of their populations subsist on a monotonous diet of cereal grains that lack essential vitamins and minerals, micronutrient deficiency is common and contributes significantly to the poor socioeconomic conditions prevalent in these countries (Timmer, 2003). Genetic engineering of staple food to fortify staple food with vitamins has come as another handy tool to face the problem. Golden rice is the name given to rice fortified with β-carotene by the media due the beautiful yellow color of grains. Scientists fortified rice with β-carotene, the precursor of vitamin A because milled rice contains essentially no provitamin A. Biosynthesis of β-carotene in plants starts with conversion of geranylgeranyl diphosphate, an early precursor of β-carotene into phytotene catalysed by the enzyme phytotene synthase. Phytotene is converted to lycopene the immediate precursor of β-carotene by the enzyme phytotene desaturase. Lycopene is finally transformed to β-carotene by lycopene cyclase (Burkhardt, 1997). Initial research confirmed that rice endosperm was capable of synthesizing geranylgeranyl diphosphate, but not phytoene. Introduction of two daffodil (*Narcissus pseudonarcissus*) genes (phytoene synthase and lycopene cyclase) and phytotene desaturase gene from the bacterium (*Erwinia uredovora*), under the control of a rice endosperm-specific glutelin promoter ensured the expression of the gene only in the rice endosperm. This reconstituted the entire pathway of β-carotene and enabled the rice endosperm to accumulate β-carotene, resulting in its characteristic golden color (Ye et al., 2000). Some of these golden rice lines were able to synthesize 1.6 μg of β-carotene per gram of grain, an amount that can really address the demand for vitamin A of rice consuming population. Recently, 'Golden Rice II' variety was reported with manifold improvement in β-carotene content (up to 37 $\mu g\ g^{-1}$). It has a more efficient maize homolog instead of daffodil phytoene synthase gene (Paine et al., 2005). Other crops engineered for higher vitamin A content are 'yellow potato' (Ducreux et al., 2005), 'orange cauliflower' (Lu et al., 2006), carrots (Baranska et al., 2006)) and tomatoes (Wurbs et al., 2006). Similar opportunities also exist to elevate the vitamin C and vitamin E content of plants through genetic engineering (Shintani and DellaPena, 1998; Herbers, 2003). A South African maize cultivar (M37W) is reported to be stably biofortified with three vitamins. The transgenic maize kernels contained 169-fold the normal amount of β-carotene, 6-fold the normal amount of ascorbate, and double the normal amount of folate after simultaneous modification of these three separate metabolic pathways. (Naqvi et al., 2009).

4.6. Edible Vaccines

Plant based edible vaccines are one of the latest advancement in vaccine technologies. It involves administering through oral route the plant materials of transgenic plants that express antigen proteins. These antigens can induce protective immunity against targeted human and animal diseases by specific immune responses. Whereas, traditional vaccination process involves administering either live vaccines or killed vaccines or attenuated vaccines or recombinant vaccines which in turn induce immune responses against the target pathogens without real infection. Production of plant based edible vaccine involves transforming plants with antigen genes to produce prorecombinant antigen in transgenic plants, and oral administration of transgenic plant products to induce targeted immune response. Some of the early reports which stimulated research in this direction are publication on production of monoclonal antibody from the tobacco plant in 1989 (Hiatt et al., 1989) and expression of hepatitis B surface antigen in transgenic plants in 1992 (Mason et al., 1992). Since then considerable efforts has been taken to develop edible vaccine resulting in numerous plant based vaccines which currently under clinical trials. Though no plant-based edible vaccines are currently commercially available there are reports that a secretory antibody vaccine is approved in the EU, a poultry vaccine against Newcastle disease is approved by the USDA, and a hepatitis B virus vaccine using a tobacco plant has been approved in Cuba (Kim and Yang, 2010). In spite of these reports some significant challenges are need to be overcome before vaccine crops become a boon in reality.

4.6.1. Plantigens

Plantigens (Plant Antigens) are those antigens produced from transgenic plant substances which cause human/animal production of antibodies. Additional advantages of antigenic molecules derived from plants besides those described earlier is that it can be administered orally without the need of purification treatments (Walmsley and Arntzen, 2003) and the production of vaccine components in plants can be achieved by both stable and transient expression (Mason et al., 2002). Toxins produced by *E. coli* and *Vibrio cholerae* cause acute watery diarrhea in human beings. The heat-labile enterotoxin (LT) from *E. coli* is a multimeric protein that is structurally, functionally and antigenically very similar to cholera toxin (CT) (Jobling et al., 1991). Potato plants were transformed with a synthetic gene of *E. coli* heat-labile enterotoxin B subunit (sLT-B) by *Agrobacterium*-mediated transformation. Mice that fed with these transgenic potato developed antibodies that were effective in inhibiting LT activity on mammalian cells. Mice that were fed with transformed potato also showed resistance to the pathogen. It is suggested with this that the transgenic potato could be used as a vaccine component or as a booster vaccine (Mason et al., 1998). The hepatitis B virus is one of the notorious viral pathogen infecting millions of people throughout the globe. Transgenic potato plants were generated with gene coding for the surface antigen HBsAg. ELISA analysis with monoclonal antibodies directed against human-serum-derived HBsAg revealed their presence in extracts from transgenic plant. Immunoaffinity chromatography also revealed the presence of 22 nm sized virus-like spherical particles. These particles exhibited properties similar to the subviral particles obtained from human serum and to the recombinant HBsAg which is formulated in the commercial vaccine produced in yeast cells. HBsAg in the particle form was found to be much more immunogenic than that in the form of the peptide alone. Mouse fed with plant derived HBsAg developed Anti-HBsAg which on recovery reacted with authentic HBsAg from human serum. However the drawback of the experiment is that the expression of HBsAg in transgenic potatoes is not sufficient for use as oral vaccine. (Mason et al., 1992; Domansky et al., 1995). Swine-transmissible gastroenteritis virus (TGEV) is an important viral pathogen of farm animals causing acute diarrhea of newborn piglets. *A. thaliana* plants were transformed with constructs encoding either the N-terminal or full length glycoprotein S of TGEV. It is now shown that mice immunized with leaf extracts from transgenic plants developed antibodies that reacted specifically to TGEV in ELISA, immunoprecipitated the virus-induced protein and neutralized the virus infectivity (Gomez et al., 1998).

4.6.2. Plantibodies

Antibodies are complex, multimeric proteins that play an essential role in animal immune system. A typical antibody protein (immunoglobulin) consists of two 'heavy' polypeptide chains and two identical 'light' polypeptide chains. Their primary function is to very specifically bind antigens initiating a cascade of immunological response. The highly specific antigen-antibody binding is because of six peptide loops, the complementarity determining regions (CDRs) that are hyper variable in amino acid sequence and length. Three of the CDRs are by domains at the N-terminus of the heavy chain and other three by the domains at the N-terminus of the light chain. These domains which provide the scaffold to display the CDRs are termed the variable heavy chain domain (V_H) and the variable light chain domain (V_L). The hypervariability of the CDRs is responsible for the great diversity of recognition reaction between the antibodies and antigens. Constant domain which is *per se* not required for the recognition reaction makes up rest of the antibody molecule. These domains also function in complement fixation, placental transfer and binding to immune system cell types.

There are many reports that declare successful production of intact antibodies and antibody fragments in transgenic plants have appeared in recent years. These antibodies of transgenic plant origin are called 'plantibodies'. The recombinant human monoclonal antibody (MAb) against hepatitis B virus (HBV) surface antigen (HBsAg) was expressed in tobacco suspension cultures. The cDNA for this antibody (CL4MAb) was derived from transformed human cell line TAPC301-CL4 and was transformed into tobacco suspension cells by *Agrobacterium*-mediated transformation. The transformed plants cells produced the CL4 derived monoclonal antibodies B294 and B303. Immunoassays showed that these two MAbs had anti-HBs relative affinity constants similar to the parental CL4MAb and B303 MAb interacted with cell surface HBsAgs and showed complement-dependent cytotoxicity similar to clinically used anti-HBs human immunoglobulins (HBIg). This study proves that MAbs to HBsAg can be produced in plants in an economical and safer way as an alternative to HBIg (Yano, 2004).

West Nile virus (WNV) is a neurotropic, enveloped virus possesing a single-stranded, 11-kilobase RNA genome with positive polarity. It causes severe neurological diseases, long-term morbidity and even death especially in aged people. At present, there is no therapeutic or vaccine is available for human use. Humanized murine mAb (Hu-E16) that binds to a highly conserved epitope on domain III (DIII) of WNV envelope (E) protein is the only effective and specific antiviral treatment for infection by WNV or other flaviviruses. This works by blocking viral fusion at concentrations that result in low occupancy of accessible sites on the virion. *N. benthamiana* was transformed with *Hu-E16* gene whose light chain (LC) and heavy chain (HC) coding sequence were optimized for use in plants. High level of expression of plant Hu-E16 was detected in transgenic tobacco within 8 days of infiltrate. It also showed high-affinity binding and neutralizing activity *in vitro* against WNV. A single dose of plant derived Hu-E16 was able to protect mice against WNV-induced mortality even 4 days after infection at rates that were indistinguishable from mammalian-cell-produced Hu-E16 (Lai et al., 2010)

5. Implications of Genetically Engineered Crops

It is estimated that around 8 billion people will be living on this earth by 2030 competing for the limited resources for living. There is an urgent need to take some tangible measures to meet the needs of the burgeoning population especially in the field of food security without destabilizing and destroying the natural resources. Transgenic crops along with important developments in other fields have the potential to address the need of the hour. Transgenic technology is believed to increase the production of food crops thus making sure food for every people on this earth and reduce many of the negative environmental impacts of farming with traditional crops. But transgenic crops which were initially received with a lot enthusiasm due to their obvious advantages are now under controversies because of skeptics, who argue against the cultivation of the transgenic plants (Dunwell, 1999). There are pros and cons of transgenic crops which are debated very hotly by the supporters and those who oppose.

5.1. Advantages of the Genetically Engineered Crops

Benefits to scientists: Conventional plant breeding has a lot of limitations which a plant breeder has to face. This procedure cannot muster a particular gene combination from the millions of crosses generated. Varieties resulting through this will not only possess most of the traits that the breeder selected for but also contain undesirable linked traits or, while one desirable gene is gained, another is lost because the genes of both parents are mixed together and re-assorted more or less randomly in the offspring (Gepts, 2002; Rommens, 2007; Singh, 2008). It needs repeated backcrossing to accentuate the expression of desired gene or to eliminate the unwanted effects of linked genes. Moreover handling more than one gene at a time is a very cumbersome task.

In contrast to this, genetic engineering permits genetic material to be inserted from unconventional sources using modern recombinant DNA techniques to obtain the desired agronomic trait. It is now well established that genes which are sourced from species, families and even kingdoms which could not previously be sources of genetic material for a particular species is inserted and expressed comfortably in the species of interest. This procedure also allows plant genome modification by removing or switching off particular genes of their genome. It also made possible to custom design a chimeric gene that does not exist in nature for better expression in target plant. The protection of 'designer genomes' by patent or other legal devices is also possible opening up the previously unexplored avenues for sublicensing genetic stock, and bringing in funds for further research and development.

Benefits to producers: Genes which were responsible for 'Green revolution" in 1960s have recently been successfully isolated and demonstrated to act in exactly the same way when used to transform other crop plant species (Peng et al., 1999). Thus transgenic crops contribute directly to the production and productivity of respective crops. Transgenic crops also increase the yield of the crop by reducing the loss incurred by weeds, insects, pathogens and other abiotic stresses directly and by lowering the cost incurred on plant protection chemicals. Marginal lands hitherto not cultivated because of higher salt content of soil and water, acidity of soils, low temperature can also be brought under regular cultivation of crops thus enlarging the cultivable land area. Also, a transgenic crop means a more predictable investment in terms of what the end result of a given growing season will yield. This is more important to poor farmers especially of developing countries whose primary income is derived from supplying a commercially important crop grown in monoculture. These factors can significantly increase the profit of farmer and ultimately the gross domestic product of the country. Increased production also means decreased poverty among world people.

Benefits to consumers: Fruits and vegetables engineered for delayed ripening ripen very slowly (however, they can be ripened by ethylene application according to the demands of the market) and this helps in exporting the fruits to longer distances without spoilage. Longer-shelf life of these fruits benefits the producer, the seller and the consumer. Plants enriched with nutrients decrease the incidence of malnutrition and deficiency diseases among poor people thus improving human heath in an affordable cost. Additionally, plant based vaccine technology has numerous advantages over vaccines from fermentation based production system. Plants can express multiple antigens in the same plant (Conrad and Fielder, 1994), can be administered very easily especially with children and can efficiently induce mucosal immune responses. As it is can easily induce mucosal immune responses which is not feasible with a conventional injection vaccine, the technology is quite useful for the development of vaccines against gastrointestinal infectious diseases (Kim and Yang, 2010).

Benefits to industry: As the crops are appropriately modified for increased synthesis of industrially wanted compound, the cost involved in processing of the raw material and recovering the product is considerably decreased. Some of the examples are potatoes with modified and increased starch content, wood with low lignin content, bio-plastics and therapeutic proteins (Andow and Zwahlen, 2006). Plant derived vaccines are free of toxins and other pathogens associated with traditional synthesis with fermentation or using animal cell lines. Moreover, the production cost is also very low and requires no expensive cold storage facilities for storage and transport.

Environmental benefits: Drastic reduction in the usage of plant protection chemicals means a lower environmental pollution. This means less damage to soil, water and air besides protecting the biodiversity of the ecosystem. Herbicide resistant crops also encourage the adoption of no-till farming, the modern concept in tillage in which no manipulation of field is performed to leave weeds and stubble undisturbed over winter. This conserves the precious fertile top soil my reducing soil erosion. It also conserves soil moisture, soil structure and soil humus. Additionally it reduces nitrate run-off decreasing the catastrophe of eutrophication in near by water bodies. *In situ* bioremediation of polluted soil, sediments, surface waters and aquifers can be done using appropriate transgenic crops which can sequester heavy metals from soils (Gleba et al., 1999) or detoxify the pollutant from the polluted site (Bizily et al., 2000,).

5.2. Risks Associated with Genetically Engineered Crops

Impact of herbicide resistant crops: Gene flow is the most touching risk arising from GM crops. The gene transfer process in decided by factors such as availability of wild relatives closely related to the transgenic crop and the mode of reproduction of the crop species. Generally there is an increased risk where the wild relatives are in close proximity to the areas where the transgenic crops are grown. Flow of herbicide resistance transgenes from transgenic crops to wild relatives can make wild plants into "super weeds" making them resistant to that particular herbicide. Jointed goatgrass (*Aegilops cylindrica)* is one of the most pernicious weeds related to wheat. It is observed that this weedy relative can acquire the herbicide-tolerant trait from Roundup Ready wheat, and can therefore thrive in crop fields unless applications of other herbicides are made (Rehman et al., 2010). Gene flow between crops can also cause "Volunteerism". It is reported from Canada that the presence of volunteer canola resistant to the herbicides glyphosate, imidazolinone and glufosinate The "triple-resistant" canola was developed through gene flow among the three canola varieties designed to resist each of the three herbicides, which were planted in close proximity to each other and has turned into a significant weed problem (Hall et al., 2000).

There is also possibility that transgenic plants *per se* can turn into weeds in ensuing generations. Evidences show that that hybridization among wild, weed and crop plants is already occurring in the nature. The emergence of Group B/2 herbicides (which work by inhibiting acetolactate synthase, ALS) resistant shattercane (*Sorghum bicolor*) a weedy relative of sorghum and the gene flow between maize and teosinte demonstrates the potential for crop relatives to become serious weeds (Zelaya and Owen 2004). It is well documented that repeated spray of single herbicide on a crop develops herbicide resistance in weed populations (Holt et al., 1993). The sulfonylureas and the imidazolinones are particularly responsible to the rapid evolution of resistant weeds and fourteen weed species have been reported to become resistant to sulfonylurea herbicides (Goldburg, 1992).

Impact of insect resistant crops: Crops expressing the Bt endotoxins can kill only the targeted insects but most crops have multitude of insect pests still have to be managed. Insecticides have to be still applied to control these pests not susceptible to the endotoxin expressed by the crop (Gould, 1994). Several field and laboratory tests have shown that many lepidopteran insects have developed resistance to Bt toxin on repeated feeding. This indicates that a major pest resistance problem can explode in near future in BT crops as the continuous expression of the toxin creates a strong selection pressure (Tabashnik 1994). Even in case of using diverse Bt genes in future to counter this pest resistance problem insects also can counter by developing multiple resistance or cross resistance (Alstad and Andow, 1995).

Impact of disease resistant crops: Reports are available that when recombination of viral genome occurs in transgenic plants under certain conditions, it produces a new viral strain with altered host range (Steinbrecher, 1996). Based on this, the risk with transgenic plants engineered for virus resistance is that DNA genome in virus resistant crops may recombine with the viral genome during replication of RNA plant viruses. This may result in unique viruses that are difficult to manage and may have serious unforeseen impact on biodiversity in the long run.

Impact on ecosystem and other non-target organisms: Although a crop uniformity may have certain economic advantages due to extensive cultivation of transgenic crops, it has two ecological drawbacks which were learnt from green revolution. First, a huge area planted to a single cultivar is very vulnerable to a new, matching strain of a pathogen or pest and second, the widespread use of a single cultivar leads to an irrevocable loss of genetic diversity (Heal, 2004).

Laboratory studies have revealed that larvae of monarch butterfly, *Danaus plexippus* reared on milkweed leaves dusted with Bt corn pollen ate less, grew more slowly and suffered around 44% mortality whereas the effect was less and mortality was nil in the case of larvae fed on leaves with non-Bt corn pollen (Losey et al., 1999). Bt crops also cause ecological imbalance in other ways. By keeping pest population at extremely low level they also starve other insects including natural enemies and parasites which depend on the pests for development and survival. Bt toxins can be incorporated into the soil through transgenic leaf litter, where they persist for 2-3 months, resisting degradation by binding to soil clay particles while maintaining toxin activity (Palm et al., 1996). Such Bt toxins later end up in the soil and water may have negative impacts on soil and aquatic invertebrates and nutrient cycling processes (Baumgarte and Tebbe, 2005). There are also reports that aphids are capable of sequestering the toxin from Bt crops and transferring it to its coccinellid predators, which in turn affect reproduction and longevity of the beneficial beetles. The potential of Bt toxins moving through food chains pose serious implications for natural biocontrol in agro-ecosystems (Birch et al., 1997).

Cultivation of herbicide resistant crops has increased the use of broad spectrum herbicides like bromoxynil and glyphosate. Bromoxynil is absorbed through skin and causes birth defects in laboratory animals, is toxic to fish, and may cause cancer in humans. Similarly, glyphosate has been reported to be toxic to beneficial predators such as spiders, mites, carabid and coccinellid beetles and to detritivores such as earthworms, as well as to aquatic organisms, including fish (Pimentel et al., 1989).

Health hazard: Experiments conducted in Scotland with rats by feeding them with transgenic potato harboring snowdrop lectin gene showed that rats developed internal organ damage. This is one of the often cited examples by people who are against the consumption of transgenic food. But it is counter argued that the experiment is too small to be conclusive (Ewen and Pusztai, 1999). Another noted case against transgenic crops is that attempts to transfer a brazil nut protein gene into soybean was stopped in US when early tests showed that people allergic to nuts reacted to the modified soy products (Nordlee et al., 1996). Other argument put forward against transgenic food is that the CaMV 35S promoter, which is often used to drive genes in transgenic crops, might escape the normal digestive breakdown and insert itself into a human chromosome where it could turn on genes (Kohli et al., 1999). But the cascade of events necessary for this development is very unlikely. Scientists also argue that the risks from consuming transgenic foods containing the CaMV 35S promoter are extremely remote. Moreover human beings and animals are consuming various transgenic foods containing this viral promoter with no documented effect on health.

Use of selectable marker is an integral part of delivering a gene package into a plant carrier. Toxicity or adverse effects of the product of these selectable marker genes on humans, wildlife, animals, beneficial insects, threatened and endangered species are yet to be studied. There is also apprehension that enteric bacteria and other pathogenic organisms can develop antibiotic resistance because of transfer of genes from transgenic plants. This is of great concern from medicinal point of view.

Regardless of the future direction of transgenic technology, the doubts for farmers, environmentalists, and the general public on the risks to human health and environment have to be addressed through rigorous scientific evaluation.

Table 1. Successful Transgene Expression in Plants

Transgenic plant	**Transgene / Source**	**Trait**	**Reference**
Fungal resistant crop			
Ricc	Endochitinase gene, *McCHIT1* (*Momordica charantia*)	Resistance to blast (*Magnaporthe grisea*) and sheath blight (*Rhizoctonia solani*)	Li et al., 2009.
Tobacco	Endochitinase gene, *chit*1 (*Metarhizium anisopliae*)	Resistance to damping off (*R. solani*)	Kern et al., 2010.
Indian mustard	Class I basic glucanase cDNA (*Lycopersicon esculentum*)	Resistance to Alternaria leaf spot (*Alternaria brassicae*)	Mondal et al., 2007.
Rice	Chitinase gene, *chi*11 (*Oryza sativa*) and β-1,3-glucanase genes (*Nicotiana tabacum*)	Resistance to sheath blight (*R. solani*)	Sridevi, 2008.
Oil seed rape	Thaumatin-like protein Hv-TLP8 NA (*Hordeum vulgare*)	Resistance to club root (*Plasmodiophora brassicae*)	Reiss, 2009.
Tomato	Defensin cDNA (*Capsicuum annum*)	Resistance to *Fusarium* sp. and *Phytophthora infestans*	Zainal et al., 2009.
Tomato	Thionin gene, *Thi2.1* (*Arabidopsis thaliana*)	Resistance to Furarium wilt (*Fusarium oxysporum* f. sp. *lycopersici*) and bacterial wilt (*Ralstonia solanacearum*)	Chan et al., 2005.
Rice	Endochitinase gene, *McCHIT1* (*Momordica charantia*)	Resistance to blast (*Magnaporthe grisea*) and sheath blight (*Rhizoctonia solani*)	Li et al., 2009.
Tobacco	Endochitinase gene, *chit*1 (*Metarhizium anisopliae*)	Resistance to damping off (*R. solani*).	Kern et al., 2010.
Indian mustard	Class I basic glucanase cDNA (*Lycopersicon esculentum*)	Resistance to Alternaria leaf spot (*Alternaria brassicae*)	Mondal et al., 2007.
Rice	Chitinase gene, *chi*11 (*Oryza sativa*) and β-1,3-glucanase genes (*Nicotiana tabacum*)	Resistance to sheath blight (*R. solani*).	Sridevi, 2008.
Oil seed rape	Thaumatin-like protein Hv-TLP8 cDNA (*Hordeum vulgare*)	Resistance to club root (*Plasmodiophora brassicae*)	Reiss, 2009.
Tomato	Defensin cDNA (*Capsicuum annum*)	Resistance to *Fusarium* sp. and *Phytophthora infestans*	Zainal et al., 2009.

Table 1. (continued)

Transgenic plant	Transgene / Source	Trait	Reference
Fungal resistant crop			
Tomato	Thionin gene, *Thi2.1* (*Arabidopsis thaliana*)	Resistance to Furarium wilt (*Fusarium oxysporum* f. sp. *lycopersici*) and bacterial wilt (*Ralstonia solanacearum*)	Chan et al., 2005.
Tobacco	Ribosome-inactivating protein (RIP) gene, *PhRIP I* (*Phytolacca heterotepala*)	Resistance to brown spot or leaf spot (*Alternaria alternate*) and grey mold (*Botrytis cinerea*)	Corrado et al., 2005.
Tomato	Stilbene synthase genes coding for trans-resveratrol (*Vitis vinifera*)	Resistance to late blight (*Phytopthera. Infestans*)	Thomzik et al., 1997.
Bread wheat	Resveratrol synthase genes, *vst1* and *vst2* (*Vitis vinifera*) and pinosylvin synthase gene, *pss* (*Pinus sylvestris*)	Resistance to stem rust (*Puccinia recondita* f. sp. *tritici*) and glume blotch (*Septoria nodorum*)	Serazetdinova et al., 2005.
Tomato	Polygalacturonase inhibitor protein gene, *pGIP* (*Pyrus communis*)	Resistance to grey mold (*Botrytis cinerea*)	Powell et al., 2000.
Rice	Trichothecene 3-*O*-acetyl- Transferase gene, *Tri101* (*Fusarium* sp.)	Resistance to *Fusarium* phytotoxin deoxynivalenol (DON)	Ohsato et al., 2007.
Peanut	Oxalate oxidase gene (*Hordeum vulgare*)	Resistance to Sclerotinia blight (*Sclerotinia minor*)	Livingstone et al., 2005.
Tobacco	Cryptogein gene, *Crypt* (*Phytophthora cryptogea*)	Resistance to downy mildew (*Peronospora parasitica* var. *nicotianae*)	Keller et al., 1999.
Banana	Human lysozyme gene, *HL*	Resistance to Panama wilt (*F. oxysporum*)	Pei et al., 2005.
Bacterial resistant crops			
Tobacco	Lytic peptide Cecropin B analogues (SB-37 and Shiva-1)	Resistance to bacterial wilt (*Pseudomonas solanacearum*) and tobacco wild fire (*P. syringae* pv. *tabaci*)	Jaynes et al., 1993 and Huang, 1997.
Apple	Lytic protein attacin E gene	Resistance to fire blight (*Erwinia amylovora*)	Norelli et al., 1994.
Tobacco	Human lysozyme gene, *HL*	Resistance to *P. syringae* pv. *tabaci.*	Nakajima et al., 1994.

Table 1. (continued)

Transgenic plant	Transgene / Source	Trait	Reference
Bacterial resistant crops			
Rice	Human lactoferrin gene, *HLF*	Bacterial seedling blight (*Burkholderia plantarii*)	Takase et al., 2005.
Tobacco	Tabtoxin resistance gene, *ttr* coding for an acetyltransferase (P. *syringae* pv. *tabaci*)	Resistance to *P. syringae* pv. *tabaci*	Anzai et al., 1989.
Tobacco	Cryptogein gene, *Crypt* (*P. cryptogea*)	Resistance to*P. syringae* pv. *tabaci*	Donghua, et el., 2004.
Rice	Receptor-like kinase gene, *Xa21* (*Oryza sativa*)	Resistance to bacterial leaf blight (*Xanthomonas oryzae* pv. *oryzae*).	Zhai et al., 2001.
Tomato	Resistance (*R*) gene, *Bs2* (*Piper nigrum*)	Resistance to bacterial black spot (*Xanthomonas campestris* pv. *vesicatoria)*	Tai et al., 1999.
Potato	Glucose oxidase gene, *GO* (*Aspergillus niger*)	Resistance to black-leg (*Erwinia carotovora* ssp. *carotovora*)	Wu et al., 1995.
Tobacco	Alpha-thionin gene (*Hordeum vulgare*)	Resistance to *P.* pv. *tabaci* and *P. syringae* pv. syringae	Carmona et al., 1993.
Tobacco and Tomato	Chimeric bacterio-opsin, *bO*	Resistance to *P. syringae* pv. *tabaci Rhizoctonia solani* and *P. syringae* pv. *tomato*	Rizhsky and Mittler, 2001.
Viral resistant crops			
Rice	Coat Protein (CP) gene of Rice tungro spherical virus (RTSV)	Resistance to RTSV	Ganesan et al., 2009.
Rice	Gene for RNA-dependent RNA polymerase of rice yellow mottle virus (RYMV)	Resistance to RYMV	Pinto et al., 1999.
Tomato	Replication-associated protein (Rep-210) of Tomato yellow leaf curl Sardinia virus (TYLCSV)	Resistance to TYLCSV	Lucioli et al., 2003.
Tobacco	Movement protein gene, *BC1* of tomato mottle Gemini virus (TMoV)	Resistance to TMoV	Duan et al., 1997.

Table 1. (continued)

Transgenic plant	Transgene / Source	Trait	Reference
Viral resistant crops			
Tobacco	Inverted repeat (IR) of Nia-protease (*Pro*) gene sequence of potato virus Y (PVY)	Resistance to PVY	Smith et al., 2000.
Mexican lime	Two different single-chain variable fragment (scFv) from the monoclonal antibodies 3DF1 and 3CA5 specific against Citrus tristeza virus (CTV) virus p25 major coat protein.	Resistance to CTV	Cervera et al., 2010.
Insect resistant crops			
Tobacco	Crystal protein genes, *Cry1Aa* and *Cry1Ab* (*Bacillus thuringiensis*)	Resistance to tobacco hornworm (*Manduca sexta*)	Vaeck et al., 1987.
Rice	Chymotrypsin inhibitor gene, *pin2* (*Solanum tuberosum*)	Resistance to yellow stem borer (*Scirpophaga incertulas*)	Ramana Rao et al., 2009.
Indian cowpea	α-amylase inhibitor-1 gene, *αAI-1* (*Phaseolus vulgaris*)	Resistance to storage pests *Callosobruchus maculatus* and *Callosobruchus chinensis*	Solleti et al., 2008.
Rice	Leaf lectin gene , *asal* (*Allium sativum)*	Resistance to sap-sucking pests brown planthopper (BPH), green leafhopper (GLH) and whitebacked planthopper (WBPH)	Yarasi et al., 2008.
Herbicide resistant crops			
Tobacco	Nitrilase gene, *bxn* that detoxify the herbicide bromoxynil (*Klebsiella ozaenae*)	Resistance to herbicide bromoxynil	Stalker et al., 1988.
Tobacco	Glyphosate-resistant EPSPS (5-enolpyruvyl shikimate-3-phosphate synthase) gene (*Pseudomonas fluorescens*)	Resistance to herbicide glyphosate	Dun et al., 2007.
Pineapple	Bialaphos resistance gene, *bar (Streptomyces hygroscopicus)*	Resistance to herbicide glufosinate	Sripaoraya et al., 2001.

Table 1. (continued)

Transgenic plant	Transgene / Source	Trait	Reference
Herbicide resistant crops			
Tobacco	Sulfonylurea resistant form of Acetolactate synthase (ALS) gene (*Arabidopsis thaliana*)	Resistant to Sulfonylurea herbicides	Odell et al., 1990.
Tomato	ACC (1-aminocyclopropane-1-carboxylic acid) deaminase gene (*Pseudomonas chlororaphis*)	Reduced ethylene production leading to delayed ripening	Klee et al., 1991.
Delayed ripening of fruits and berries			
Papaya	ACC (1-aminocyclopropane-1-carboxylic-acid) oxidase in gene sense orientation (*Carica papaya*)	Suppression of ACC oxidase leading to delayed ripening	Lopez-Gomez et al., 2009.
Coffee	*ACC oxidase* gene in antisense orientation	Delayed ripening of berries	Ribas et al., 2006
Nutritionally enhanced food crops			
Rice	Ferritin gene (*Glycine max*)	Enhanced iron content	Goto et al., 1999.
Rice	Nicotianamine synthase gene, *OsNAS3* (*O. sativa*)	Enhanced iron and zinc content	Lee et al., 2009.
Oil seed rape	β-ketoacyl-[ACP] synthase genes *KAS I* and *KAS II*, (*Cuphea pullcherima*) and D9-desaturase (*Carthamus tinctorius*)	Reduced levels of saturated fatty acids	Scarth and Tang, 2006.
Wheat	cDNA of legumin gene, *legA* (*Pisum sativum*)	Enhanced legumin content in grains	Stoger, et al., 2001.
Potato	Human milk protein β-casein gene	Enhanced levels of protein	Chong et al., 1997.
Potato	Albumin gene, *AmA1* (*Amaranthus hypochondriacus*)	Increased protein content with enhanced levels of several essential amino acids	Chakraborty et al., 2000.
Potato	R1-enzyme gene in antisense orientation	Reduced phosphate content in starch	Lorberth et al., 1998.
Rice	Phytoene synthase gene, *psy*, Lycopene synthese gene, *lyc*, (*Narcissus pseudonarcissus*) and Phytotene desaturase gene, *crtl* (*Erwinia uredovora)*	Enhanced levels of β-carotene (1.6 μg g^{-1}) (Golden rice)	Ye et al., 2000.

Table 1. (continued)

Transgenic plant	Transgene / Source	Trait	Reference
Nutritionally enhanced food crops			
Rice	Phytoene synthase gene, *psy*, *mays*) and Phytotene desaturase gene, *crt1* (*Erwinia uredovora)*	(*Zea* Enhanced levels of β-carotene (37 μg g^{-1}) (Golden rice II)	Paine et al., 2005.
Maize	Phytoene synthase cDNA, *psy1* (*Zea mays*), Carotene desaturase, gene, *crtI* (*Pantoea ananatis)*, dehydroascorbate reductase cDNA, *dhar* (*Oryza sativa*), and GTP cyclohydrase (GCH1) gene, *folE* (*E. coli*).	Enhanced levels of β-carotene, ascorbate and folate	Naqvi et al., 2009.
Edible vaccines			
Potato	Synthetic gene of heat-labile enterotoxin B subunit, *sLT-B*, *E. coli*	Inhibiting of heat-labile enterotoxin (LT) from *E. coli*	Mason et al., 1998
Potato	Gene coding for the surface antigen HBsAg	Mouse fed with plant derived HBsAg developed Anti-HBsAg which on recovery reacted with authentic HBsAg from human serum.	Mason et al., 1992 and Domansky et al., 1995.
Arabidopsis thaliana	N-terminal or full length glycoprotein S of Swine transmissible gastroenteritis virus TGEV	Mice immunized with leaf extracts from transgenic plants developed antibodies against TGEV	Gomez et al., 1998.
Tobacco	cDNA for the antibody (CL4MAb) from transformed human cell line (TAPC301-CL4)	Poduction of the CL4 derived monoclonal antibodies B294 and B303	Yano, 2004.
Tobacco	Humanized murine mAb gene, *Hu-E16* (coding sequence of light chain (LC) and heavy chain (HC) optimized for plants)	A single dose of plant derived Hu-E16 protects mice against West Nile virus (WNV) induced mortality	Lai et al., 2010.

REFERENCES

Abbo, S., Lev-Yadun. S. & Ladizinsky. G. (2001). Tracing the wild genetic stocks of crop plants, *Genome, 44,* 309-310.

Adams, K. L. & Wendel, J. F. (2005). Polyploidy and genome evolution in plants. *Curr. Opin. Plant Biol.,* 8, 135-141.

Anand, A & Mysore, K. S. (2006). *Agrobacterium* biology and crown gall disease. In S. S. Gnanamanickam. (Ed.), *Plant-Associated Bacteria.* (359-384). Netherlands: Springer Science...

Alimohammadi. M. & Bagherieh-Najjar B. M. (2009). *Agrobacterium*-mediated transformation of plants: basic principles and influencing factors. *Afr. J. Biotechnol., 8,* 5142-5148.

Alstad, D. N. & Andow, D. A. (1995). Managing the evolution of insect resistance to transgenic plants. *Science, 268,* 1894-1896.

Andow, D. A. & Zwahlen, C. (2006). Assessing environmental risks of transgenic plants. *Ecol. Lett., 9,* 196-214.

Andrews, S. C., Arosio, P., Bottke, W., Briat, J. F., Von Darl, M., Harrison, P. M., Laulhere, J. P., Levi, S., Lobreaux, S. & Yewdall, S. J. (1992). Structure, function and evolution of ferritins. *J. Inorg. Biochem., 47,* 161-174.

Anzai, H., Yoneyama, K. & Yamaguchi, I. (1989). Transgenic tobacco resistant to a bacterial disease by the detoxification of a pathogenic toxin. *Mol. Gen. Genet., 219,* 492-494.

Bagasra, O. & Prilliman, K. R. (2004). RNA interference: the molecular immune system. *J. Mol. Histol., 35,* 545-553.

Baranska. M., Baranski, R., Schulz, H. & Nothnagel, T. (2006). Tissue-specific accumulation of carotenoids in carrot roots. *Planta, 224,* 1028-1037.

Baumgarte, S. & Tebbe, C. C. (2005). Field studies on the environmental fate of the Cry1Ab Bt-toxin produced by transgenic maize (MON810) and its effect on bacterial communities in the maize rhizosphere, *Mol. Ecol., 14,* 2539-2551.

Baumgartner. B & Chrispeels, M. J. (1976). Partial characterization of a protease inhibitor which inhibits the major endopeptidase present in the cotyledons of mung beans, *Plant Physiol., 58,* 1-6.

Bendahmane, M., Chen, I., Asurmendi, S., Bazzini, A. A., Szecsi, J. & Beachy, R. N. (2007). Coat protein-mediated resistance to TMV infection of *Nicotiana tabacum* involves multiple modes of interference by coat protein. *Virology, 366,* 107-116.

Bhat, S. R. & Chopra, V. L. (2006). Choice of technology for herbicide-resistant transgenic crops in India: examination of issues. *Curr. Sci., 91,* 435-438.

Binns, A. N. & Thomashow, M. F. (1988). Cell biology of *Agrobacterium* infection and transformation of plants. *Ann. Rev. of Microbiol., 42,* 575-606.

Birch, R. G. (1997). Plant transformation: problems and strategies for practical application. *Annu. Rev. Plant Physiol. Plant Mol. Biol., 48,* 297-326.

Bizily, S. P., Rugh, C. L. & Meagher, R.B. (2000). Phytodetoxification of hazardous organomercurials by genetically engineered plants. *Nat. Biotechnol., 18,* 213- 217.

Bleecker, A. B., Somerville, C. & Kende, H. (1988). Insensitivity to ethylene conferred by a dominant mutation in *Arabidopsis thaliana. Science, 241,* 1086-1089.

Bleibaum, JL; Genez, A; Fayet-Faber, J; McCarter, DW; Thompson. GA. Modifications in palmitic acid levels via genetic engineering of β-ketoacyl-ACP synthases. In: *Abstracts of the national plant lipid symposium*. Minneapolis, Minnesota; 1993; Poster A18.

Blevins, T., Rajeswaran, R., Shivaprasad, V. P., Beknazariants, D., Si-Ammour, A., Park, H. S., Vazquez, F., Robertson, D., Meins, F., Hohn, T. & Pooggin, M. (2006). Four plant Dicers mediate viral small RNA biogenesis and DNA virus induced silencing. *Nucleic Acids Res. 34,* 6233-6246.

Bohlmann, H. (1994). The role of thionins in plant protection. *Crit. Rev. Plant Sci., 13,* 1-16.

Bol, J. F., Linthorst, H. J. M. & Cornelissen, B. J. C. (1990). Plant pathogenesis-related proteins induced by virus infection. *Annu. Rev. Phytopathol., 28,* 113-138.

Bouzar, H. & Moore, L. W. (1987). Isolation of different *Agrobacterium* biovars from a natural oak savanna and tallgrass prairie. *Appl. Environ. Microbiol., 53,* 717-721.

Brady, C. J. (1987). Fruit ripening. *Ann. Rev. Plant Physiol., 38,* 155-178.

Bravo, A., Gill, S. S., & Soberón, M. (2007). Mode of action of *Bacillus thuringiensis* Cry and Cyt toxins and their potential for insect control. *Toxicon., 49,* 423-435.

Broekaert, W. F., Cammue, B. P. A., DeBolle, M. F. C., Thevissen, K., DeSamblanx, G. W. & Osborn, R.W. (1997). Antimicrobial peptides from plants. *Crit. Rev. Plant Sci., 16,* 297-323.

Brown, H. M. (1990). Mode of action, crop selectivity, and soil relations of the sulfonylurea herbicides, *Pestic. Sci., 29,* 263-281.

Bucher, E., Lohuis, D., van Poppel, P. M., Geerts-Dimitriadou, C., Goldbach, R. & Prins, M. (2006). Multiple virus resistance at a high frequency using a single transgene construct. *J. Gen. Virol., 87,* 3697-3701.

Burkhardt, P. K., Beyer, P., Wunn, J., Kloti, A., Armstrong, G. A., Schledz, M., Von Lintig, J. & Potrykus, I. (1997). Transgenic rice (*Oryza sativa*) endosperm expressing daffodil (*Narcissus pseudonarcissus*) phytoene synthase accumulates phytoene, a key intermediate of provitamin A biosynthesis. *Plant J., 11,* 1071-1078.

Burr, T. J. & Katz, B. H. (1984). Grapevine cuttings as potential sites of survival and means of dissemination of *Agrobacterium tumefaciens*. *Plant Dis., 68,* 976-978.

Callaway, A., Giesman-Cookmeyer, D., Gillock, E. T., Sit, T. L., & Lommel, S. A. (2001). The multifunctional capsid proteins of plant RNA viruses. *Annu. Rev. Phytopathol., 39,* 419-460.

Canto, T. & Palukaitis, P. (1999). Replicase-mediated resistance to cucumber mosaic virus does not inhibit localization and/or trafficking of the viral movement protein. *Mol. Plant Microbe Interact., 12,* 743-747.

Cao, Y. J., Wei. Q., Liao, Y., Song, H. L., Li, X., Xiang, C. B. & Kuai B. K. (2009). Ectopic overexpression of *AtHDG11* in tall fescue resulted in enhanced tolerance to drought and salt stress. *Plant Cell Rep., 28,* 579-588.

Carlini, C. R. & Grossi-de-Sa, M. F. (2002). Plant toxic proteins with insecticidal properties. A review on their potentialities as bioinsecticides. *Toxicon., 20,* 1515-1539.

Carmona, M. J., Molina, A., Fernandez, J. A., Lopez-Fando, J. J. & Garcia-Olmedo, F. (1993). Expression of the alpha-thionin gene from barley in tobacco confers enhanced resistance to bacterial pathogens. *Plant J., 3,* 457-462.

Carpita, N. C. & Gibeaut, D. M. (1993). Structural models of primary cell walls in flowering plants: consistency of molecular structure with the physical properties of the walls during growth. *Plant J., 3,* 1-30.

Carr, J. P. & Zaitlin, M. (1991). Resistance in transgenic plants expressing a nonstructional gene sequence of tobacco mosaic virus is a consequence or markedly reduced virus replication. *Mol. Plant Microbe Interact., 4,* 579-585.

Carr, J. P., Gal-On, A., Palukaitis, P. & Zaitlin, M. (1994). Replicase mediated resistance to cucumber mosaic virus in transgenic plants involves suppression of both virus replication in the inoculated leaves and long-distance movement. *Virology, 199,* 439-447.

Carr, J. P., Marsh, L. E., Lomonossoff, G. P., Sekiya, M. & Zaitlin, M. (1992). Resistance to tobacco mosaic virus induced by the 54-kDa gene sequence requires expression of the 54-kDa protein. *Mol. Plant Microbe Interact., 5,* 397-404.

Castle, L. A., Siehl, D. L., Gorton, R., Patten, P. A., Chen, Y. H., Bertain, S., Cho, H. J., Wong, N. D., Liu, D. & Lassner, M. W. (2004). Discovery and directed evolution of a glyphosate tolerance gene. *Science, 304,* 1151-1154.

Cervera, M., Esteban, O., Gil, M., Gorris, M. T., Martínez, M. C., Pena, L. & Cambra M. (2010). Transgenic expression in citrus of single-chain antibody fragments specific to Citrus tristeza virus confers virus resistance. *Transgenic Res.,* (in press).

Chakraborty, S., Chakraborty, N., Datta, A. (2000). Increased nutritive value of transgenic potato by expressing a nonallergenic seed albumin gene from *Amaranthus hypochondriacus. Proc. Natl. Acad. Sci. U.S.A., 97,* 3724-3729.

Chan, Y. L., Prasad, V., Sanjaya, Chen, K, H., Liu, P.C., Chan, M. T. & Cheng, C, P. (2005). Transgenic tomato plants expressing an *Arabidopsis* thionin (*Thi2.1*) driven by fruit-inactive promoter battle against phytopathogenic attack. *Planta, 221,* 386-393.

Chilton M. D., Drummond, M. H., Merio, D. J,, Sciaky, D., Montoya A. L., Gordon, M. P. & Nester, E. W. (1977). Stable incorporation of plasmid DNA into higher plant cells: the molecular basis of crown gall tumorigenesis, *Cell, 11,* 263-271.

Chong, D. K. X., Roberts, W., Arakawa, T., Illes, K., Bagi, G., Slattery, C. W. & Langridge, W. H. R. (1997). Expression of human milk protein β-casein in transgenic potato plants. *Transgenic Res., 6,* 289-296.

Chopra, V. L. (2000). *Plant breeding: Theory and practice* (second edition). New Delhi. IBH Pub. Co. Pvt. Ltd.

Christou. P. & Twyman. R. M. (2004). The potential of genetically enhanced plants to address food insecurity. *Nutr. Res. Rev., 17,* 23-42.

Cohen, S. N., Chang, A. C. Y., Boyer, H. W. & Helling, R. B. (1973). Construction of biologically functional bacterial plasmids *in vitro. Proc. Natl. Acad. Sci. U.S.A., 70,* 3240-3244.

Collard. B. C. & Mackill, D. J. (2008). Marker-assisted selection: an approach for precision plant breeding in the twenty-first century. *Philos. Trans. R. Soc. Lond. B Biol. Sci., 363,* 557-572.

Conrad, U. & Fielder, U. (1994). Expression of engineered antibodies in plant cell. *Plant Mol. Biol., 26,* 1023-1030.

Corrado, G., Bovi, P. D., Ciliento, R., Gaudio, L., Di Maro, A., Aceto, S., Lorito, M. & Rao, R. (2005). Inducible expression of a *Phytolacca heterotepala* ribosome-inactivating protein leads to enhanced resistance against major fungal pathogens in tobacco. *Phytopathology, 95,* 206-215.

Dale. J. W & Schantz, M. V. (2007) *From genes to genomes: concepts and applications of DNA technology* (second Edition), West Sussex, John Wiley & Sons Ltd.

Dangl, J. L., Dietrich, R. A. & Richberg, M. H. (1996). Death don't have no mercy: cell death programs in plant-microbe interactions. *Plant Cell, 8,* 1793-1807.

De Block, M., Herrera-Estrella, L., Van Montagu, M., Schell, J., Zambryski, P. (1984). Expression of foreign genes in regenerated plants and their progeny. *EMBO J., 3,* 1681-1689.

De Haan, P., Gielen, J. J., Prins, M., Wijkamp, I. G., van Schepen, A., Peters, D., van Grinsven, M. Q. & Goldbach, R. (1992). Characterization of RNA-mediated resistance to tomato spotted wilt virus in transgenic tobacco plants. *Biotechnology, 10,* 1133-1137.

De Jaeger, G., De Wilde, C., Eeckhout, D., Fiers, E. & Depicker, A. (2000). The plantibody approach: expression of antibody genes in plants to modulate plant metabolism or to obtain pathogen resistance *Plant Mol. Biol., 43,* 419-428.

De Leo, F., Volpicella, M., Licciulli, F., Liuni, S., Gallerani, R. & Ceci. L. R. (2002). Plant-PIs: a database for plant protease inhibitors and their genes. *Nucleic Acids Res., 30,* 347-348.

De Veau, E. J. I., Gross, K. C., Huber, D. J. & Watada, A. E. (1993). Degradation and solubilization of pectin by *β*-galactosidases purified from avocado mesocarp. *Physiol. Plant, 87,* 279-285.

DeCleene, M. & DeLey, J. (1976). The host range of crown gall. *Bot. Rev., 42,* 389-466.

Desjardins, A. E. (2009). From yellow rain to green wheat: 25 years of trichothecene biosynthesis research, *J. Agric. Food Chem., 57,* 4478-4484.

Ding, S. W. & Voinnet, O. (2007). Antiviral immunity directed by small RNAs. *Cell, 130,* 413-426.

Domansky, N., Ehsani, P., Salmanian, A-H. & Medvedeva. T. (1995). Organ-specific expression of hepatitis B surface antigen in potato, *Biotechnol. Lett., 17,* 863-866.

Domoney, C., Welham, T., Sidebottom, C. & Firmin, J. L. (1995). Multiple isoforms of *Pisum* trypsin inhibitors result from modification of two primary gene products. *FEBS Lett., 360,* 15-20.

Dong, X. (1998). SA, JA, ethylene and disease resistance in plants. *Curr. Opin. Plant Biol., 1,* 316-323.

Donghua, J., Zejian, G., Xujun, C., Zhiqiang, C. & Zhong, Z. (2004). Study on tobacco disease resistances mediated by the elicitor gene cryptogein from *Phytophthora cryptogea*. *Chin. J. Agr. Biotechnol., 1,* 125-131.

Duan, J. J., Lundgren, J. G., Naranjo. S. & Marvier. M. (2010). Extrapolating non-target risk of *Bt* crops from laboratory to field, *Biol. Lett., 6,* 74-77.

Duan, Y. P., Powell, C. A., Webb, S. E., Purcifull, D. E & Hiebert, E. (1997). Geminivirus resistance in transgenic tobacco expressing mutated BC1 protein, *Mol. Plant Microbe Interact.,* 10, 617-623.

Ducreux, L. J., Morris, W. L., Hedley, P. E., Shepherd. T., Davies, H. V., Millam, S. & Taylor, M. A. (2005). Metabolic engineering of high carotenoids potato tubers containing enhanced levels of beta-carotene and lutein. *J. Exp. Bot., 56,* 81-89.

Duke, SO; Baerson, SR; Rimando, AM. Herbicides: glyphosate. In: Plimmer JR, Gammon DW, Ragsdale NN,. *Encyclopedia of Agrochemicals*. New York.: John Wiley & Sons; 2003; 708-869.

Dun, B. Q., Wang, X. J., Lu, W., Zhao, Z. L., Hou, S. N., Zhang, B. M., Li, G. Y., Evans, T. C. Jr., Xu, M. Q. & Lin. M. (2007). Reconstitution of glyphosate resistance from a split

5-enolpyruvyl shikimate-3-phosphate synthase gene in *Escherichia coli* and transgenic tobacco. *App. Environ. Microbiol., 73,* 7997-8000.

Dunwell, J. M. (1999). Transgenic crops: the next generation, or an example of 2020 vision, *Ann. Bot., 84,* 269-277.

Dutton, M. V. & Evans, C. S. (1996). Oxalate production by fungi: its role in pathogenicity and ecology in the soil environment. *Can. J. Microbiol., 42,* 881-895.

Eckes. P., (1992). Inhibition of fruit ripening by antisense-RNA technology, *Angew. Chem. Int. Ed. Engl, 31,* 175-177.

Ehrenfeld, N., Romano, E., Serrano, C. & Arce-Johnson, P. (2004). Replicase mediated resistance against potato leafroll virus in potato Desirée plants. *Biol. Res., 37,* 71-82.

Eisemann, C. H., Donaldson, R. A., Pearson, R. D., Cadagon, L. C., Vuocolo, T. & Pellam, R. L. (1994). Larvicidal action of lectins on *Lucilia cuprina*; mechanism of action. *Entomol. Exp. Appl., 72,* 1-11.

Escobar. A. M. & Dandekar. A. M. (2003). *Agrobacterium tumefaciens* as an agent of disease. *Trends Plant Sci., 8,* 380-386.

Ewen S. W. B. & Pusztai, A. (1999). Effect of diets containing genetically modified potatoes expressing *Galanthus nivalis* lectin on rat small intestine, *Lancet, 354,* 1353-1354.

Farrand, S. K., Van Berkum, P. B. & Oger, P. (2003). *Agrobacterium* is a definable genus of the family Rhizobiaceae. *Int. J. Syst. Evol. Microbiol., 53,* 1681-1687.

Fedtke. C., & Duke, S. O. (2004). Herbicides. In B. Hock, & E. F. Elstner, (Eds.), *Plant Toxicology.* (247-330). New York: Marcel Dekker Inc....

Fire, A., Xu, S., Montgomery, M. K., Kostas, S., Driver, S. A. & Mello, C. C. (1998). Potent and specific genetic interference by double-stranded RNA in *Caenorhabditis elegans. Nature, 391,* 806-11.

Fischer, R. L. & Bennett, A. B. (1991). Role of cell wall hydrolases in fruit ripening. *Annu. Rev. Plant Physiol. Plant Mol. Biol. 42,* 675-703.

Flor, H. H. (1971). Current status of the gene-for-gene concept. *Annu. Rev. Phytopathol., 9,* 275-296.

Franco, O. L., Rigden, D. J., Melo, F. R. & Grossi-de-Sa, M. F. (2002). Plant α-amylase inhibitors and their interaction with insect α-amylases. Structure, function, and potential for crop protection. *Eur. J. Biochem., 269*, 397-412.

Franco, O. L., Rigden, D. J., Melo, F. R., Bloch, C. Jr., Silva, C. P. & Grossi-de-Sa, M. F. (2000). Activity of wheat α-amylase inhibitors towards bruchid α-amylases and structural explanation of observed specificities. *Eur. J. Biochem., 267,* 1466- 1473.

Franz, J. E., Mao. M. K. & Sikorski J. A. (1997). Glyphosate: a unique global herbicide. *ACS Monograph, 189,* 617-642.

Freyssinet, G.,& Cole, D. J. (1999). Herbicide tolerance in crops: A commercial reality. In A. Altman, Z. Meira Z, & I. Shamay (Eds.), *Plant biotechnology and in vitro biology in the 21st century.* (481-485). Dordrecht, Kluwer academic publishers.

Gamage, T. V., & Rehman, M. S. (1999). Post harvest handling of foods of plant origin. In M. S. Rehman, *Handbook of food preservation* (11-46). New York: Marcel Dekker Inc..

Ganesan, U., Suri, S. S., Rajasubramaniam, S., Rajam, M. V. & Dasgupta, I. (2009). Transgenic expression of coat protein gene of Rice tungro bacilliform virus in rice reduces the accumulation of viral DNA in inoculated plants. *Virus genes, 39,* 113-119.

Gepts, P. A. (2002). Comparison between crop domestication, classical plant breeding, and genetic engineering, *Crop Sci., 42,* 1780-1790.

Giesbach, R. J. (1994). Improved plant transformation through electrophoresis. *Plant Sci., 102,* 81-89.

Gleba, D., Borisjuk, N. V., Borisjuk, L. G., Kneer, R., Poulev, A., Sarzhinskaya, M., Dushenkov, S., Logendra, S., Gleba, Y. Y., Raskin, I. (1999). Use of plant roots for phytoremediation and molecular farming. *Proc. Natl. Acad. Sci. U.S.A., 96,* 5973-5977.

Goldburg, R. J. (1992). Environmental concerns with the development of herbicide-tolerant plants. *Weed Tech., 6,* 647-652.

Golemboski, D.B., Lomonossoff, G.P. & Zaitlin, M. (1990). Plants transformed with a tobacco mosaic virus nonstructural gene sequence are resistant to the virus. *Proc. Natl. Acad. Sci. U.S.A., 87,* 6311-6315.

Gomez, N., Carrillo, C., Salinas, J., Parra, F., Borca, M. V. & Escribano, J. M. (1998). Expression of immunogenic S polypeptides from transmissible gastroenteritis coronovirus in transgenic plants. *Virology, 249,* 352-358.

Gomez-Galera, S., Rojas, E., Sudhakar, D., Zhu, C., Pelacho, A. M., Capell, T. & Christou, P. (2010). Critical evaluation of strategies for mineral fortification of staple food crops, *Transgenic Res., 19,* 165-180.

Good, X., Kellogg, A. J., Wagoner, Langhoff, D., Matsumura. W & Bestwick, R. K. (1994). Reduced ethylene synthesis by transgenic tomatoes expressing S-adenosylmethionine hydrolases. *Plant Mol. Biol., 26,* 781-790.

Goodman, R. N. & Novacky, A. J. (1994). *The hypersensitive response reaction in plants to pathogens: A resistance phenomenon.* St. Paul, Minnesota, N.American phytopathological society press.

Goto, F., Yoshihara, T., Shigemoto, N., Toki, S. & Takaiwa, F. (1999). Iron fortification of rice seed by the soybean ferritin gene, *Nat. Biotechnol., 17,* 282-286.

Gould, F. (1994). Potential and problems with high-dose strategies for pesticidal engineered crops. *Biocontrol Sci. Technol., 4,* 451-461.

Grayer, R. J. & Kokubun, T. (2001). Plant fungal interactions: the search for phytoalexins and other antifungal compounds from higher plants. *Phytochemistry, 56,* 253-263.

Green, J. M. (2009). Evolution of glyphosate-resistant crop technology. *Weed Sci., 57,* 108-117.

Greenberg, J. T., Guo, A., Lessig, D. F., & Ausubel, F. M. (1994). Programmed cell death in plants: a pathogen-triggered response activated coordinately with multiple defense functions. *Cell, 77,* 551-564.

Gutierrez, C. (2000). Geminiviruses and the plant cell cycle. *Plant Mol. Biol., 43,* 763-772.

Hall, L., Topinka. K., Huffman, J., Davis. L. & Good. A. (2000). Pollen flow between herbicide-tolerant *Brassica napus* is the cause of multiple-resistant *B. napus* volunteers. *Weed Sci., 48,* 688-694.

Hamilton C. M., Frary, A., Lewis, C. & Tanksley, S. D. (1996). Stable transfer of intact high molecular weight DNA into plant chromosomes. *Proc. Natl. Acad. Sci. U.S.A., 93,* 9975-9979.

Hamilton, A. J., Lycett, G. W., & Grierson, D. (1990). Antisense gene that inhibits synthesis of the hormone ethylene in transgenic plants. *Nature, 346,* 284-287.

Hammond-Kosack, K. E. & Jones, J. D. G. (1997). Plant disease resistance genes. *Annu. Rev. Plant Physiol. Plant Mol. Biol., 48,* 575-607.

Hanley-Bowdoin, L., Settlage, S. B., Orozco, B. M., Nagar, S. & Robertson, D. (2000). Geminiviruses: models for plant DNA replication, transcription, and cell cycle regulation. *Crit. Rev. Biochem. Mol. Biol., 35,* 105-140.

Hansen, G. & Wright, M. S. (1999). Recent advances in the transformation of plants. *Trends Plant Sci., 4,* 226-231

Hatziloukas, E., & Panopoulos, N. J. (1992). Origin, structure, and regulation of *argK*, encoding the phaseolotoxin-resistant ornithine carbamoyltransferase in *Pseudomonas syringae* pv. *phaseolicola*, and functional expression of *argK* in transgenic tobacco. *J. Bacteriol., 174,* 5895-5909.

Hay, J. V. (1990). Chemistry of sulfonylurea herbicides. *Pestic. Sci., 29,* 247-261.

Heal, G. M., Walker, B., Levin, S., Arrow, K., Dasgupta, P., Daily, G., Ehrlich, P., Maler, K-G, Kautsky, N., Lubchenco, J., Schneider, S. & Starrett, D. (2004). Genetic diversity and interdependent crop choice in agriculture. *Resour. Energy Econ., 26,* 175-184.

Heath, M. C. (1998). Apoptosis, programmed cell death and the hypersensitive response. *Eur. J. Plant Pathol., 104,* 117-124.

Herbers, K. (2003). Vitamin production in transgenic plants, *J. Plant Physiol., 160,* 821-829.

Heyer, A. G., Lloyd, J. R. & Kossmann, J. (1999). Production of modified polymeric carbohydrates. *Curr. opin. Biotechnol., 10,* 169-174.

Hiatt, A., Cafferkey, R. & Bowdish, K. (1989). Production of antibodies in transgenic plants. *Nature, 342,* 76-78.

Hilder, V. A., Gatehouse, A. M. R., Sheerman, S. E., Barker, R. F. & Boulter, D. A. (1987). Novel mechanism of insect resistance engineered into tobacco. *Nature, 300,* 160-163.

Hoerlein, G. (1994). Glufosinate (phosphinothricin), a natural amino acid with unexpected herbicidal properties. *Rev. Environ. Contam. Toxicol. 138,* 73-145.

Holt, J. S., Powles, S. B. & Holtum, J. A. M. (1993). Mechanisms and agronomic aspects of herbicide resistance. *Annu. Rev. Plant Physiol. Plant Mol. Biol., 44,* 203-229.

Horsch, R. B., Fraley, R. T., Rogers, S. G., Sanders, P. R., Lloyd. A. & Hoffmann. N. (1984). Inheritance of functional foreign genes in plants. *Science, 223,* 496-498.

Houlb, E. B. (2001). The arms race is ancient history in *Arabidopsis*, the wildflower. *Nat., Rev. Genet.*, 2, 516-527.

Huang, Y., Nordeen, R. O., Di, M., Owens, L. D. & McBeath J. H. (1997). Expression of an engineered cecropin gene cassette in transgenic tobacco plants confers disease resistance to *Pseudomonas syringae* pv. *tabaci*. *Phytopathology, 87,* 494-499.

Hulbert, S. H., Webb, C. A., Smith, S. M. & Sun, Q. (2001). Resistance gene complexes: evolution and utilization. *Annu. Rev. Phytopathol., 39,* 285-312.

Hunt, M. D., & Ryals, J. A. (1996). Systemic acquired resistance signal transduction. *Crit. Rev. Plant Sci., 15,* 583-606.

Jaynes, J. M., Nagpala, P., Destefano-Beltran, L., Huang, J. H., Kim, J. H., Denny, T. & Cetiner, S. (1993). Expression of a cecropin B lytic peptide analog in transgenic tobacco confers enhanced resistance to bacterial wilt caused by *Pseudomonas solanacearum*. *Plant Sci.,* 89, 43-53.

Jobling M. G. & Holmes, R. K. (1991). Analysis of structure and function of the B subunit of cholera toxin by the use of site-directed mutagenesis. *Mol. Microbiol., 5,* 1755-1767.

Jouanin, L., Bonade-Bottin, M., Girard, C., Morrot, G. & Giband, M. (1998). Transgenic plants for insect resistance. *Plant Sci., 131,* 1-11.

Karesch, H., Bilang, R., Scheid, O. M., & Potrykus, I. (1991). Direct gene transfer to protoplasts of *Arabidopsis thaliana*. *Plant Cell Rep., 9*, 575-578.

Keller, H., Pamboukdjian, N., Ponchet, M., Poupet, A., Delon, R., Verrier, J. L., Roby, D. & Ricci, P. (1999). Pathogen-induced elicitin production in transgenic tobacco generates a hypersensitive response and nonspecific disease resistance. *Plant Cell, 11*, 223-235.

Kern, M. F., Maraschin, S de F., Vom Endt, D., Schrank, A., Vainstein, M. H. & Pasquali, G. (2010). Expression of a chitinase gene from *Metarhizium anisopliae* in tobacco plants confers resistance against *Rhizoctonia solani*. *Appl. Biochem. Biotechnol., 160*, 1933-1946.

Kevany, B. M., Taylor, M. G. & Klee, H. J. (2008). Fruit-specific suppression of the ethylene receptor *LeETR4* results in early-ripening tomato fruit. *Plant Biotechnol. J., 6*, 295-300.

Khan, E. U. & Liu J. H. (2009). Plant biotechnological approaches for the production and commercialization of transgenic crops. *Biotechnol. Biotechnol. Equip., 23*, 1281-1288.

Kikkert, J. (1993). The Biolistic PDS-1000/He device. *Plant Cell Tiss. Org. Cult., 33*, 221-226.

Kikot, G. E., Hours, R. A. & Alconada, T. M. (2009). Contribution of cell wall degrading enzymes to pathogenesis of *Fusarium graminearum*: a review, *J. Basic Microbiol., 49*, 231-241.

Kim, T. G. & Yang, M. S. (2010). Current trends in edible vaccine development using transgenic plants, *Biotechnol. Bioprocess Eng., 15*, 61-65.

Kitajima, S. & Sato, F. (1999). Plant pathogenesis-related proteins: molecular mechanisms of gene expression and protein function. *J. Biochem., 125*, 1-8.

Klee, H. J., Hayford, M. B., Kretzmer, K. A., Barry, G. F., & Kishore, G. M. (1991). Control of ethylene synthesis by expression of a bacterial enzyme in transgenic tomato plants. *The Plant Cell, 3*, 1187-1193.

Klein, R. M. & Link, G. K. K. (1955). The etiology of crown gall. *Q. Rev. Biol., 30*, 207-277.

Kohli, A., Griffiths, S., Palacios, N., Twyman, R. M., Vain, P., Laurie, D.A. & Christou, P. (1999). Molecular characterization of transforming plasmid rearrangements in transgenic rice reveals a recombination hotspot in the CaMV 35S promoter and confirms the predominance of microhomology mediated recombination. *Plant J., 17*, 591-601.

Koiwa, H., Bressan, R. A. & Hasegawa. P. M. (1997). Regulation of protease inhibitors and plant defense. *Trends Plant Sci.. 2*, 379-384.

Kosslak, R. M., Chamberlin, M. A., Palmer, R. G. & Bowen, B. A. (1997). Programmed cell death in the root cortex of soybean root necrosis mutants. *Plant J., 11*, 729-745.

Kossmann, J., Abel, G. J., Springer, F., Lloyd, J. R. & Willmitzer. L. (1999). Cloning and functional analysis of a cDNA encoding a starch synthase from potato (*Solanum tuberosum* L,) that is predominantly expressed in leaf tissue. *Planta, 208*, 503-511.

Lai, H., Engle, M., Fuchs, A., Keller, T., Johnson, S., Sergey Gorlatov, S., Diamond, M. S. & Chen, Q. (2010). Monoclonal antibody produced in plants efficiently treats West Nile virus infection in mice. *Proc. Natl. Acad. Sci. U.S.A., 107*, 2419-2424.

Lay, F. T. & Anderson, M. A. (2005). Defensins-components of the innate immune system in plants. *Curr. Protein Pept. Sci., 6*, 85-101.

Lee, S., Jeon, U. S., Lee, S. J., Kim, Y. K., Persson, M., Hustedb, S., Schjorringb, J. K., Kakeic, Y., Masudac, H., Nishizawac, N. K. & An, G. (2009). Iron fortification of rice seeds through activation of the nicotianamine synthase gene. *Proc. Natl. Acad. Sci. U.S.A., 106*, 22014-22019.

Lee, Y. H., Yoon, I. S., Suh, S. C. & Kim, H. I. (2002). Enhanced resistance in transgenic cabbage and tobacco expressing a glucose oxidase gene from *Aspergillus niger*. *Plant Cell Rep., 20,* 857-863.

Leibbrandt, N. B. & Snyman, S. J. (2003). Stability of gene expression and agronomic performance of a transgenic herbicide-resistant sugarcane line in South Africa, *Crop Sci., 43,* 671-677.

LeVine, H. (2006). *Genetic engineering: A reference handbook* (second edition). Santa Barbara, California: ABC-CLIO.

Li. P., Pei. Y. Sang. X., Ling, Y., Yang, Z. & He. G. (2009). Transgenic indica rice expressing a bitter melon (*Momordica charantia*) class I chitinase gene (*McCHIT1*) confers enhanced resistance to *Magnaporthe grisea* and *Rhizoctonia solani*. *Eur. J. Plant Pathol., 125,* 533-543.

Lin, H., Jian, M., Liang, L. Y., Pei, W. J., Liu. X. Z. & Zhang, H. Y. (2010). Production of polyploids from cultured shoot tips of *Eucalyptus globulus* Labill by treatment with colchicine, *Afr. J. Biotechnol., 9,* 2252-2255.

Lioret, C. (1956). Sur la mise en evidence d'un acide amine non identifie particulier aux tissus de 'crown-gall'. *Bull. Soc. Fr. Physiol., 2,* 76.

Lis, H. & Sharon N. (1998). Lectins: carbohydrate-specific proteins that mediate cellular recognition. *Chem. Rev., 98,* 637-674.

Liu, Y. J., Cheng, C. S., Lai, S. M., Hsu, M. P., Chen, C. S. & Lyu, P. C. (2006). Solution structure of the plant defensin VrD1 from mung bean and its possible role in insecticidal activity against bruchids. *Proteins, 63,* 777-786.

Liu, Z., Du. L. & Wan. B. (2005). Pathogenesis-related proteins in higher plants. *Nat. Prod. Res. Dev., 17,* 229-234.

Livingstone, D. M., Hampton, J. L., Phipps, P. M. & Grabau, E. A. (2005). Enhancing resistance to *Sclerotinia minor* in peanut by expressing a barley oxalate oxidase gene, *Plant Physiol., 137,* 1354-1362.

Longstaff, M., Brigneti, G., Boccard, F., Chapman, S. & Baulcombe, D. (1993). Extreme resistance to potato virus X infection in plants expressing a modified component of the putative viral replicase. *EMBO J., 12,* 279-386.

Lopatin, M. I. (1939). Influence of bacterial root canker on the development of the cherry tree in the orchard. *Plant Prot., 18,* 167-173.

Lopez-Gomez, R., Cabrera-Ponce, J. L., Saucedo-Arias, L. J., Carreto-Montoya, L., Villanueva-Arce, R., Diaz-Perez, J. C., Gomez-Lim, M. A. & Herrera-Estrella, L. (2009). Ripening in papaya fruit is altered by ACC oxidase cosuppression. *Transgenic Res., 18,* 89-97.

Lorberth, R., Ritte, G., Willmitzer, L & Kossmann, J. (1998). Inhibition of a starch-granule-bound protein leads to modified starch and repression of cold sweetening. *Nat. Biotechnol., 16,* 473-477.

Losey, J. E., Raynor, L. S. & Carter. M. E. (1999). Transgenic pollen harms monarch larvae. *Nature, 399,* 214.

Lu, S., Van Eck, J., Zhou, X., Lopez, A. B., O'Halloran, D. M., Cosman, K. M., Conlin, B. J., Paolillo, D. J., Garvin, D. F., Vrebalov, J., Kochian, L. V., Küpper, H., Earle, E. D., Cao. J. & Li, L. (2006). The cauliflower *Or* gene encodes a DnaJ cysteine-rich domain-containing protein that mediates high levels of β-carotene accumulation. *Plant Cell, 18,* 3594-3605.

Lucioli, A., Noris, E., Brunetti, A., Tavazza, R., Ruzza, V., Castillo, A. G., Bejarano, E. R., Accotto, G. P. & Tavazza, M. (2003). *Tomato yellow leaf curl Sardinia virus* rep-derived resistance to homologous and heterologous geminiviruses occurs by different mechanisms and is overcome if virus-mediated transgene silencing is activated. *J. Virol.*, 6785-6798.

Manfredini, C., Sicilia, F., Ferrari, S., Pontiggia, D., Salvi, G., Caprari, C., Lorito, M. & De Lorenzo, G. (2005). Polygalacturonase-inhibiting protein 2 of *Phaseolus vulgaris* inhibits BcPG1, a polygalacturonase of *Botrytis cinerea* important for pathogenicity, and protects transgenic plants from infection. *Physiol. Mol. Plant Pathol., 67,* 108-115.

Mason, H. S., Haq, T. A., Clements, J. D. & Arntzen, C. J. (1998). Edible vaccine protects mice against *Escherichia coli* heat-labile enterotoxin (LT): potatoes expressing a synthetic LT-B gene. *Vaccine, 16,* 1336-1343.

Mason, H. S., Lam, D. M. & Arntzen, C. J. (1992). Expression of hepatitis B surface antigen in transgenic plants. *Proc. Nat. Acad. Sci. U.S.A., 89,* 11745-11749.

Mason, H. S., Warzecha, H., Mor, T. & Arntzen, J. (2002). Edible plant vaccines: applications for prophylactic and therapeutic molecular medicine. *Trends Mol. Med., 8,* 324-329.

Matzke, M. A., Matzke. A. J., Pruss, G. J. & Vance V. B. (2001). RNA-based silencing strategies in plants. *Curr. Opin. Genet. Dev., 11,* 221-227.

Maule, A. (1991). Virus movement in infected plants. *Crit. Rev. Plant Sci., 9,* 457-473.

McCabe, D. & Christou, P. (1993). Direct DNA transfer using electrical discharge particle acceleration (ACCELL™ technology). *Plant Cell Tiss. Org. Cult., 33,* 227-236.

McDowell, J. M. & Dangl, J. L. (2000). Signal transduction in the plant immune response. *Trends Biochem. Sci., 25,* 79-82.

McGurl, B., Mukherjee, S., Kahn, M. & Ryan, C. A. (1995). Characterization of two proteinase inhibitor (ATI) cDNAs from alfalfa leaves (*Medicago sativa* var. Vernema): the expression of ATI genes in response to wounding and soil microorganisms. *Plant Mol. Biol., 27,* 995-1001.

Melchers, L. S. & Stuiver, M. H. (2000). Novel genes for disease resistance breeding. *Curr. Opin. Plant Biol., 3,* 147-152.

Mentag. R, Luckevich, M., Morency, M. J. & Seguin. A. (2003). Bacterial disease resistance of transgenic hybrid poplar expressing the synthetic antimicrobial peptide D4E1, *Tree Physiol,* 23, 405-411.

Mitterbauer, R., Poppenberger, B., Raditschnig, A., Lucyshyn, D., Lemmens, M., Glossl, J. & Adam, G. (2004). Toxin-dependent utilization of engineered ribosomal protein L3 limits trichothecene resistance in transgenic plants. *Plant Biotechnol. J., 2,* 329-340.

Molina, A., Ahl Goy, P., Fraile, A., Sanchez-Monge, R. & Garcia-Olmedo, F. (1993). Inhibition of bacterial and fungal plant pathogens by thionins of types I and II. *Plant Sci., 92,* 169-177.

Mondal, K. K., Bhattacharya, R. C., Koundal, K. R. & Chatterjee, S. C. (2007). Transgenic Indian mustard (*Brassica juncea*) expressing tomato glucanase leads to arrested growth of *Alternaria brassicae. Plant Cell Rep., 26,* 247-252.

Mosqueda, G., Van den Broeck, G., Saucedo, O., Bailey, A. M., Alvarez-Morales, A. Herrera-Estrella L. (1990). Isolation and characterization of a gene from *Pseudomonas syringae* pv. *phaseolicola* encoding a phaseolotoxin-insensitive ornithine carbamoyltransferase. *Mol. Gen. Genet., 222,* 461-466.

Mourgues, F., Brisset, M. N. & Chevreau, E. (1998). Strategies to improve plant resistance to bacterial diseases through genetic engineering. *Trends Biotechnol., 16,* 203-210.

Muhitch, M. J., McCormick. S. P., Alexander, N. J. & Hohn, T.M. (2000). Transgenic expression of the *TRI 101* or *PDR 5* gene increases resistance of tobacco to the phytotoxic effects of the trichothecene 4,15-diacetoxyscirpenol. *Plant Sci., 157,* 201-207.

Muthukrishnan, S., Liang, G. H., Trick, H. N. & Gill, B. S. (2001). Pathogenesis-related proteins and their genes in cereals. *Plant Cell Tiss. Org. Cult., 64,* 93-114.

Nakajima, H., Muranaka, T., Ishige. F., Akutsu, K. & Oeda, K. (1994). Fungal and bacterial disease resistance in transgenic plants expressing human lysozyme, *Plant Cell Rep., 16,* 674-679.

Naqvi, S., Zhu, C., Farre, G., Ramessar, K., Bassie, L., Breitenbach, J., Perez Conesa, D., Ros, G., Sandmann, G., Capell, T. & Christou, P. (2009). Transgenic multivitamin corn through biofortification of endosperm with three vitamins representing three distinct metabolic pathways. *Proc. Natl. Acad. Sci. U.S.A., 106,* 7762-7767.

Nester, E. W., Gordon, M. P., Amasino, R. M. & Yanofsky, M. F. (1984). Crown gall: a molecular and physiological analysis. *Ann. Rev. Plant Physiol., 35,* 387-413.

Neuhaus. G. & Spangenberg. G. (1990). Plant transformation by microinjection techniques. *Physiol. Plant, 79,* 213-217.

Newell. C. A. (2000). Plant transformation technology: developments and applications, *Mol. Biotechnol., 16,* 53-65.

Nicholson, R. L. & Hammerschmidt, R. (1992). Phenolic compounds and their role in disease resistance. *Annu. Rev. Phytopathol., 30,* 369-389.

Nielsen, K. & Boston. R. S. (2001). Ribosome-inactivating proteins: a plant perspective. *Annu. Rev. Plant Physiol. Plant Mol. Biol., 52,* 785-816.

Nordlee, J. A., Taylor, S. L., Townsend, J. A., Thomas, L. A. & Bush, R. K. (1996). Identification of a Brazil-nut allergen in transgenic soybeans. *N. Engl. J. Med., 334,* 688-692.

Norelli, J. L., Aldwinckle, H. S., Destefano Beltran, L. & Jaynes, J. M. (1994). Transgenic 'Mailing 26' apple expressing the attacin E gene has increased resistance to *Erwinia amylovora. Euphytica, 77,* 123-128.

Odell. J. T., Caimi, P. G., Yadav, N. S. & Mauvais. J. C. (1990). Comparison of increased expression of wild-type and herbicide-resistant acetolactate synthase genes in transgenic plants, and indication of posttranscriptional limitation on enzyme activity, *Plant Physiol., 94,* 1647-1654.

Ohsato, S., Ochiai-Fukuda, T., Nishiuchi. T., Takahashi-Ando, N., Koizumi, S., Hamamoto, H., Kudo, T., Yamaguchi, I. & Kimura, M. (2007). Transgenic rice plants expressing trichothecene 3-*O*-acetyltransferase show resistance to the *Fusarium* phytotoxin deoxynivalenol, *Plant Cell Rep., 26,* 531-538.

Opabode, J. T. (2006). *Agrobacterium*-mediated transformation of plants: emerging factors that influence efficiency. *Biotechnol. Mol. Bio. Rev., 1,* 12-20.

Padgette, S. R., Re, D. B., Barry, G. F. Eichholtz, D. E., Delannay, X., Fuchs, R. L., Kishore, G. M., & Fraley, R. T. (1996). New weed control opportunities: Development of soybeans with a Roundup ready gene. In S. O. Duke (Ed.), *Herbicide resistant crops: Agricultural, environmental, economic, regulatory, and technical aspects* (53-84). Boca Raton, Florida: CRC Press,

Paine, J. A., Shipton, C. A., Chaggar, S., Howells, R. M., Kennedy. M. J., Vernon. G., Wright. S. Y., Hinchliffe, E., Adams, J. L., Silverstone, A. L. & Drake, R. (2005). Improving the nutritional value of Golden Rice through increased pro-vitamin A content. *Nat. Biotechnol.* 23, 482-487.

Palm, C. P., Schaller, D. L., Donegan, K. K., & Seidler, R. J. (1996). Persistence in soil of transgenic plants produced *Bacillus thuringiensis* var. *kurstaki*-endotoxin. *Can. J. Microbiol., 42, 1258-1262.*

Park, S. W., Vepachedu, R., Sharma, N. & Vivanco, J. M. (2004). Ribosome-inactivating proteins in plant biology, *Planta, 219,* 1093-1096.

Paszkowski, J., Shillito, R, D., Saul, M., Mandak, V., Hohn, T., Honn, B. & Potrykus, I. (1984). Direct gene transfer to plants. *EMBO J., 3,* 2717-2722.

Payan, F. (2004). Structural basis for the inhibition of mammalian and insect α-amylases by plant protein inhibitors. *Biochim. Biophys. Acta, 1696,* 171-180.

Pei, X. W., Chen, S. K., Wen, R. M., Ye, S., Huang, J. Q., Zhang, Y. Q., Wang, B. S., Wang, Z. X. & Jia, S. R. (2005). Creation of transgenic bananas expressing human lysozyme gene for panama wilt resistance, *J. Integr. Plant Biol., 47,* 971-977.

Pelegrini, P. B., Murad, A. M., Grossi-de-Sa, M. F., Mello, L. V., Romeiro, L. A. S., Noronha, E. F., Caldas, R. A. & Franco, O. L. (2006). Structure and enzyme properties of *Zabrotes subfasciatus* α-amylase. *Arch. Insect Biochem. Physiol., 61,* 77-86.

Peng, J., Richards, D. E., Hartley, N. M., Murphy, G. P., Devos, K. M., Flintham, J. E., Beales, J., Fish, L. J., Worland, A. J., Pelica, F., Sudhakar, D., Christou, P., Snape, J. W., Gale, M. D. & Harberd, N. P. (1999). 'Green revolution' genes encode mutant gibberellin response modulators. *Nature, 400,* 256-261.

Pimentel, D., Hunter, M. S., LaGro, J. A., Efroymson, R. A., Landers, J. C., Mervis, F.T., McCarthy, C.A. & Boyd, A.E. (1989). Benefits and risks of genetic engineering in agriculture. *Bioscience, 39,* 606-614.

Pinto, Y. M., Kok, R. A. & Baulcombe, D. C. (1999). Resistance to rice yellow mottle virus (RYMV) in cultivated African rice varieties containing RYMV transgenes. *Nat. Biotechnol. 17,* 702-707.

Pline-Srnic, W, A. (2006). Physiological mechanisms of glyphosate resistance. *Weed Technol., 20,* 290-300.

Powell, A. L., Van Kan. J., Ten Have, A., Visser, J., Greve, L. C., Bennett A. B. & Labavitch, J. M. (2000). Transgenic expression of pear PGIP in tomato limits fungal colonization. *Mol. Plant Microbe Interact., 13,* 942-950.

Powell-Abel, P., Nelson, R. S., De, B., Hoffmann, N., Rogers, S. G., Fraley, R. T. & Beachy, R. N. (1986). Delay of disease development in transgenic plants that express the tobacco mosaic virus coat protein gene, *Science, 232,* 738-743.

Prasanna, V., Prabha, T. N. & Tharanathan, R. N. (2007). Fruit ripening phenomena-an overview. *Crit. Rev. Food Sci. Nutr., 47,* 1-19.

Primrose, S. B. & Twyman, R. M. (2006). *Principles of gene manipulation and genomics* (seventh edition), Malden, Massachusetts, Blackwell Publishing.

Qi, B., Fraser, T., Mugford, S., Dobson, G., Sayanova, O., Butler, J., Napier, J. A., Stobart, A. K. & Lazarus, C. M. (2004). Production of very long chain polyunsaturated omega-3 and omega-6 fatty acids in plants. *Nat. Biotechnol., 22,* 739-745.

Ramalho, L. F., Costa, F. T., Souza, L. A., Pelegrini, P. B., Grossi-de-Sa, M. F., Neto, S. M., Bloch, C., Jr., Laumann, R. A., Noronha, E. F. & Franco, O. L. (2007). Isolation of a novel *Carica papaya* α-amylase inhibitor with deleterious activity toward *Callosobruchus maculatus. Pest. Biochem. Physiol., 87*, 255-260.

Ramana Rao, M. V., Behera, K. S., Baisakh, N., Datta, S. K. & Rao, G. J. N., (2009). Transgenic indica rice cultivar 'Swarna' expressing a potato chymotrypsin inhibitor *pin2* gene show enhanced levels of resistance to yellow stem borer. *Plant Cell Tiss. Organ Cult., 99,* 277-285.

Ranjekar, P. K., Patankar, A., Gupta, V., Bhatnagar. R., Bentur. J. & Ananda Kumar, P. (2003). Genetic engineering of crop plants for insect resistance, *Curr. Sci., 84,* 321-329

Rao, A. Q., Bakhsh, A., Kiani, S., Shahzad, K., Shahid, A. A., Husnain, T. & Riazuddin, S. (2009). The myth of plant transformation. *Biotechnol. Adv., 27,* 753-763.

Rehman, M., Hansen, J. L., Mallory-Smith, C. A., Hang. A, Burton, C. & Zemetra, R. S. (2010). Determining the pollen parent of field-grown backcross progenies of wheat (*Triticum aestivum* L.) × jointed goatgrass (*Aegilops cylindrica* Host) hybrids using genomic *in situ* hybridization (GISH), *Crop Sci., 50,* 1474-1479.

Reiss, E., Schubert, J., Scholze, P., Kramer, R. & Sonntag. K. (2009). The barley thaumatin-like protein Hv-TLP8 enhances resistance of oilseed rape plants to *Plasmodiophora brassicae. Plant Breeding, 128,* 210-212.

Ribas, A. F., Pereira, L. F. P. & Vieira, L. G. E. (2006). Genetic transformation of coffee. *Braz. J. Plant Physiol., 18,* 83-94.

Rizhsky. L. & Mittler. R. (2001). Inducible expression of bacterio-opsin in transgenic tobacco and tomato plants. *Plant Mol. Biol., 46,* 313-323.

Romeis, J., Meissle. M. & Bigler, F. (2006). Transgenic crops expressing *Bacillus thuringiensis* toxins and biological control. *Nat. Biotechnol., 24,* 63-71.

Rommens, C. & Kishore, G. M. (2000). Exploiting the full potential of disease-resistance genes for agricultural use. *Curr. Opin. Biotechnol., 11,* 120-125.

Rommens, C. M. (2007). Intragenic crop improvement: combining the benefits of traditional breeding and genetic engineering. *J. Agric. Food Chem., 55,* 4281-4288.

Ryan, CA. Insect-induced chemical signals regulating natural plant protection responses. In: Denno, RF., Mcclure MS,. *Variable plants and herbivores in natural and managed systems.* New York: Academic Press; 1989; 43-60.

Salmon, V., Legrand, D., Slomianny, M. C., el Yazidi, I., Spik, G., Gruber, V., Bournat, P., Olagnier, B., Mison, D., Theisen, M. & Merot, B. (1998). Production of human lactoferrin in transgenic tobacco. *Protein Expr. Purif., 13,* 127-135.

Sanford, J. C. (1988). The biolistic process-a new concept in gene transfer and biological delivery. *Trends Biotechol., 6,* 299-302.

Sanford, J. C., & Johnston, S. A. (1985). The concept of parasite derived resistance-deriving resistance genes from the parasite's own genome. *J. Theor. Biol., 113,* 395-405.

Sanford. J. C., (2000). The development of the biolistic process. *In Vitro Cell Dev. Biol., Plant, 36,* 303-308.

Scarth, R. & Tang, J. (2006). Modification of *Brassica* oil using conventional and transgenic approaches. *Crop. Sci., 46,* 1225-1236.

Schaller. G. E. & Bleecker. A. B. (1995). Ethylene-binding sites generated in yeast expressing the *Arabidopsis ETR1* gene. *Science, 270,* 1809-1811.

Schnepf. H. E. (1995). *Bacillus thuringiensis* toxins: regulation, activities and structural diversity, *Curr. Opin. Biotechnol., 6,* 305-312.

Schroth, M. N., McCain A. H., Foottt J. H. & Huisman, O. C. (1988). Reduction in yield and vigor of grapevine caused by crown gall disease. *Plant Dis., 72,* 241-246.

Serazetdinova, L., Oldach, K. H. & Lorz, H. (2005). Expression of transgenic stilbene synthases in wheat causes the accumulation of unknown stilbene derivatives with antifungal activity. *J. Plant Physiol., 162,* 985-1002.

Sharon, N. & Lis, H. (1989). Lectins as cell recognition molecules. *Science, 246,* 227-234.

Shelton, A. M., Zhao, J. Z. & Roush, R. T. (2002). Economic, ecological, food safety, and social consequences of the deployment of Bt transgenic plants. *Annu. Rev. Entomol., 47,* 845-81.

Shintani, D. & DellaPena, D. (1998). Elevating the vitamin E content of plants through metabolic engineering. *Science, 282,* 2098-2100.

Shu, Q. Y. & Lagoda, P. J. L. (2007). Mutation techniques for gene discovery and crop improvement. *Mol. Plant Breed., 5,* 193-195.

Sindhu Singh, A., Zheng, Z., & Murai, N. (1997). The pea seed storage protein legumin was synthesized, processed, and accumulated stably in transgenic rice endosperm. *Plant Sci., 130,* 189-196.

Singh. B. D. (2008). *Plant breeding: principles and methods* (eighth edition). New Delhi, Kalyani Publishers.

Sitbon, F., Hennion, S., Little C. H. A. & Sundberg, B. (1999). Enhanced ethylene production and peroxidase activity in IAA overproducing transgenic tobacco plants is associated with increased lignin content and altered lignin composition. *Plant Sci., 141,* 165-173.

Sivakumar, S., Mohan, M., Franco, O. L. & Thayumanavan, B. (2006). Inhibition of insect pest α-amylases by little and finger millet inhibitors. *Pest. Biochem. Physiol., 85,* 155-160.

Slabaugh, M. B., Leonard, J. M. & Knapp. S. J. (1998). Condensing enzymes from *Cuphea wrightii* associated with medium-chain fatty acid biosynthesis. *Plant J., 13,* 611-620.

Smith, E. F. & Townsend, C.O. (1907). A plant tumor of bacterial origin. *Science, 25,* 671-673.

Smith, E. F., Brown, N. A. & Townsend, C. O. (1911). Crown gall of plants: its cause and remedy. *USDA Bur. Plant Ind. Bull., 213,* 1-215.

Smith, N. A., Singh, S. P., Wang, M. B., Stoutjesdijk, P. A., Green, A. G. & Waterhouse P. M. (2000). Total silencing by intron-spliced hairpin RNAs. *Nature, 407,* 319-320.

Snyman, S. J., Leibbrandt, N. & Botha. F. C. (1998). Buster resistant sugarcane. *Proc. S. Afr. Sug. Technol. Ass., 72,* 138-139.

Solleti. S., Bakshi. S., Purkayastha, J., Panda, S. K. & Sahoo, L. (2008). Transgenic cowpea (*Vigna unguiculata*) seeds expressing a bean α-amylase inhibitor 1 confer resistance to storage pests, bruchid beetles. *Plant Cell Rep., 27,* 1841-1850.

Song, W. Y., Wang, G. L., Chen, L. L., Kim, H. S., Pi, L. Y., Holsten, T., Gardner, J., Wang, B., Zhai, W. X., Zhu, L. H., Fauquet, C. & Ronald, P. (1995). A receptor kinase-like protein encoded by the rice disease resistance gene, *Xa21. Science, 270,* 1804-1806.

Song. X., Gu. Y. & Qin. G. (2007). Application of a transformation method via the pollen-tube pathway in agriculture molecular breeding. *Life Sci. J., 4,* 77-79.

Sridevi, G., Parameswari, C., Sabapathi, N., Raghupathy, V. & Karuppannan, V. (2008). Combined expression of chitinase and β -1,3-glucanase genes in indica rice (*Oryza sativa* L.) enhances resistance against *Rhizoctonia solani. Plant Sci., 175,* 283-290.

Sripaoraya, S., Marchant, R., Power, J. B. & Davey, M. R. (2001). Herbicide-tolerant transgenic pineapple (*Ananas comosus*) produced by microprojectile bombardment. *Ann. Bot., 88,* 597-603.

Srivastava, A. & Raj, S. K. (2008). Coat protein-mediated resistance against an Indian isolate of the *Cucumber mosaic virus* subgroup IB in *Nicotiana benthamiana, J. Biosci.,* 33, 249-257.

Stalker, D. M., Mcbride, K. E. & Malyj, D. L. (1988). Herbicide resistance in transgenic plants expressing a bacterial detoxification gene. *Science, 242,* 419-423.

Stec, B. (2006). Plant thionins-the structural perspective. *Cell. Mol. Life Sci., 63,* 1370-1385.

Steinbrecher, R. A. (1996). From green to gene revolution: the environmental risks of genetically engineered crops. *The Ecologist, 26,* 273-282.

Stintzi, A., Heitz, T., Prasad, V., Wiedemann-Merdinoglu, S., Kauffmann, S., Geoffroy, P., Legrand, M. & Fritig, B. (1993). Plant 'pathogenesis-related' proteins and their role in defense against pathogens. *Biochemie., 75,* 687-706.

Stoger, E., Parker, M., Christou. P. & Casey, R. (2001). Pea legumin overexpressed in wheat endosperm assembles into an ordered paracrystalline matrix. *Plant Physiol., 125,* 1732-1742

Stokes, T. (2001). Gene transformation gets acupuncture. *Trends Plant Sci., 6,* 244-251.

Stram, Y. & Kuzntzova, L. (2006). Inhibition of viruses by RNA interference. *Virus Genes 32,* 299-306.

Sundar, I. K. & Sakthivel, N. (2008). Advances in selectable marker genes for plant transformation. *J. Plant Physiol., 165,* 1698-1716.

Tabashnik, B. E. (1994). Genetics of resistance to *Bacillus thuringiensis. Annu. Rev. of Entomol., 39,* 47-79.

Tabashnik, B. E., Carriere, Y., Dennehy, T. J., Morin, S., Sisterson, S. M., Roush, R. T., Shelton. A. M. & Zhao, J. Z. (2003). Insect resistance to transgenic Bt crops: lessons from the laboratory and field. *J. Econ. Entomol., 96,* 1031-1038.

Tai, T. H., Dahlbeck, D., Clark, E. T., Gajiwala, P., Pasion, R., Whalen, M. C., Stall, R. E. & Staskawicz, B. J. (1999). Expression of the *Bs2* pepper gene confers resistance to bacterial spot disease in tomato. *Proc. Natl. Acad. Sci. U.S.A., 23,* 14153-14158.

Takase, K. & Hagiwara, K. (1998). Expression of human alpha-lactalbumin in transgenic tobacco. *J. Biochem., 123,* 440-444.

Takase, K., Hagiwara, K., Onodera, H., Nishizawa, Y., Ugaki, M., Omura, T., Numata, S., Akutsu, K., Kumura, H. & Shimazaki, K. (2005). Constitutive expression of human lactoferrin and its N-lobe in rice plants to confer disease resistance. *Biochem. Cell Biol., 83,* 239-249.

Takken, F. L. W. & Joosten, M. H. A. J. (2000). Plant resistance genes: their structure, function and evolution. *Eur. J. Plant Pathol., 106,* 699-713.

Tavladoraki, P., Benvenuto, E., Trinca, S., Demartinis, D., Cattaneao, A. & Galeffi, P. (1993). Transgenic plants expressing a functional single chain Fv antibody are specifically protected from virus attack. *Nature, 366,* 469-472.

Taylor, N. J. & Fauquet. C. M. (2002). Microparticle bombardment as a tool in plant science and agricultural biotechnology. *DNA Cell Biol., 21,* 963-977.

Templeton, M. D., Reinhardt, L. A., Collyer, C. A., Mitchell, R. E. and Cleland, W. W. (2005). Kinetic analysis of the L-ornithine transcarbamoylase from *Pseudomonas savastanoi* pv. *phaseolicola* that is resistant to the transition state analogue (R)-N^{δ}-(N'-sulfodiaminophosphinyl)-L-ornithine. *Biochemistry, 44,* 4408-4415.

Theil, E. C. (1987). Ferritin: structure, gene regulation, and cellular function in animals, plants and microorganisms. *Annu. Rev. Biochem., 56,* 289-315.

Theil, E. C. (1990). Regulation of ferritin and transferrin receptor mRNAs. *J. Biol. Chem., 265,* 4771-4774.

Theologis, A., Zarembinski, T, I., Oeller, P. W., Liang, X., & Abel, S. (1992). Modification of fruit ripening by suppressing gene expression, *Plant Physiol., 100,* 549-551.

Thomma, B., Eggermont, K., Penninckx, I., Mauch-Mani, B., Vogelsang, R., Cammue, B. P. A. & Broekaert, W. F. (1998). Separate jasmonate-dependent and salicylate-dependent defense-response pathways in *Arabidopsis* are essential for resistance to distinct microbial pathogens. *Proc. Natl. Acad. Sci. U.S.A., 95,* 15107-15111.

Thompson, C., Dunwell, J. M., Johnstone, C. E., Lay, V., Schmitt, M., Watson, H. & Nisbet, G. (1995). Degradation of oxalic acid by transgenic oilseed rape plants expressing oxalate oxidase. *Euphytica, 85,* 169-172.

Thomzik, J. E., Stenzel, K., Stocker, R., Schreier, P. H., Hain, R., & Stahl, D. J. (1997). Synthesis of a grapevine phytoalexin in transgenic tomatoes (*Lycopersicon esculentum* Mill.) conditions resistance against *Phytophthora infestans. Physiol. Mol. Plant Pathol., 51,* 265-278.

Timmer, C. P. (2003). Biotechnology and food systems in developing countries. *J. Nutr., 133,* 3319-3322.

Trieu, A. T., Burleigh, S. H., Kardailsky, I. V., Maldonado-Mendoza, I. E., Versaw, W.K., Blaylock, L. A., Shin, H., Chiou, T. J., Katagi, H., Dewbre, G. R., Weigel, D. & Harrison, M. J. (2000). Technical Advance: transformation of *Medicago truncatula* via infiltration of seedlings or flowering plants with *Agrobacterium. Plant J., 22,* 531-541.

Tripathi, L., Tripathi, J. N. & Tushemereirwe, W. K. (2004). Strategies for resistance to bacterial wilt disease of bananas through genetic engineering, *Afr. J. Biotechnol.,* 3, 688-692.

Tsaftaris, A. (1996). The development of herbicide-tolerant transgenic crops. *Field Crops Res., 45,* 115-123.

Tu, J., Datta, K., Khush, G. S., Zhang, Q. & Datta, S. K. (2000). Field performance of *Xa21* transgenic indica rice (*Oryza sativa* L.), IR72. *Theor. Appl. Genet., 101,* 15-20.

Tzfira. T. & Citovsky. V. (2006). *Agrobacterium*-mediated genetic transformation of plants: biology and biotechnology. *Curr. Opin. Biotechnol., 17,* 147-154.

Vaeck, M., Reynaerts, A., Hofte, H., Jansens, S., De Beuckeleer, M., Dean, C., Zabeau, M., Montagu, M. V. & Leemans. J. (1987). Transgenic plants protected from insect attack. *Nature, 328,* 33-37.

Van Loon, L. C. & Van Kammen, A. (1970). Polyacrylamide disc electrophoresis of the soluble leaf proteins from *Nicotiana tabacum* var. "Samsun" and "Samsun NN". II. Changes in leaf protein constitution after infection with tobacco mosaic virus. *Virology., 40,* 199-211.

Van Wert. S. L. & Saunders. J. A. (1992). Electrofusion and electroporation of plants, *Plant Physiol., 99,* 365-367.

Veluthambi. K, Gupta. A. K. & Sharma. A. (2003). The current status of plant transformation technologies, *Curr. Sci., 84,* 368-380.

Verberne, M. C., Verpoorte, R., Bol, J. F., Mercado-Blanco, J. & Linthorst, H. (2000). Overproduction of salicylic acid in plants by bacterial transgenes enhances pathogen resistance. *Nat. Biotechnol., 18,* 779-783.

Vijayan, M. & Chandra, N. (1999). Lectins. *Curr. Opin. Struct. Biol., 9,* 707-714.

Visser, R. G. F., Suurs, L. C. J. M., Bruinenberg, P. M., Bleeker, I. & Jacobsen, E. (1997). Comparison between amylose-free and amylose containing potato starches. *Starch/ Stärke, 49,* 438-443.

Voinnet, O. (2005). Induction and suppression of RNA silencing: insights from viral infections. *Nat. Rev. Genet., 6,* 206-220.

Walmsley, A. M. & Arntzen, C. J. (2003). Plant cell factories and mucosal vaccine. *Curr. Opin. Biotechnol., 14,* 145-150.

Wargovich, M. J. (2000). Anticancer properties of fruits and vegetables. *HortScience,35,* 573-575.

Waugh, R., Leader, D. J., McCallum, N. & Caldwell, D. (2006). Harvesting the potential of induced biological diversity. *Trends Plant Sci., 11,* 71-79.

Wei, L., Guangqin, G. & Guochang, Z. (2000). *Agrobacterium*-mediated transformation: state of the art and future prospect. *Chin. Sci. Bull.,* 45, 1537-1540.

Williamson, V. M. (1999). Plant nematode resistance genes. *Curr. Opin. Plant Biol., 2,* 327-331

Wu, G., Shortt, B. J., Lawrence, E. B., Leon, J., Fitzsimmons, K. C., Levine, E. B., Raskin, I. & Sha. D. M. (1997). Activation of host defense mechanisms by elevated production of H_2O_2 in transgenic plants. *Plant Physiol., 115,* 427-435.

Wu, G., Shortt, B. J., Lawrence, E. B., Levine, E. B., Fitzsimmons, K. C. & Shah, D. M. (1995). Disease resistance conferred by expression of a gene encoding H_2O_2-generating glucose oxidase in transgenic potato plants. *Plant Cell, 7,* 1357-1368.

Wurbs, D., Ruf, S. & Bock, R. (2007). Contained metabolic engineering in tomatoes by expression of carotenoid biosynthesis genes from the plastid genome. *Plant J., 49,* 276-288.

Yang, S. F. & Hoffman, N. E. (1984). Ethylene biosynthesis and its regulation in higher plants. *Annu. Rev. Plant Physiol., 35,* 155-189.

Yano, A., Maeda, F. & Takekoshi, M. (2004). Transgenic tobacco cells producing the human monoclonal antibody to hepatitis B virus surface antigen. *J. Med. Virol., 73,* 208-215.

Yao, K., De Luca, V. & Brisson, N. (1995). Creation of a metabolic sink for tryptophan alters the phenylpropanoid pathway and the susceptibility of potato to *Phytophthora infestans. Plant Cell, 7,* 1787-1799.

Yarasi, B., Sadumpati, V., Immanni, C. P., Vudem. D. R. & Khareedu, V. R. (2008). Transgenic rice expressing *Allium sativum* leaf agglutinin (ASAL) exhibits high-level resistance against major sap-sucking pests, *BMC Plant Biol., 8,* 102.

Ye, X., Al-Babili, S., Kloti, A., Zhang, J., Lucca, P., Beyer, P. & Potrykus, I. (2000). Engineering the provitamin A (beta-carotene) biosynthetic pathway into (carotenoid-free) rice endosperm. *Science, 287,* 303-305.

Yoder, O. C. & Turgeon B. G. (1996). Molecular-genetic evaluation of fungal molecules for roles in pathogenesis to plants. *J. Genet., 75,* 425-440.

Young, J. M., Kuykendall, L. D., Martinez-Romero, E., Kerr, A. & Sawada, H. (2001). A revision of *Rhizobium* Frank 1889, with an emended description of the genus, and the inclusion of all species of *Agrobacterium* Conn 1942 and *Allorhizobium undicola* de

Lajudie *et al.* 1998 as new combinations: *Rhizobium radiobacter, R. rhizogenes, R. rubi, R. undicola* and *R. vitis. Int. J. Syst. Evol. Microbiol., 51,* 89-103.

Zainal, Z., Marouf, E., Ismail, I & Fei, C. K. (2009). Expression of the *Capsicuum annum* (Chili) defensin gene in transgenic tomatoes confers enhanced resistance to fungal pathogens. *Am. J. Plant Physiol., 4,* 70-79.

Zelaya, I. A. & Owen, D. K. M. (2004). Evolved resistance to acetolactate synthase-inhibiting herbicides in common sunflower *(Helianthus annuus)*, giant ragweed *(Ambrosia trifida)*, and shattercane *(Sorghum bicolor)* in Iowa. *Weed Sci., 52,* 538-548.

Zhai, W. X., Wang, W.M., Zhou, Y.L., Li, X.B., Zheng, X.W., Zhang, Q., Wang, G.L. and Zhu, L.H. (2001). Breeding bacterial blight-resistant hybrid rice with the cloned bacterial blight resistance gene *Xa21. Mol. Breed., 8,* 285-293.

Zhai, W., Li, X., Tian, Z., Zhou, Y., Pan, X., Cao, S., Zhao, X., Zhao, B., Zhang, Q. & Zhu, L. (2000). Introduction of a rice blight resistance gene, *Xa21*, into five Chinese rice varieties through an *Agrobacterium*-mediated system. *Sci. China C Life Sci., 43,* 361-368.

Zhu, C., Naqvi, S., Gomez-Galera, S., Pelacho, A. M., Capell. T. & Christou, P. (2007). Transgenic strategies for the nutritional enhancement of plants. *Trends Plant Sci., 12,* 548-555.

Ziegler, A. & Torrance, L. (2002). Applications of recombinant antibodies in plant pathology. *Mol. Plant Pathol., 3,* 401-407.

Zupan, J. R. & Zambryski. P. C. (1995). Transfer of T-DNA from *Agrobacterium* to the plant cell. *Plant Physiol., 107,* 1041-1047.

In: Genetically Engineered Crops
Editors: Lindsay M. Grover
ISBN: 978-1-61122-526-6

Chapter 2

THE APPLICATION OF SITE-SPECIFIC RECOMBINATION SYSTEMS FOR BIOSAFETY AND GENOME MANIPULATION IN THE PRODUCTION OF TRANSGENIC PLANTS

Yuan-Yeu Yau * ***and Ludmila Tyler***
University of California-Berkeley, Albany, California, USA

ABSTRACT

By enabling the development of new crop varieties with increased nutritional content, disease resistance, and stress tolerance, plant biotechnology plays an important role in addressing the food needs of the world's population. According to ISAAA Brief 39-2008, the area of land used globally for cultivating transgenic crops grew from 114.3 million hectares in 2007 to 125 million hectares in 2008. In 2007, the direct global farm income benefit from biotech crops was $10.1 billion (Brooke, 2009). So far, the majority of transgenic crops have been produced either by *Agrobacterium*-mediated transformation or by biolistic transformation. The gene controlling a trait of interest is usually delivered along with a selectable marker gene (such as an antibiotic-resistance gene) for later selection of the transgenic plants from the bulk of non-transformed plants. After serving this purpose, the selectable marker gene is no longer needed. However, this approach for producing transgenic crops has raised concerns from both the regulatory and public sectors over the presence of selectable marker genes in the food supply or their possible escape into the environment through the outcrossing of transgenic crops to related species. Therefore, research in plant transgenesis continues to focus on ways to generate transgenic crops free of selectable marker genes. Several approaches for producing selectable-marker-free transgenic crops have been reported. Among these approaches, one of the most powerful is the application of microbial site-specific recombination systems to the transformation technology. This article will discuss the promise of transgenic breeding for future agriculture, the public concerns over genetically modified plants, and the utilization of site-specific recombination systems (including the

* Corresponding author: Plant Gene Expression Center, USDA-ARS & Department of Plant & Microbial Biology, University of California-Berkeley, 800 Buchanan St., Albany, CA 94710

novel ParA system recently described by our laboratory) for the production of marker-free transgenic plants. The use of site-specific recombination systems to tailor controlled transgene expression in plants and to genetically engineer plants for breeding programs will be discussed, as well.

Conventional Breeding and the Benefits of Transgenic Crop Production

The principles of genetics have been used for years to improve plant performance, especially yield and quality. In its early form, the application of genetics occurs through rough, non-systematic phenotypical or "visual" selections usually conducted by the farmers. Local farmers select and save the best-looking or highest-yielding plants and the largest seeds for future growing seasons. Plant lines resulting from this method are called "landraces" in the scientific community.

As genetics becomes better understood, advanced breeding methods with more complicated and systematic strategies are gradually developed and employed to increase yield in various crops. These methods include tailoring the selection schemes for specific crops. Conventional breeding is born. Plant breeders try to generate elite varieties in breeding programs by accumulating favorable alleles from within the gene pool or from cross-compatible wild relatives (Poehlman and Sleper, 1995). The practice usually involves sexually crossing two or more lines and selecting the lines with superior genotypes, which combine the desired traits and high heritability.

The most prominent, successful example of conventional breeding is the Green Revolution, which started in the 1940s and continued into the 1950s and 1960s. The Green Revolution is often attributed to the American scientist and Nobel laureate (Peace Prize, 1970) Dr. Norman E. Borlaug. He originally worked to improve wheat by crossbreeding new disease-resistant (*e.g.* resistant to wheat rust), high-yielding, and dwarf wheat varieties in Mexico. The effort was a huge success. Mexico was, for the first time, able to produce more wheat than its own citizens needed and exported wheat by the 1960s. The successful case of wheat in Mexico helped the breeding technologies spread worldwide, to be adopted by many countries, including the US and nations in Asia.

The improvements in wheat were quickly replicated in other grains. For example, the development and release of a high-yielding, semi-dwarf variety of rice known as IR-8 from the International Rice Research Institute (IRRI) in the Philippines helped avert massive famine in India in the early 1960s. The IR-8 'magic rice' spread throughout Asia in the following decades, leading to the rice 'Green Revolution.' IR-8 was bred to include a "short stature" gene. The reduced height of IR-8 plants caused them to exhibit lodging-resistance, responsiveness to fertilizer inputs, and improved yields through increases in the harvest index (grain-straw ratio) (Khush, 1999). The short stature of IR-8 rice is now known to be due to a mutant gibberellin-synthesis gene, *sd1* (Sasaki *et al.*, 2002). This gene encodes an oxidase enzyme which is involved in the biosynthesis of the plant growth hormone gibberellin (GA). GA is one of the most important determinants of plant height. Because of the Green Revolution, global food production expanded dramatically, preventing hunger for millions of people (Borlaug, 2002). This type of traditional plant breeding is still commonly practiced today around the world.

Although it is an important tool in increasing plant performance and has improved many traits in different crops in the past, conventional breeding has some limitations. These limitations arise from the fact that genes for traits of interest can only be transmitted to the next generation by sexually crossing plants which bear these genes: ***(1)*** Crossing is difficult for plant species with primarily asexual modes of reproduction. Although some techniques can overcome this obstacle, they tend to be challenging and have limited applicability. Protoplast fusion, for example, requires a suitable method for protoplast production and is therefore not available for many plant species. ***(2)*** Sometimes even in species which reproduce sexually, a crossing barrier prevents crosses between different species from being successful. ***(3)*** When two plants are crossed, 50% of each parental genome will go into the next generation, which means that many traits – including undesirable ones – will exist in the offspring. A prolonged process called backcrossing can be employed to recover the elite line's genomic makeup. However, a backcrossing program usually last several generations. This translates into years for *annual* crops (where the plant usually germinates, grows, flowers, and dies within a year), let alone for *biennial* or *perennial* crops. For biennial crops, a plant takes two years to complete its biological lifecycle: the plant grows vegetatively during the first year, then undergoes a *vernalization* process during a period of cold temperatures (winter), flowers in the spring of the following year, sets seeds, and finally dies. For perennial crops, a plant grows more than two years via a root-stock.

After recombinant DNA technology became available during the first half of the 1970s (Cohen *et al.*, 1973), genetic engineering emerged as a powerful new tool for modifying genomes, including those of plants. Genetic engineering and genetic transformation overcome the limitations of conventional breeding. Genetic engineering physically "removes" a gene from an organism – virtually any organism including bacteria – by cloning the gene (PCR-amplifying it and cutting and pasting it into a DNA vector). Genetic transformation transfers the gene into another organism (*e.g.* a plant) to create a transgenic organism with a new trait. Because sexual crossing is not needed, this method circumvents the "crossing barrier" which occurs between plants in conventional breeding programs. In addition to this advantage, the merits of transgenic breeding include, but are not limited to, the following: (1) Genes for new traits can be acquired from different species, including bacteria and animals, and (2) a novel gene can be created for new use(s) in a technique called "gene shuffling." Various gene fragments can be "shuffled" and engineered together to form a novel gene and then genetically transformed into plants for new function(s). For example, Wulff *et al.* have generated a great variety of novel *Hcr9* (homologues of *C. fulvum* resistance gene *Cf-9*) shufflants *in vitro* by using members of the conserved *Hcr9* gene family for a gene shuffling experiment. Restriction-enzyme-cut fragments of DNA were used to generate a shufflant library. Some of these shufflants showed novel specificities and were found to act as auto-activators of the hypersensitive response, a plant defense mechanism, in *Nicotiana* species (Wulff *et al.*, 2004).

PRODUCTION OF TRANSGENIC PLANTS

1. Vector Construction

To start a genetic transformation experiment, recombinant DNA technology is used to build a construct based, for example, on a bacterial plasmid. For *Agrobacterium*-mediated transformation of plants, a gene of interest and a selectable marker gene are inserted between a pair of specialized DNA sequences, the left border (LB) and right border (RB) of the *Agrobacterium tumefaciens* transferred DNA (T-DNA). Constructs built without LB and RB sequences can be delivered into the plant nuclear genome by several methods described below, including the gene gun method and polyethylene glycol (PEG)-mediated protoplast transformation methods.

2. Genetic Transformation of Plants

There are several ways to genetically transform plants. They are: (1) *Agrobacterium*-mediated genetic transformation, (2) biolistic or particle bombardment, nicknamed "gene-gun" transformation, (3) electroporation of tissues or protoplasts, (4) microinjection, (5) polyethylene glycol (PEG)-mediated protoplast transformation, (6) virus-mediated transformation, and (7) whisker-mediated transformation. *Agrobacterium tumefaciens* is a soil-borne pathogen which infects many higher plant species and causes Crown Gall disease. *A. tumefaciens* contains a plasmid called the Ti, or tumor-inducing, plasmid. The bacterium has the ability to transfer and integrate a small part of the Ti plasmid (the T-DNA) into the genome of an infected plant. The proteins and mechanisms involved in the transfer and integration of T-DNAs have been discussed in great detail in many publications. An understanding of these processes has allowed scientists to engineer T-DNAs in order to design vectors for delivering genes of interest into higher plants. Particle bombardment uses a high-pressure gas (*e.g.* helium) to accelerate DNA-coated gold microparticles (0.6-1.6 μm in diameter) into plant tissues in a particle bombardment device such as the *Bio-Rad* PDS-1000/He™ system (*Bio-Rad* laboratories, Inc.). Electroporation uses pulses of high-voltage electricity to induce transient pores in cell membranes and thus allow the entry of DNA molecules. Microinjection injects DNA molecules into a plant cell using a finely drawn micropipette needle. PEG-mediated protoplast transformation involves digesting plant cell walls with enzymes (cellulase and macerozyme) to produce protoplasts before PEG-treating the protoplasts for genetic transformation. Viral transformation proceeds by packaging the desired genetic material into a suitable plant virus and allowing this modified virus to infect the plants. Whisker-mediated genetic transformation involves the use of silicon carbide (SiC) whiskers (microfibers, 10-80 μm long and 0.6 μm in diameter) to deliver the DNA molecules into a plant's genome. The first report of plant tissue transformation utilizing this last method was published by Kaeppler *et al.* (1990). Transformed plant cells were generated by adding a DNA/fiber mixture to packed cells from maize or tobacco suspension cultures and vortexing for 60 seconds.

Among these methods, *Agrobacterium*-mediated and gene-gun-mediated genetic transformation are the most popular for plants. *Agrobacterium*-mediated transformation had most often been used for dicots (plants with two cotyledons), while the gene-gun method was mostly used for monocots (rice, wheat, and other plants with a single cotyledon), although, nowadays, improved protocols allow several monocots (such as rice, wheat, maize, barley and sorghum) to be routinely transformed with *Agrobacterium* (see review by Cheng *et al.,* 2004).

3. Selection of Transgenic Plants

Because the transformation efficiency is less than optimal for many plant species, one of the most important steps in the production of transgenic plants is the ability to distinguish a few transgenic plants from the bulk of non-transgenic plants. To achieve this, researchers usually deliver a construct containing both the gene of interest and a selectable marker gene. Once this construct is integrated into the plant genome, the protein expressed from the selectable marker gene can, for example, detoxify a specific antibiotic or herbicide. Using this mechanism, scientists can select the genetically transformed plant cells on a medium containing the antibiotic or herbicide. Non-transformed plants are susceptible to the selection agent and begin to die, while a resistant transgenic plant expressing the selectable marker gene will survive (Fig. 1). Approximately fifty marker genes have reportedly been used for transgenic plant research or crop development (Miki and McHugh 2004). According to Miki and McHugh, the top three most commonly employed selectable marker genes are a kanamycin resistance gene, a hygromycin resistance gene, and a phosphinothricin resistance gene.

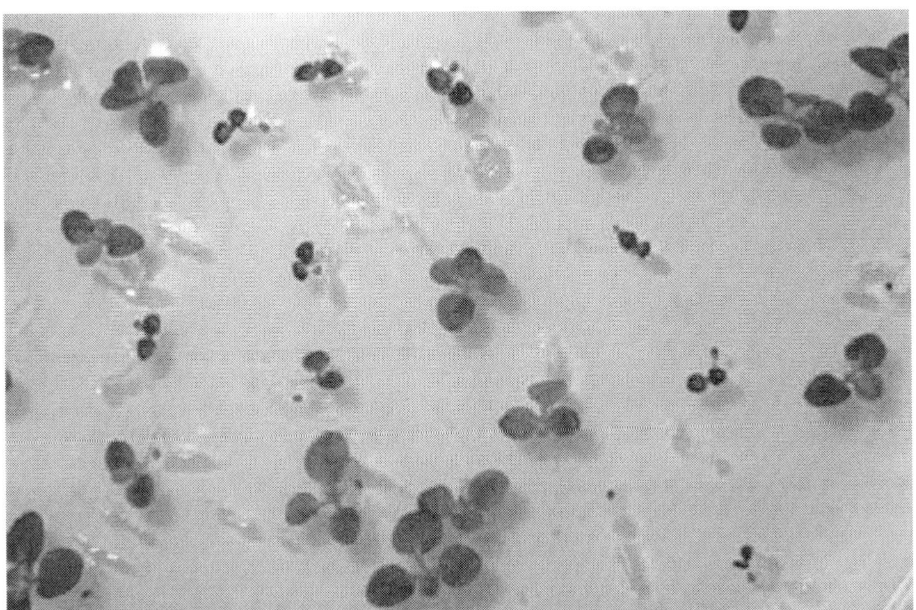

Figure 1. Tobacco seedlings on selection medium, MS + hygromycin (20 μg/mL). Resistant plantlets grow well in the medium, whereas susceptible plantlets remain small and fail to establish root systems.

The kanamycin resistance gene is derived from the bacterial amino-glycoside 3'-phosphotransferase II (APH [3'] II, E.C. 2.7.1.95) gene, also known as the neomycin phosphotransferase II (*npt*II) gene. The NPTII enzyme catalyses the ATP-dependent phosphorylation of the 3'-hydroxyl group of the amino-hexose portion of certain amino-glycosides including neomycin, kanamycin and geneticin. First, the *npt*II gene from *Escherichia coli* is used to construct a binary vector for plant genetic transformation. Then, transformed plant tissues are selected by placing the tissues on a medium, such as MS (Murashige-Skoog) medium, containing kanamycin. Transgenic tissues expressing the *npt*II gene will survive kanamycin selection and grow in the medium. The concentration of kanamycin used depends on the plant species. For model plants such as *Arabidopsis* and tobacco, the concentration of kanamycin in the medium is usually 50 μg/mL for *Arabidopsis* transgenic seed selection and 100 μg/mL for tobacco transgenic seed selection or 200 μg/mL for tobacco transformed-leaf-tissue selection.

Hygromycin B is an aminocyclitol antibiotic inhibitor of protein synthesis with a broad spectrum of activity against prokaryotes and eukaryotes. The *E. coli* gene *aphIV* [also called *hph* or *hpt*, coding for hygromycin phosphotransferase (HPT, E.C. 2.7.1.119)] confers resistance on prokaryotic and eukaryotic cells, including cells from higher plants. In the absence of the *aphIV* gene, hygromycin B is very toxic to plants. Sterile, liquid hygromycin can be purchased from a commercial supplier and added directly to the medium for transgenic plant selection.

Members of the genus *Streptomyces* produce hundreds of antibiotics, one of which is *bialaphos* (also known as PPT, for phosphinothricin, or glufosinate). Bialaphos inhibits glutamine synthetase (GS), the key enzyme in the nitrogen assimilation pathway. The toxicological characteristics of glufosinate have made it a popular herbicide, and it has been commercialized under the names Basta® and Liberty® (http://www.patentlens.net/daisy/Phosph/ext/navaggregator/navaggregator).

Antibiotic-producing *Streptomyces* strains have evolved mechanisms to avoid the toxicity of their own products. Such strains often contain modifying enzymes which can inactivate the antibiotics, such as an enzyme that detoxifies glufosinate. Thompson *et al.* (1987) have isolated the gene encoding this enzyme from *Streptomyces hygroscopicus*. The gene – named *bar* (for bialaphos resistance) – encodes a phosphinothricin acetyl transferase (PAT). Numerous researchers have used the *bar* gene as a selectable marker, cloning it into DNA constructs for plant genetic transformation. Transgenic plants expressing PAT can survive and grow in medium supplemented with the bialaphos compound. Scientists can also select *bar*-gene-expressing transgenic plants by spreading the bialaphos solution directly onto the seedlings or leaves of the plants.

Concerns About the Use of Selectable Marker Genes

Many questions have been raised regarding the safety of transgenic crops, both for the consumer and the environment. The fact that selectable marker genes remain in the genomes of genetically modified plants has triggered both regulatory and public concerns, which include: (1) If plant products consumed by humans as food contain antibiotic resistance genes, will these products increase the incidence of antibiotic-resistant bacterial strains? (2) If seeds or pollen grains contain herbicide-resistance genes, will spreading of the transgenic seeds and crossing to wild relatives via the transgenic pollen ("*gene flow*") contaminate the environment by creating new, aggressive weeds which cannot be eliminated by herbicide application? Gene flow of an herbicide-resistance transgene has been studied in Canada in herbicide-resistant canola fields (Warwick *et al.*, 2003). In this study, researchers observed that gene flow occurred from commercial, herbicide-resistant *B. napus* to natural, wild populations of *B. rapa*, if the plants were grown close to each other. However, the researchers also found that barrier crops can be planted as "absorbers" to prevent gene flow.

Due to the concerns described above, before the release of genetically modified plants from laboratories for field trials and commercialization, intensive studies and tests regarding potential impacts on human health and the environment need to be done, and regulatory processes need to be set. Although studies with mice revealed that *npt*II degraded rapidly in a simulated gastric environment and was unlikely to cause an allergic response (Fuchs *et al.*, 1993), and although gene flow can be carefully handled and monitored to prevent it, much uncertainty still exists. Besides, studies of the impact of other marker genes, such as an herbicide-resistance gene, on human or animal health are rare, and the outcomes remain unclear. These risks may, however, be alleviated by removing the selectable marker genes once they are no longer needed. Therefore, the production of selectable-marker-gene-free transgenic crops is becoming an extremely attractive alternative. The holy grail of current transgenic plant research has been to develop strategies that can remove the vector backbone and selectable marker genes, after their use, from the final transgenic product. In addition, removing the selectable marker gene would allow researchers to re-use the same selectable marker, without the need to utilize multiple markers. Consequently, several approaches have been developed to generate selectable-marker-gene-free transgenic crops.

Strategies for Producing Marker-Gene-Free Transgenic Plants

As reviewed by Hohn *et al.* (2001) and Darbani *et al.* (2007), there are several approaches to producing marker-free transgenic plants. These approaches are:

(I) Avoid using selectable marker genes in the transformation process for producing genetically modified (GM) crops. Theoretically, because the transformed plants carry a transgene, it should be possible to distinguish them from un-transformed plants using molecular tools such as PCR. However, a project like this would be extremely labor-intensive, time-consuming, and costly. Without selection, the majority of the screened plants would be un-transformed. This approach is impractical even in today's laboratories with the availability of automated analysis and with a PCR protocol which directly uses plant tissues for PCR amplification, eliminating the need for isolating genomic DNA. (Kits for performing PCR from plant material include the Phire® Plant Direct PCR kit sold by Biolabs, www.finnzymes.com.)

(II) Use a "co-transformation and segregation"-based strategy to remove selectable marker genes. One example of this approach is to build the selectable marker gene and the transgene of interest (such as a trait gene) into two separate constructs which are co-transformed into plant tissues by either *Agrobacterium*-mediated or biolistic genetic transformation. Resistant transgenic plants surviving selection are screened for the presence of the gene of interest. By selecting for the transgene of interest and not the marker gene in the segregating population of the subsequent generation, one can obtain transgenic lines with the transgene of interest but without the marker gene. Under this strategy for producing marker-free plants, co-transformation using *Agrobacterium* – rather than particle bombardment – stands out as the more-often-used method; this is likely due to the simplicity and efficiency of *Agrobacterium*-mediated transformation. Many variations of this approach, referred to as "2 T-DNA" transformation, have been described (Huang *et al.*, 2004).

The co-bombardment method is analogous to the "2 T-DNA" *Agrobacterium*-mediated transformation method. However, there are only a handful of case studies utilizing the co-bombardment method to produce marker-free transgenic plants. This may be because bombardment tends to deliver multiple copies of transgenes into a single locus to form a complex insertion. Although studies on producing marker-free plants through bombardment are limited, unlinked integrations of transgenes have been reported, and a low percentage of marker-free plants were recovered after segregating away the unlinked marker gene. By using "cassette DNA (not the whole plasmid)" and "limiting the amount of input DNA," Prakash *et al.* (2009) devised a novel method for producing marker-free transgenic corn through co-bombardment. The so-called "cassettes" were DNA fragments isolated from plasmids via appropriate restriction-enzyme digestions. The bombardments were performed with 2.5 ng of the *npt*II-gene cassette per shot and 15 ng of the gene-of-interest cassette per shot, amounts which are much lower than the 0.6-1 μg/shot suggested by the manufacturer. With this approach, the researchers identified 28 marker-free corn plants from the progeny of 103 R_0 plants containing both co-transformed constructs.

There are disadvantages to the "co-transformation and segregation"-based approach: (1) It is time-consuming and labor-intensive. (2) This method can only be applied to sexually propagated plants whose genomes undergo recombination and segregation, not to vegetatively propagated plants. (3) In addition, if the two constructs become closely linked to each other when they are introduced into plants, then the chances of finding two separated transgenes (the marker gene and the transgene of interest) in the segregating population will be dramatically reduced.

(III) Employ a transposon-based strategy for marker-gene removal. The maize transposable element *Ac* has the ability to move to new locations within a genome (Fedoroff, 1989). According to a previous study, about 10% of the *Ac* elements that are excised do not reinsert and therefore disappear from the genome or reinsert into a sister chromatid that is subsequently lost during somatic segregation. Ebinuma *et al.* (1997) employed the *Ac* element to remove the isopentenyl transferase gene (*ipt*), which was used for visually distinguishing transgenic plants from non-transgenic plants by identifying the *ipt* "shooty" phenotype. In the experiment, a T-DNA was constructed to contain an *Ac* element linked to a 35S promoter-*ipt* gene-terminator marker-gene cassette, and a gene-of-interest cassette. The idea is that when the *Ac* element transposes, the *ipt* gene transposes along with the *Ac* element and may eventually be lost with the *Ac* element, resulting in a T-DNA which only contains the gene-of-interest cassette. By six months post infection, the researchers had obtained marker-free transgenic tobacco plants – from which *Ac* had disappeared – from 4.8% of the transgenic clones.

(IV) Remove selectable marker genes via intrachromosomal homologous recombination of two bacteriophage λ *attP* sites. Usually, using intrachromosomal homologous recombination (ICR) between two homologous sequences to induce transgene deletion in plants is not efficient, due to the low frequency of ICR. According to previous studies, ICR between closely linked repeated sequences occurs at a rate of approximately 10^{-6} (Puchta and Hohn, 1996). However, Peter Meyer's group has devised an ICR strategy which is based on the recombination of the *attP* region of bacteriophage λ but which does not require the presence of recombinase and other helper proteins (Zubko *et al.*, 2000). In this study, a construct was built to contain two 250-bp λ phage attachment sites (or *attP*) as homologous repeats flanking a selectable marker gene. The construct was transformed into tobacco plants, and higher rates of ICR-induced selectable-marker-gene deletion were detected in different parts of two of eleven lines. There were no detectable deletion events in the other nine lines. This result was attributed to position effect, in this case, the influence which the integration region of the individual transgene has on the efficiency of recombination. In parallel, the researchers observed that ICR is not always associated with precise homologous recombination between the two *attP* regions; larger deletions (in which regions outside the *attP* sequences were lost, as well) were also detected among the deletion events in the two lines. Illegitimate recombination was proposed as the underlying mechanism for the majority of the deletion events. It is unclear why the rates of ICR-induced DNA-fragment (selectable-marker-gene) excision were higher than expected. Puchta (2000) has suggested several possibilities, including that the increased rates may be caused by T-DNAs integrating into genomic double-stranded breaks (DSBs), and the only known way to increase homologous – and simultaneously, illegitimate – recombinations by several orders of magnitude is to induce DSBs. However, except for this paper, apparently no other laboratories have reported using a similar approach to remove selectable markers. It would be interesting to explore if other *attP*-like sites from other site-specific recombination systems can be used for this same purpose.

(V) Utilize zinc finger nucleases (ZFNs) for marker gene removal. The ZFNs are fairly new molecular tools for manipulating plant genomes, including mutating and adding genes. Genomic integrity is very important for maintaining a healthy cell, and cells must repair damaged DNA with high fidelity. One of the most dangerous DNA lesions is a DSB, which can lead to cell death or to mutation. When eukaryotic chromosomes suffer DSBs, the cellular DNA repair systems are activated to repair these breaks through both homology-dependent (homologous recombination or HR) and non-homology-dependent (nonhomologous end-joining or NHEJ) mechanisms. Repair of DSBs by NHEJ generates localized mutations by causing small deletions and insertions at the break site.

The ZFNs, originally termed hybrid restriction enzymes, are fusions of zinc-finger-based DNA recognition modules to endonuclease domains (Kim *et al.*, 1996). These enzymes are engineered to contain a zinc finger DNA-binding domain (composed of 3 zinc fingers) and the non-sequence-specific DNA-cleavage domain from the *Fok*I type II restriction endonuclease (Fn domain) (Porteus and Carroll, 2005). The cleavage domain of ZFNs must dimerize in order to cut DNA; therefore, efficient cleavage requires two zinc-finger-binding sites in close proximity. Because the zinc finger domain can be engineered to recognize a wide range of novel DNA sequences, a specific ZFN can be designed, produced, introduced and directed to a specific genomic sequence for cleavage. ZFNs have been widely used by genomic engineers to introduce DSBs at endogenous loci in eukaryotic systems to generate gene mutations (due to a small deletion and insertion of sequences at the breaks by the NHEJ repair mechanism) and to achieve gene replacement (through the homologous recombination repair mechanism). For instance, Petolino *et al.* (2010) demonstrated that a *GUS* reporter gene flanked by two ZFN cleavage sites was deleted from a stably transformed plant by crossing it with a second plant expressing a corresponding ZFN gene. In this study, ZFN binding sites flanking the targeted gene were pre-integrated into the plant genome; however, since ZFNs can be designed to bind and cleave a wide range of endogenous DNA sequences, these results constitute a first test of a general strategy for targeted gene deletions (Petolino *et al.*, 2010).

(VI) Exploit site-specific recombination to remove selectable marker genes. Site-specific recombination systems, which are discussed extensively below, have gained in popularity over the years, as a method for marker gene removal. These genetic systems contain three basic components— the recombinase protein and two recombination sites, which are either identical or dissimilar, for example *attP* or *attB*. These components are essential for site-specific integration or excision between the two molecules which contain the recombination sites. Researchers have cloned the DNA sequences for these three components and used them to build constructs for various applications in transgenic plant production. Dale and Ow (1991) conducted pioneering work to demonstrate that the Cre-*lox* site-specific recombination system was capable of excising unnecessary selectable marker genes from transgenic plants. In the two decades since then, numerous papers have been published using Cre-*lox* or other newly-identified site-specific recombination systems for marker gene removal in plants. The common feature of these studies is the production of transgenic plants with a selectable marker gene flanked by two recombination sites oriented in the same direction. Upon expression of the recombinase, a site-specific recombination event loops out the marker gene placed between the recombination sites.

The Sources of Site-Specific Recombination Systems

Site-specific recombination systems are commonly found in prokaryotes and lower eukaryotes such as the unicellular yeasts. The original biological functions of site-specific recombination systems are to integrate or excise bacteriophage into or out of the host genome, to maintain plasmid copy number during replication, to "switch" gene expression, to alter cell-surface components in some bacteria, to generate the final products of transposon transposition, and to accelerate the replication process by inverting the convergent replication forks in autonomous plasmids (Stark *et al.*, 1992).

Based on evolutionary and mechanistic relatedness, the recombinase super-family is composed of two fundamental groups, the *tyrosine* recombinases and the *serine* recombinases, named after the active amino acid residues that form covalent protein-DNA linkages in the reaction intermediates (Grindley *et al.*, 2006). Virtually all of the site-specific recombination systems fall into these two groups. Serine recombinases are subdivided into the small recombinases, which catalyze only intramolecular strand exchange (excision and/or inversion), and large recombinases, which catalyze both inter-and intramolecular strand exchange (integration and excision) (Smith and Thorpe, 2002).

Site-specific recombination is characterized by the following three features: (1) recombination occurs at specific DNA sequences (known as "recognition sites") on the interacting DNA molecules, (2) recombination is conservative, with the reaction resulting in no gain or loss of nucleotides, and (3) strand exchange occurs in small regions (with homologous DNA sequences) within the recognition sites.

The simplest site-specific recombination systems comprise two identical recognition sites and a single-polypeptide recombinase protein that catalyzes the recombination event (Sadowski, 1993). These systems include Cre-*lox* from the *E. coli* bacteriophage P1 (Sauer and Henderson, 1988), FLP-*FRT* from the 2-μ cycle yeast *Saccharomyces cerevisiae* (Golic and Lindquist, 1989), and R-*RS* from *Zygosaccharomyces rouxii* (Onouchi *et al.*, 1991); Cre, FLP and R are the recombinases, and *lox*, *FRT,* and *RS* are the respective DNA recognition sites. More complex site-specific recombination systems may have recombination sites that are dissimilar in sequence and may require different recombinases and accessory protein(s) (Grindley *et al.*, 2006).

MODES OF ACTION OF SITE-SPECIFIC RECOMBINATION

Depending on the initial arrangement of recognition sites on the involved molecule(s), site-specific recombination can lead to four possible outcomes: translocation, integration, excision, and inversion of DNA molecules (Figure 2 and Figure 3). Recombination between recognition sites *in trans* causes a reciprocal translocation of two linear DNA molecules, while co-integration occurs if at least one of the two molecules is circular (Ow, 2002). Excision of intervening DNA results from recombination between two recognition sites oriented in the same direction *in cis*. Inversion events result from recombination between two recognition sites oriented in opposite directions (Grindley *et al.*, 2006).

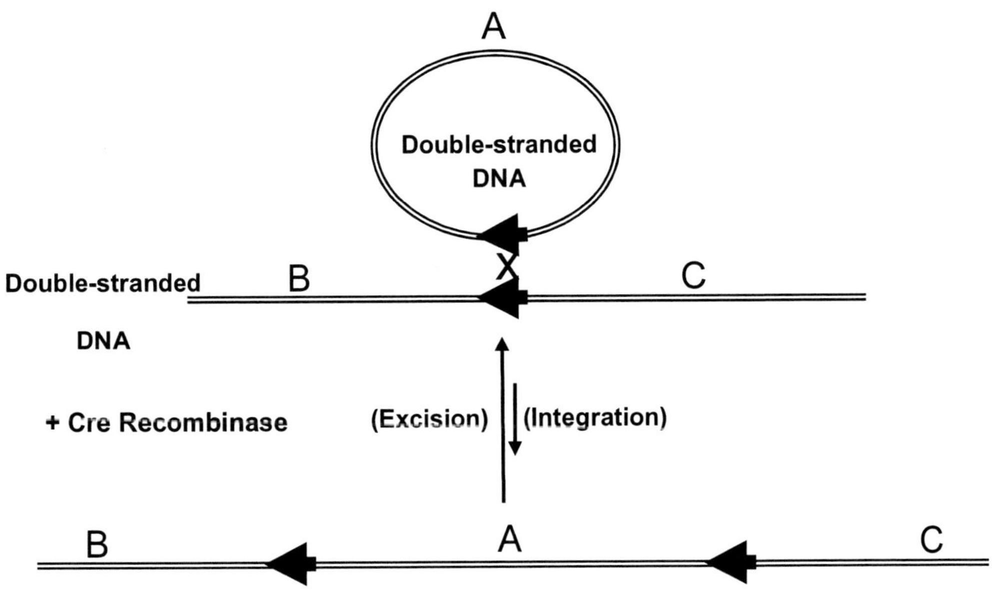

Figure 2. Cre-protein-mediated integration and excision reactions between two DNA molecules carrying *loxP* sites (short, dark arrows). The reactions are freely reversible, but excision is kinetically favored over integration.

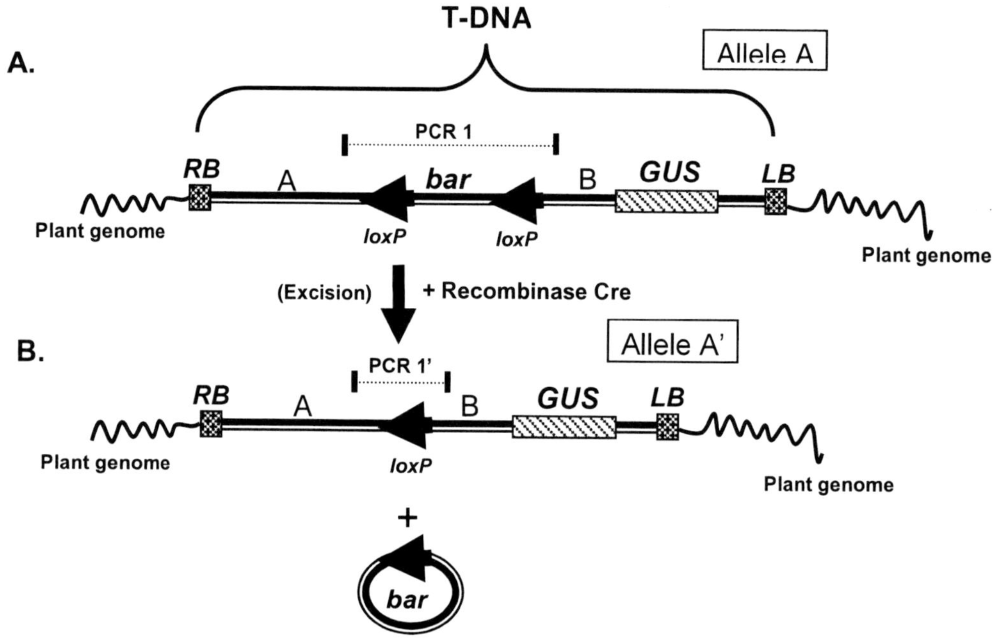

Figure 3. Cre-mediated removal of a selectable marker gene. A T-DNA fragment containing a selectable marker gene, *bar*, and a reporter gene, *GUS*, is integrated into the plant genome (allele A) (A). The *bar* gene is flanked by a pair of *loxP* sites (in the same orientation, arrowheads). The *bar* gene is excised from the T-DNA after a site-specific recombination event catalyzed by the Cre recombinase, yielding the new allele A' (B). The plants with allele A' can be identified and distinguished from plants with allele A by using PCR. (Contrast the sizes of the products for PCR1 *vs*. PCR1'.). RB: right border of the T-DNA, LB: left border of the T-DNA, A: DNA fragment "A," B: DNA fragment "B," *bar*: *bar* selectable marker gene, *GUS*: β-glucuronidase reporter gene.

The simplest site-specific recombination systems, Cre-*lox*, FLP-*FRT* and R-*RS*, belong to the small tyrosine group. Within these systems, the genetic cross-over between the two recognition sites regenerates a single site which is identical in nucleic acid sequence to the parental sites and releases the intervening DNA fragment. Because of the identical nature of the two recognition sites, the recombination reactions (co-integration and excision) are fully reversible, although the intramolecular reaction (excision) is kinetically favored over the intermolecular reaction (co-integration) (Figure 2).

A second group of site-specific recombination systems includes β-*six* (Diaz *et al.*, 2001), the bacterial γδ (Tn1000) transposon-encoded resolvase/invertase sub-family site-specific DNA recombination system γδ-*res* (Schwikardi and Dröge, 2000), CinH-*RS2* (Kholodii G, 2001; Thomson and Ow, 2006), and ParA-*MRS* (Gerlitz *et al.*, 1990; Thomson *et al.*, 2009); β, γδ, CinH, and ParA are small serine recombinases, and *six, res*, *RS2*, and *MRS* are the respective DNA recognition sites. The natural function of the γδ-*res* system is to resolve, through recombination, the co-integrate structure that is generated during transposition (Stark *et al.*, 1992). Recombination mediated by these small serine recombinases also regenerates a single product site of the same sequence as the parental sites. However, as mentioned earlier, the small serine recombinases catalyze intra-molecular excision; due to conformational stain, they cannot facilitate intermolecular integration.

A third group of site-specific recombination systems is represented by phiC31 (Thomason *et al.*, 2001), TP901-1 (Stoll *et al.*, 2002), R4 (Olivares *et al.*, 2001), and Bxb1 (Kim *et al.*, 2003; Keravala *et al.*, 2006; Thomson and Ow, 2006). These large serine recombinases (to distinguish from the small serine site-specific recombinases) act on two recognition sites, *attB* and *attP* (*b*acterial and *p*hage *att*achment sites) that differ in sequence, to yield hybrid sites known as *attL* and *attR*. These recombinases can facilitate an excision, inversion or integration reaction but cannot reverse the reaction, because the recognition-site sequences of *attB* and *attP* change to the sequences of *attL* and *attR*. The hybrid sites *attL* and *attR* are no longer substrates for the recombinase proteins. Reversing the reaction is only possible through the addition of a second protein, an excisionase (Ghosh *et al.*, 2006; Thorpe *et al.*, 2000). The phiC31 integrase is a 68-kDa, single-polypeptide, bacteriophage serine recombinase. It catalyzes site-specific recombination to allow integration of the *Streptomyces* phage phiC31 into the host chromosome (Smith and Thorpe, 2002).

Cre, FLP and R are also called *bi-directional* recombinases because the recombination reactions (co-integration and excision) are fully and freely reversible. These particular recombinases are capable of excision, inversion, and integration. phiC31 is an example of a *uni-directional* site-specific recombination system. Once co-integration occurs between two molecules through *attB* and *attP* site-specific recombination, the integrated molecule cannot be freely excised. To date, only members of the tyrosine recombinase family have shown bi-directional activity, while both tyrosine and serine recombinase families have shown uni-directional functionality. Furthermore, within the serine recombinase family, there are members which specialize in deletion reactions and cannot catalyze inversion or integration reactions.

A Brief History of the Study of the Cre-lox Site-Specific Recombination System and its Application in Eukaryotic DNA Excision

Cre-*lox* is the best-known and most widely investigated site-specific recombination system, as well as the foremost system for site-specific recombination studies in eukaryotic cells. Dr. Brian Sauer is credited with discovering this system (Sauer, 1987), but many workers from diverse fields have found creative ways to utilize Cre-*lox* recombination in manipulating mammalian and plant genomes. Cre stands for "**c**auses **re**combination," and *loxP* stands for "**lo**cus of crossing over in **P**1." The native *loxP* site is composed of an 8-bp spacer region flanked by two identical 13-bp inverted repeats (arm regions). P1 is maintained in the *E. coli* genome as a single-copy plasmid. The role of the Cre recombinase is to resolve dimers of P1 that arise after replication, in order to allow partitioning of the two P1 molecules at cell division.

In 1981, Sternberg and his colleague found that "a site (called *loxP*) must be present in both partners in the recombination in order for recombination to occur, and a P1 gene (called *cre*), whose product is necessary for recombination, must also be present" (Sternberg and Hamilton, 1981).In 1982, Hoess *et al.* reported the detailed nucleotide sequence of the *loxP* site, began to clarify which DNA sequences are necessary for *loxP* recombination, and implicated a region of dyad symmetry as being important (Hoess *et al.*, 1982). In 1983,

Abremski *et al.* (1983) demonstrated that when Cre recombinase was used *in vitro*, the products of recombination between *loxP* sites on a supercoiled DNA substrate were present as free supercoiled circles, *i.e.* topologically unlinked products. In 1984, Abremski and his colleague described the purification and characterization of the Cre recombinase protein. In an *in vitro* study, they observed that intramolecular recombination (two *loxP* sites on the same DNA molecule) is considerably more efficient than intermolecular recombination (Abremski and Hoess, 1984). In 1985, Hoess and his co-worker reported the mechanism of strand cleavage and exchange in the Cre-*lox* site-specific recombination system, locating the point at which strand exchange takes place with one radioactively labeled and one unlabeled *loxP* site (Hoess and Abremski, 1985). In 1987, Sauer was the first to demonstrate that the prokaryotic Cre-*lox* site-specific recombination system of phage P1 could function in an efficient manner in a lower eukaryote, the yeast *Saccharomyces cerevisiae*. This study demonstrated that a prokaryotic recombinase (Cre) can enter a eukaryotic nucleus and efficiently and accurately catalyze excision and recombination at specific DNA sites (*loxP* sites) to remove the fragment between them (Sauer, 1987). This finding opened the door for scientists to start using prokaryotic site-specific recombination systems for eukaryotic genome manipulation. In 1988, Sauer and Henderson showed that Cre can be functionally expressed in cultured mouse cells to catalyze excisive recombination of extrachromosomal DNA (Sauer and Henderson, 1988). In 1989, Sauer and Henderson showed that the Cre protein can promote DNA recombination at *loxP* sites stably resident on a mammalian chromosome, resulting in the release of a 2.5 kb yeast DNA fragment flanked by two directly repeated *loxP* sites (Sauer and Henderson, 1989). In 1990, David Ow's group was the first to demonstrate the functioning of the Cre-*lox* system in plants (Dale and Ow, 1990). The group then reported, in 1991, that the Cre protein can precisely and conservatively (without loss or alteration of nucleotides in the recombination sites) catalyze excision events in the tobacco genome to remove a selectable marker gene (Dale and Ow, 1991). Since then, numerous publications have employed the Cre-*lox* site-specific recombination system for mammalian and plant genome manipulation, including excising selectable marker genes from transgenic plants. In parallel, other site-specific recombination systems were discovered and tested in eukaryotic cells.

Applications of Site-Specific Recombination Systems for Plant Genome Manipulation

1. Site-Specific Recombination Systems Are Used to Delete Selectable Marker Genes in Plants.

The most straightforward application of site-specific recombination in plant cells is the removal of undesired DNA segments, especially unwanted selectable marker genes from transgenic plants. After the discovery that the Cre-*lox* system functioned well in plants (Dale and Ow, 1990), site-specific recombination systems have been used in a number of different ways to remove selectable marker genes, for example in transgenic tobacco (Dale and Ow 1991; Bayley *et al.*, 1992; Gleave *et al.*, 1999) and rice (Hoa *et al.*, 2002). This approach for removing unneeded selectable marker genes has the benefit, as described earlier in this chapter, of reducing concerns raised by regulatory groups and consumers about the introduction of new genes. Site-specific excision has also allowed recycling of the selectable marker genes for multiple rounds of genetic modification in plants (Dale and Ow, 1991). For example, the kanamycin resistance gene (*npt*II) is by far the most popular selectable marker used in transgenic crop production. However, it can be difficult to use the same selectable marker to introduce a second gene of interest into a transgenic plant that already contains an *npt*II resistance gene in the genome. Although other selectable markers exist, only a limited number are of practical use. Therefore, it is desirable to develop a system for removing the selectable marker gene after there is no need for its presence, thus permitting re-use of the marker and alleviating safety concerns.

1.1. Constitutively Expressing a Recombinase Gene for Selectable-Marker-Gene Excision

In initial studies using site-specific recombination systems to remove unwanted DNA marker genes in plants, two main strategies were employed: line crossing (target line x recombinase-expressing line) and re-transformation (two rounds of transformation). (1) For line crossing, target lines are developed by transforming plants with a T-DNA construct containing a gene of interest and a selectable marker gene which is flanked by recognition sites. The target lines are then screened via Southern analysis for single-copy transgenes and for precise construct structure (no rearrangements, truncations, or similar events occuring after transformation). In the meantime, recombinase-expressing lines are developed by genetically transforming plants and screening for lines which express the recombinase at high levels. The characterized target lines are then crossed to the characterized recombinase-expressing lines. The F_1 seeds derived from the cross are harvested, germinated and screened for the presence of both the target gene and the recombinase gene. Progeny of the F_1 plants are then tested for germinal transmission of the final product of excision. (2) For the re-transformation strategy, after the target lines are generated and screened, they are subjected to another round of genetic transformation, this time with a recombinase-expressing gene construct (for which a different selectable marker is needed). The re-transformed lines are screened for the presence of the recombinase-expressing gene. Transgenic lines with the gene are allowed to set seeds, and plants derived from these seeds are tested for germinal transmission of the marker-gene-excised, final transgenic product. The un-needed recombinase gene can then be removed from the marker-free lines by genetic segregation. Although marker-free transgenic crop plants can be produced in laboratories using these two processes, there are some disadvantages: the crossing strategy involves generating and then crossing target lines and recombinase-expressing lines, selecting for complete excision which is heritable, and finally segregating away the recombinase gene. This strategy is restricted to sexually propagated species, and some crops, such as potato, are only propagated asexually. For other crops, long generation times make crossing schemes impractical. On the other hand, the re-transformation method is lengthy. There are also reports that overexpression of the recombinase gene may cause abnormalities in the transgenic plants (Coppoolse *et al.*, 2003). Similarly, undesired effects following the overexpression of site-specific recombinases in mammalian cells have been documented (Buchholz, 2009). Loonstra *et al.* (2001) reported that Cre expression caused numerous chromosomal aberrations and markedly reduced proliferation in cultured mammalian cells. Also, Naiche *et al.* (2007) observed that Cre activity causes widespread apoptosis and lethal anemia during embryonic development, and Liu *et al.* (2009) found that phiC31 integrase induces a DNA damage response and chromosomal rearrangements in human adult fibroblasts.

1.2. Transiently Expressing a Recombinase Gene for Selectable-Marker-Gene Excision

There are ways to avoid the possibility of causing aberrant developmental phenotypes due to a constitutively overexpressed, stably integrated recombinase gene, as well as to bypass the need for crossing the recombinase gene in and out. These approaches include: (1) The recombinase can be transiently expressed from a DNA construct which, instead of inserting into the plant genome, is eventually degraded and removed from the plant cells. (2) The recombinase gene can be placed under an inducible promoter and arranged with the marker gene in such a way that, when the recombinase gene is activated, it can auto-excise its own coding sequence, as well as that of the marker gene. This section (1.2) discusses transient expression, while the next section (1.3) examines inducible expression.

Agrobacterium T-DNA vectors and plant virus vectors can serve as vehicles for delivering recombinase genes into plant cells for transient, high-level expression of recombinase proteins, without integration of the recombinase genes into the genome. There have been reports of successful, site-specific-recombinase-mediated selectable-marker-gene excisions through transient expression of Cre, with both *Agrobacterium* T-DNA vectors (Kopertekh and Schiemann, 2005) and plant-RNA-virus-based vectors (Kopertekh *et al.*, 2004; Jia *et al.*, 2006) being used as vehicles to deliver the *cre* gene.

Virus-based vectors are particularly suitable for temporary recombinase expression in plants, since the foreign genes delivered by viruses into the infected plant and amplified by viral replicase do not integrate into the plant genome (Koprowski and Yusibov, 2001). Jia and co-workers described the use of a movement-function-improved m30B viral vector [including a modifying movement protein (MP) gene from a well-developed plant virus vector, the Tobacco Mosaic Virus (TMV) 30B vector], to transiently express Cre recombinase and thus remove a selectable marker gene from transgenic tobacco. *Agrobacterium* carrying the TMV vector was pressure-infiltrated, with a needle-less syringe, into plant tissues containing an excision construct consisting of *loxP* sites flanking the *npt*II selectable marker. The authors obtained selectable-marker-free transgenic plants with 34% excision efficiency. The progeny inherited the marker-free genotype from the regenerated parental plants and were also virus-free, since TMV is not transmitted to seeds. A further examination by Southern analysis confirmed that no stably integrated *cre* gene was detected in the genomes of the marker-free transgenic plants. Other plant-virus-based vectors have also reportedly been used to delete marker genes via transient *cre* gene expression. This includes using a Potato Virus X (PVX)-based vector to transiently express Cre to remove a selectable marker gene from transgenic *Nicotiana benthamiana* (Kopertekh *et al.*, 2004).

Transient recombinase expression via biolistic methods was also demonstrated in plants. To achieve Cre-mediated site-specific integration in corn, Srivastava and a co-worker (Srivastava and Ow, 2001) co-bombarded a *cre*-gene construct and an integration vector into callus from a target line. The data showed that 60% of the single-copy-transgene-inserted plants lacked the recombinase gene, indicating that the *cre* gene expression was transient.

Additionally, polyethylene glycol (PEG)-mediated transformation can deliver a recombinase gene on a plasmid into protoplasts for transient expression of the recombinase protein (Albert and Ow, 2000; Yau *et al.*, 2010). Yau *et al.* used this method to integrate a gene of interest into a pre-determined locus in the tobacco genome. For this study, the authors co-transformed two constructs, the "integration construct" containing a Bxb1 *attB* site and the *bxb1* (or Bxb1-recombinase-expressing) construct, into protoplasts derived from a tobacco target line. The target line had previously been genetically transformed to embed a Bxb1 *attP* site into the genome. The *bxb1* construct provides recombinase activity for the *attB*-containing integration plasmid to recombine with the chromosomal *attP* site. By selecting only for the selectable marker carried on the integration construct, several putative integration lines were obtained. The integration lines were examined further to determine whether they contained the Bxb1 recombinase gene. No *bxb1* gene was detected, indicating that none had stably integrated into the genome and that the Bxb1 recombinase protein which facilitated integration of the integration construct into the embedded *attP* site in the target line was transiently expressed.

1.3. Induced Expression of a Recombinase Gene for Selectable-Marker-Gene Excision

In order to further control the excision process to remove the selectable marker at a certain time, recombinase genes driven by various inducible promoters have been developed. These include *heat-shock-inducible* promoters and *chemical-inducible* promoters. By activating the promoters with inducers (heat or chemicals), the expression of the recombinase gene can be tightly controlled. To shorten and simplify the process of excising selectable markers, various groups have designed a one-step auto-excision strategy for DNA removal using inducible promoters. In this strategy, the marker gene, the recombinase gene, and the gene(s) of the interest are cloned into a single construct. The recombinase gene under the control of an inducible promoter and the marker gene are placed into a cassette flanked by two recognition sites oriented in the same direction, while the gene of interest is placed outside the region flanked by the recognition sites. After transformation and molecular analysis, the transgenic plants are treated with inducers to initiate expression of the recombinase (*e.g.* Cre) to delete the region between the recognition sites (*e.g. loxP*); the recombinase and the selectable marker gene are both removed. Recent studies using this method have been reviewed by Gidoni *et al.* (2008). For example, a heat-shock-regulated, Cre-*lox*-mediated, site-specific excision of extraneous DNA in transgenic tobacco plants was reported (Liu *et al.*, 2005). In this study, the *Arabidopsis* heat-shock-responsive promoter *HSP81-1* (Takahashi *et al.,* 1992) was used to control the expression of Cre recombinase. In nearly all cases, the selectable markers (*bar* and *npt*II), the Cre gene, and other unneeded components located between two 34-bp *loxP* sites were efficiently removed after heat-shock treatments: Testing a total of 115 leaves (from 10 lines) revealed that the excision frequency was 95.7%. This strategy of using a heat-shock-inducible promoter to control timely expression of the recombinase gene has also been applied to *Arabidopsis* (Hoff *et al.*, 2001), maize (Zhang *et al.*, 2003), tobacco (Wang *et al.*, 2005), potato (Cuellar *et al.*, 2006) and aspen (Fladung *et al.*, 2010) for site-specific recombination events.

Chemically regulated, site-specific DNA excision to remove the selectable marker gene in transgenic plants via the site-specific recombination systems was also reported in *Arabidopsis*) [by Cre-*lox*] (Zuo *et al.*, 2001), aspen [by FLP-*FRT*] (Matsunaga *et al.*, 2002), rice [by Cre-*lox*] (Sreekala *et al.*, 2005), tomato [by Cre-*lox*] (Zhang *et al.*, 2006; Zhang *et al.*, 2009) and tobacco [by FLP-*FRT*] (Woo *et al.*, 2009). In the study by Zuo *et al.*, an estrogen-receptor-based transactivator termed XVE [for the bacterial repressor LexA (X), the acidic transactivating domain of VP16 (V), and the regulatory region of the human estrogen receptor

(E) (Zuo *et al.*, 2000)] and the Cre-*lox* system were used in combination. In transgenic plants, β-estradiol-activated XVE stimulated expression of the Cre protein under the control of the target promoter; this resulted in Cre-mediated site-specific recombination events and deletion of the unnecessary DNA. The XVE system was subsequently adopted by several groups for producing marker-free transgenic rice (Sreekala *et al.*, 2005) and tomato (Zhang *et al.*, 2006; Zhang *et al.*, 2009). Woo *et al.* (2009) reported auto-excision of selectable marker genes from transgenic tobacco via a stress-inducible FLP/*FRT* site-specific recombination system. In this system, an oxidative-stress-inducible promoter [a peroxidase (POD) promoter] was fused to the FLP recombinase gene. Hydrogen peroxide (H_2O_2) was used to induce the recombinase gene, and the successful excision of selectable marker genes via hydrogen peroxide (H_2O_2)-regulated site-specific recombination was observed.

Because heat-shock treatments and chemicals are required for recombinase activation for marker gene deletion, these treatments may be limited to certain plant species and may present complications for the transformation process (Li *et al.*, 2007). Other problems associated with these inducible systems include low promoter induction, which can lead to incomplete transgene excision, and background activity, which is due to unwanted leaky expression and leads to unintended excision. Leakiness in promoter activity among transgenic lines is common and has been reported by several laboratories which used chemical- or heat-shock-inducible promoters for recombinase gene expression (Gidoni *et al.*, 2008).

To prevent potential leakiness of the promoter, a 'double-lock' method was employed by Joubès *et al.* (2004) to tightly control a heat-shock-inducible promoter (*HSP18.2*) and thus avoid premature activation of the inducible system. This strategy combines the Cre-*lox* site-specific recombination system and sequence encoding a mammalian glucocorticoid receptor (GR) [specifically, the hormone-binding domain of the rat GR] in a single construct for selectable-marker-gene removal. The *cre* recombinase gene is fused to the GR sequence, and this entire coding region is driven by a heat-shock-inducible promoter. Hence, the ligand-inducible recombinase is under the control of two inducers: Heat shock induction of the promoter provides regulation at the transcriptional level, and glucocorticoid activation of the Cre-GR fusion provides regulation at the post-transcriptional level. In the absence of glucocorticoids, the Cre-GR protein remains in the cytosol, but when chemical inducer is added, the fusion protein moves into the nucleus. To accomplish selectable marker deletion, stably transformed tobacco calli were first treated at 37°C for 2 hours to activate the *cre* gene. Subsequently, the calli were treated with a glucocorticoid compound, dexamethasone, for 24 hours to trigger translocation of the GR-fused Cre protein into the nucleus, where Cre excised sequences flanked by *loxP*. The researchers found that, in 93% of the calli, the DNA fragment (*egfp*) between the two *loxP* sites was excised through Cre-*lox*-mediated recombination. No deletion was detected when only chemical induction or only heat treatment was used, confirming the effectiveness of the "double-lock" approach.

1.4. Genetically Programmed Recombinase Gene Expression for Selectable Marker Excision

Although inducible promoters offer control over the timing of recombinase expression, the need for external induction and problems with efficacy make inducible promoters less than ideal. One issue concerns how to efficiently apply inducing chemicals to transgenic plants on the farm. Application methods used in the laboratory – where plant tissue or callus can be conveniently placed on medium supplemented with the inducer(s) in a sterile Petri dish – are not feasible in the field. Besides, the chemicals used for induction could negatively impact other plants on the farm. Pre-mature recombination events, due to recombinases expressed from leaky promoters, are another concern. Chimerism, and possibly cost, has also hampered widespread field applications of such advanced approaches. Therefore, it is more desirable to have an auto-excision event catalyzed by recombinase proteins which are driven by a genetically controlled, endogenous trigger.

Several labs have made progress in this direction: They have successfully utilized pollen-specific or tissue-specific promoters to drive recombinase gene expression for the genetically controlled excision of selectable markers (Mlynárová *et al.*, 2006; Li *et al.*, 2007; Luo *et al.*, 2007; Verweire *et al.*, 2007; Frédéric *et al.*, 2009). When a pollen-specific promoter is used to regulate recombinase gene expression, selectable marker gene removal becomes an integral part of the biology of pollen maturation. Mlynárová *et al.* (2006) have used the tobacco NTM19 promoter to drive the Cre recombinase gene (*cre*), in order to auto-excise a *loxP*-flanked cassette embedded in transformed plants. The activity of the NTM19 promoter, which is highly specific to the microspore, is first detected in early uni-nucleate microspores and is absent from all vegetative tissues. The *loxP*-flanked cassette, in this case, was composed of the "NTM19 promoter-*cre* gene" DNA fragment and a selectable marker gene, connected to a gene of interest outside the *loxP* sites. Once the NTM19 promoter was activated during early pollen development, Cre-mediated site-specific recombination excised the DNA fragment located between the *loxP* sites. The system was highly efficient, with a failure rate of only 2 out of the 16,800 seeds which were derived from the selectable-marker-free tobacco pollen. Additionally, expression of the *cre* gene was tightly regulated; leakiness of the promoter was not observed (Mlynárová *et al.*, 2006). Verweire *et al.* (2007) also successfully demonstrated the use of *Arabidopsis* germline-specific promoters from the *APETALA1* and *SOLO DANCERS* genes to achieve genetically programmed, Cre-recombinase-mediated auto-excision. Marker-free progeny plants were efficiently obtained in this study, as well. More recently, Frédéric *et al.* (2009) evaluated seven different germline-related promoters for their suitability in regulating Cre expression in transgenic plants. Five out of the seven promoters were able to drive efficient Cre-mediated gene excision. The data also showed that use of these promoters resulted in much less variation in recombination frequency than previously reported for the 35S promoter.

2. Site-Specific Recombination Systems Are Used for Resolution-Based Single-Copy Transgenic Line Production

Agrobacterium-mediated transformation and particle bombardment are the two most widely used methods for plant transformation. Unfortunately, these methods of gene delivery lead to the random integration of transgenic material throughout the plant genome, and the number of transgenes that insert into the host chromosome cannot be determined in advance. Complex, multi-copy patterns in the genome add another concern, as they are often linked to gene silencing (Meyer and Saedler, 1996). Furthermore, evidence has indicated that DNA integration is often fragmented or concatenated at a single locus. When transgenic plants are produced through *Agrobacterium*-mediated transformation, the integration of multiple T-DNA copies into a single locus as direct or inverted repeats is fairly common (De Buck *et al.*,

1999). Moreover, biolistic (gene-gun) transformation tends to result in more complex integrations than *Agrobacterium*-mediated transformation does (Tobias *et al.*, 2007). It is possible to obtain a concatenated DNA array consisting of 3 to 500 or more transgene "monomers" at a single integration site (Dorer, 1997). These concatenated transgenes, especially when arranged as inverted repeats, are unstable due to the interaction of the repeated, homologous sequences within the genome; these circumstances can lead to variegated transgene expression, as seen in transgenic mice (Dobie *et al.*, 1997); transgene silencing, such as in transgenic plants (Wang and Waterhouse, 2000; De Buck *et al.*, 2001); and genetic instability from deleted sequences (Collick *et al.*, 1996). The expression level of the transgene does not necessarily correspond to the number of copies in the integrated array and must be determined experimentally by assaying mRNA and/or protein levels.

The formation of concatenated transgenes is a common problem in genomic manipulation. Therefore, it is desirable to develop a method for resolving these concatemers. Although the site-specific Cre-*lox* recombination system is mainly used to remove selectable marker genes, exposure of the modified genome to Cre recombinase has also been utilized to reduce the tandem arrays of transgenes at a single locus to a single integrated copy. This technique was originally shown to function in wheat by Srivastava and colleagues (1999). For this procedure, the transgenes or T-DNAs need to be engineered to contain one or two *loxP* sites. When a transgenic plant with a complex array of T-DNAs at a locus (regardless of the orientation and number) is crossed to a Cre-expressing plant, site-specific recombination between the outermost *loxP* sequences in the same orientation should resolve the multiple copies into a single T-DNA. Further exploiting the features of the Cre-*lox* system, Srivastava and Ow (2001) conducted an experiment to generate single-copy primary transformants. The authors hypothesized that co-bombarding a transgene flanked by oppositely-oriented *loxP* sites into maize calli along with a Cre-expressing construct would result in low copy numbers of the transgene insertion in the majority of the primary transformants, because Cre would resolve the concatemerized transgene copies before or after they were inserted into the plant genome. Indeed, 85% of the transformants had 1-2 transgene insertions. In another example, De Buck *et al.* (2007) designed a specialized T-DNA construct that contains two inversely-oriented *loxP* sites located inside and immediately adjacent to the T-DNA right and left borders, in order to produce a single-copy transgene insertion at a locus. This transgene reduction technique ensures more control over the system by resolving complex patterns of integration and allowing direct comparison between sites of integration and gene expression.

3. Site-Specific Recombination Systems Are Used in Breeding Programs for Hybrid Seed Production

Hybrids are the offspring derived from intercrossing between two inbred lines or among more than two inbred lines, *e.g.* four in the case of "double-crosses." In many crop species, hybrids are desirable because of superior agronomic characteristics, especially in terms of yield. Hybrids of most crops yield 10-30% more than their pure inbred parental lines (Mayo, 1980). This phenomenon is called "*heterosis*" or "hybrid vigor." Heterosis is the opposite of *inbreeding depression*, which occurs with increasing homozygosity. Some hypotheses, including the dominance and overdominance hypotheses, have been offered to explain the genetic basis of heterosis. The dominance hypothesis attributes the superiority of hybrids to the suppression of undesirable or deleterious recessive alleles from one parent by dominant alleles from the other parent. However, the heterosis effect disappears after the first (F_1) generation, so it is pointless for farmers to save seeds produced by a hybrid crop.

Since hybrid vigor is so powerful, breeders have continuously tried to develop new methods for producing hybrid seeds or crops to increase yield. This includes developing better inbred varieties, conducting intensive field trials of inbred varieties, and devising novel crossing schemes for producing hybrids. Corn and rice hybrids have been the most successful. However, because many important food crops are self-pollinating, producing hybrid seed often requires methods for promoting outcrossing. For example, for hybrid corn production, two inbred varieties can be planted in a field, in alternating rows. The tassels (male part) of one variety are removed to ensure that the second variety is the only pollen source in the field. Unfortunately, de-tasseling is labor-intensive and costly.

Another valuable method is to utilize cytoplasmic male-sterile (CMS) lines which produce no pollen and can serve as the female parent for hybrid crossing. For instance, breeders have sought and exploited CMS lines to produce hybrid rice. A male-fertility-restoring gene is usually crossed into the male parent used in hybrid seed production, so that the hybrid plants will be able to self-pollinate *(http://vivi.cornell.edu/* impact/individual l/vivo/individual16535). However, there are several drawbacks to CMS systems, such as, very few CMS lines are available, and it is hard to obtain good maintainer and restorer lines. Therefore, seeking alternative methods for producing hybrid crops has become one of the most important goals for crop breeders.

Luo *et al.* (2000) have incorporated FLP/*FRT* site-specific recombination in designing a sterility-fertility restoration system for hybrid seed production. In this study, male-sterile lines were produced by transforming an antisense *Arabidopsis bcp1* gene construct into *Arabidopsis*. The *bcp1* gene is essential for pollen fertility in *Arabidopsis*. At the same time, the researchers also generated FLP-recombinase-expressing lines. All the progeny of crosses between the sterile lines and the FLP-recombinase-expressing lines produced normal siliques and seeds. These results demonstrated the utility of site-specific recombination for developing hybrids.

Gils *et al.* (2008) have also employed a phiC31-based system to develop a two-component hybrid seed production system to effectively generate and maintain the male-sterile parent, produce hybrid seeds, and fully restore fertility in the hybrid plants. Utilizing excision mediated by the phiC31 recombinase, the researchers exploited the functional interaction of two transgenic loci that are in the same position on two homologous chromosomes. Two transgenic fragments (A1 and A2) in the two loci were generated from a single DNA construct containing the A1-A2 components. The DNA construct (A1-A2), built with the phiC31 recombination sites *attP* and *attB*, was transformed into *Arabidopsis*. The transformed lines were then crossed to plants expressing phiC31 recombinase. After site-specific excision, A1-only or A2-only progeny were generated. By re-crossing these progeny, the researchers obtained the male-sterile lines and the maintainer lines.

4. Novel Site-Specific Recombination Systems Can Potentially Be Exploited for Plant Genome Manipulations

Although there are currently several microbial site-specific recombination systems available for research uses, some of the systems are patented by the private sector. It is therefore worthwhile to identify other efficient site-specific recombination systems for genome modification *in planta*. The new-found site-specific recombination systems can also be used together with already-existing systems to expand the toolkit for plant genome manipulations. For example, a plant can be engineered to contain both Cre and phiC31 site-specific recombination components in the genome for site-specific excision and integration purposes (Srivastava and Ow, 2004).

Recently, our group, under the leadership of David W. Ow, reported that four members of the small serine resolvase subfamily, CinH, ParA, Tn1721, and Tn5053, can catalyze site-specific excision in a lower eukaryote, yeast. Compared to Cre-*lox*, the estimated deletion rates for CinH, ParA, Tn1721, and Tn5053 were 91.8%, 93.3%, 54.8% and 72.6%, respectively (Thomson and Ow, 2006). ParA is a small serine resolvase from a plasmid operon, *parCBA*, which is responsible for the maintenance of the broad host range plasmids RK2 and RP4; ParA catalyzes the recombination of two identical 133-bp recognition sites known as *MRS*. The ParA site-specific recombination system has been used in genetic transformations to remove selectable marker genes from *Arabidopsis* (Thomson *et al.*, 2009). The ParA recombinase excised the *MRS*-flanked DNA, and the excision event was detected in subsequent generations in the absence of the ParA enzyme, indicating germinal transmission of the excision event. We also tested the phiC31 system for its ability and efficacy in removing a selectable marker gene via site-specific recombination in *Arabidopsis*. In our experiments, the phiC31 system has proven highly effective in removing – through site-specific excision – a transgenic, target sequence embedded in the *Arabidopsis* genome and in generating stably excised progeny plants that carry only the recombined target DNA of interest in the absence of the recombinase gene (Thomson *et al.*, 2010a). Likewise, Kempe *et al.* (2010) reported that the phiC31 site-specific recombinase could excise a transgene from wheat chromosomes, and that the recombinant loci were inherited by the subsequent generation. The other site-specific recombination systems, including Bxb1, CinH and β, are also being investigated for applications in the genomic engineering of crop plants (Gronlund *et al.* 2007; Thomson *et al.*, 2010b). Gronlund *et al.* (2007) reported using the β/*six* system to delete a spacer region that is flanked by directly oriented *six* recognition sites. However, the β/*six* system requires an accessory protein, a host factor, for recombination.

Final Remarks and Future Perspectives

Genetic transformation techniques have been used to engineer disease resistance and other traits into plants, generating GM crops. The acceptance of biotech crops has increased steadily, and they have becoming a trend in farming. The "ISAAA Brief 39-2008: Executive Summary" from the International Service for the Acquisition of Agri-Biotech Applications (ISAAA) reported the following statistics for 2008: (1) the number of countries electing to grow biotech crops grew to 25, up from 6 in 1996, (2) the global hectarage of biotech crops increased to 125 million, and (3) the number of farmers growing GM crops reached 13.3 million globally. Amazingly, in 2008, 85% of the United States' maize crop, measured in hectares, was GM. Biotech crops have contributed to food security and the alleviation of poverty and hunger, as well as reducing impacts on the environment. In China, the cultivation of *Bt* cotton by small farmers has reduced insecticide use by 60% (ISAAA Brief 39-2008). Biotechnology cannot, and is not intended to, replace conventional breeding. Instead, biotechnology is complementary to conventional breeding; together, the two approaches promote progress in agriculture.

However, marker genes left behind in the process of generating transgenic plants have caused concerns among regulatory agencies and consumers. Producing marker-gene-free crops will alleviate these concerns. As this review outlines, many approaches have been employed to delete marker genes in transgenic plants, with good results. In particular, over the past 20 years, site-specific recombination systems have become onc of the most powerful tools for marker gene deletion. These systems have proven effective for manipulating plant genomes but will require fine-tuning to transition from the laboratory environment to farmers' fields.

REFERENCES

Abremski, K., Hoess, R. (1984) Bacteriophage P1 site-specific recombination: purification and properties of the Cre recombinase protein. *J Biol Chem* 259: 1509-1514.

Abremski, K., Hoess, R., Sternberg, N. (1983) Studies on the properties of P1 site-specific recombination: evidence for topologically unlinked products following recombination. *Cell* 32: 1301-1311.

Albert, H., Ow, D.W. (2000) Cre-*lox* directed integration of transgenes into the tobacco genome. *Plant Molecular Biology Manual* N3: 1-13.

Abremski, K., Hoess, R., Sternberg, N. (1983) Studies on the properties of P1 site-specific recombination: evidence for topologically unlinked products following recombination. *Cell* 32: 1301-1311.

Bayley, C.C., Morgan, M., Dale, E.C., Ow, D.W. (1992) Exchange of gene activity in transgenic plants catalyzed by the Cre-*lox* site-specific recombination system. *Plant Mol Biol* 18: 353-361.

Borlaug, N.E. (2002) Feeding a world of 10 billion people: the miracle ahead. *In Vitro Cell, Dev Biol-Plant* 38: 221-228.

Brookes, G., Barfoot, P. (2009) Global impact of biotech crops: Income and production effects, 1996-2007. *AgBioForum* 12: 184-208.

Buchholz, F. (2009) Engineering DNA processing enzymes for the postgenomic era. *Current Opinion in Biotechnology* 20: 383-389.

Cheng, M., Lowe, B.A., Spencer, T.M., Ye, X., Armstrong, C.L. (2004) Factors influencing *Agrobacterium*-mediated transformation of monocotyledonous species. *In Vitro Cell, Dev Biol-Plant* 40: 31-45.

Cohen, S.N., Chang, A.C.Y., Boyer, H.W., Helling, R.B. (1973) Construction of biologically functional bacterial plasmids *in vitro*. *Proc Natl Acad Sci USA* 70: 3240-3244.

Collick, A., Drew, J., Penberth, J., Bois, P., Luckett, J., Scaerou, F., Jeffreys, A., Reik, W. (1996) Instability of long inverted repeats within mouse transgenes. *EMBO J* 15: 1163-1171.

Coppoolse, E.R., de Vroomen, M.J., Roelofs, D., Smit, J., van Gennip, F., Hersmus, B.J.M., Nijkamp, H.J.J., van Haaren, M.J.J. (2003) Cre recombinase expression can result in phenotypic aberrations in plants. *Plant Mol Biol* 51: 263-279.

Cuellar, W., Gaudin, A., Solòrzano, D., Casas, A., Nopo, L., Chudalayandi, P., Medrano, G., Kreuze, J., Ghislain, M. (2006) Self-excision of the antibiotic resistance gene *nptII* using a heat inducible Cre-loxP system from transgenic potato. *Plant Mol Biol* 62: 71-82.

Dale, E., Ow, D.W. (1990) Intra- and intermolecular site-specific recombination in plant cells mediated by bacteriophage P1 recombinase. *Gene* 91: 79-85.

Dale, E., Ow, D.W. (1991) Gene transfer with subsequent removal of the selection gene from the host genome. *Proc Natl Acad Sci USA* 88: 10558-10562.

Darbani, B., Eimanifar, A., Stewart, Jr., C.N., Camargo, W.N. (2007) Methods to produce marker-free transgenic plants. *Biotechnol J* 2: 83-90.

De Buck, S., Jacobs, A., Van Montagu, M., Depicker, A. (1999) The DNA sequences of T-DNA junctions suggest that complex T-DNA loci are formed by a recombination process resembling T-DNA integration. *Plant J* 20: 295-304.

De Buck, S., Van Montagu, M., Depicker, A. (2001) Transgene silencing of invertedly repeated transgenes is released upon deletion of one of the transgene involved. *Plant Mol Biol* 46: 433-445.

Diaz, V., Servert, P., Prieto, I., Gonzalez, M.A., Martinez, A.C., Alonso, J.C., Bernad, A. (2001) New insights into host factor requirements for prokaryotic beta-recombinase-mediated reactions in mammalian cells. *J Biol Chem* 276: 16257-16264.

Dobie, K., Mehtali, M., McClenaghan, M., Lathe, R. (1997). Variegated gene expression in mice. *Trends Genet* 13: 127-130.

Dorer, D. (1997) Do transgene arrays form heterochromatin in vertebrates? *Transgenic Res* 6: 3-10.

Ebinuma, H., Sugita, K., Matsunaga, E., Yamakado, M. (1997) Selection of marker-free transgenic plants using the isopentyl transferase gene. *Proc Natl Acad Sci USA* 94: 2117-2121.

Fedoroff, N.V. (1989) in *Mobile DNA*, eds. Douglas, E.B. and Martha, M.H. (Am. Soc. Microbiol., Washington, DC), pp. 375-411.

Fladung, M., Becker, D. (2010) Elimination of marker genes and targeted integration via FLP/*FRT* recombination system from yeast in hybrid aspen (*Populus tremula* L. x *P. tremuloides* Michx.). *Tree Genet Genomics* 6: 205-217.

Frédéric, V.E., Verweire, D., Claeys, M., Depicker, A., Angenon, G. (2009) Evaluation of seven promoters to achieve germline directed Cre-*lox* recombination in *Arabidopsis thaliana*. *Plant Cell Rep* 28: 1509-1520.

Fuchs, R.L., Ream, J.E., Hammond, B.G., Naylor, M.W., Leimgruber, R.M., Berberich, S.A. (1993) Safety assessment of the neomycin phosphotransferase II (NPTII) protein. *Biotechnology* 11: 1543-1547.

Gerlitz, M., Hrabak, O., Schwab, H. (1990) Partitioning of broad-host-range plasmid RP4 is a complex system involving site-specific recombination. *J Bacteriol* 172: 6194-6203.

Ghosh, P., Wasil, L.R., Hatfull, G.F. (2006) Control of phage Bxb1 excision by a novel recombination directionality Factor. *PLoS Biol* 4: e186. doi:10.1371/journal.pbio.0040186

Gidoni, D., Srivastava, V., Carmi, N. (2008) Site-specific excisional recombination strategies for elimination of undesirable transgenes from crop plants. *In Vitro Cell Dev Biol- Plant* 44: 457-467.

Gils, M., Marillonnet, S., Werner, S., Grutzner, R., Giritch, A., Engler, C., Schachschneider, R., Klimyuk, V., Gleba, Y. (2008) A novel hybrid seed system for plants. *Plant Biotechnol* 6: 226-235.

Gleave, A.P., Mitra, D.S., Mudge, S.R., Morris, B.A.M. (1999) Selectable marker-free transgenic plants without sexual crossing: transient expression of *cre* recombinase and use of a conditional lethal dominant gene. *Plant Mol Biol* 40: 223–235.

Golic, K.G., Lindquist, S. (1989) The FLP recombinase of yeast catalyzes site-specific recombination in the *Drosophila* genome. *Cell* 59: 499-509.

Grindley, N.D., Whiteson K.L., Rice P.A. (2006) Mechanisms of site specific recombination. *Annu Rev Biochem* 75: 567-605.

Gronlund, J.T., Stemmer, C., Lichota, J., Merkle, T., Grasser, K.D. (2007) Functionality of the β/*six* site-specific recombination system in tobacco and *Arabidopsis*: a novel tool for genetic engineering of plant genomes. *Plant Mol Biol* 63: 545-556.

Hoa, T.T., Bong, B.B., Huq, E., Hodges, T.K. (2002) Cre/*lox* site-specific recombination controls the excision of a transgene from the rice genome. *Theor Appl Genet* 104: 518-525.

Hoess, R.H., Abremski, K. (1985) Mechanism of strand cleavage and exchange in the Cre-*lox* site specific recombination system. *J Mol Biol* 181: 351-362.

Hoess, R.H., Ziese, M., Sternberg, N. (1982) P1 site-specific recombination: Nucleotide sequence of the recombining sites. *Proc Natl Acad Sci USA* 79: 3398-3402.

Hoff, T., Schnorr, K.M., Mundy, J.A. (2001) Recombinase-mediated transcriptional induction system in transgenic plants. *Plant Mol Biol* 45: 41-49.

Hohn, B., Levy, A.A., Puchta, H. (2001) Elimination of selection markers from transgenic plants. *Curr Opin in Biotechnol* 12: 139-143.

Huang, S., Gilbertson, L.A., Adams, T.H., Malloy, K.P., Reisenbigler, E.K., Birr, D.H., Snyder, M.W., Zhang, Q., Luethy, M.H. (2004) Generation of marker-free transgenic maize by regular two-border *Agrobacterium* transformation vectors. *Transgenic Res* 13: 451-461.

Jia, H., Pang, Y., Chen, X., Fang, R. Removal of the selectable marker gene from transgenic tobacco plants by expression of Cre recombinase from a tobacco mosaic virus vector through Agroinfection. *Transgenic Res* 15: 375-384.

Joubès, J., De Schutter, K., Verkest, A., Inzé, D., De Veylder, L. (2004) Conditional, recombinase-mediated expression of genes in plant cell cultures. *Plant J* 37: 889-896.

Kaeppler, H. F., Gu, W., Somers, D. A., Rines, H.W., Cockburn, A.F. (1990) Silicon carbide fiber-mediated DNA delivery into plant cells. *Plant Cell Rep* 9: 415-418.

Kempe, K., Rubtsova, M., Berger, C., Kumlehn, J., Schollmeier, C., Gils, M. (2010) Transgene excision from wheat chromosomes by phage phiC31 integrase. *Plant Mol Biol* 72: 673–687.

Keravala, A., Groth, A.C., Jarrahian, S., Thyagarajan, B., Hoyt, J.J., Kirby, P.J., Calos, M.P. (2006) A diversity of serine phage integrases mediate site-specific recombination in mammalian cells. *Mol Gen Gent* 276: 135-146.

Kholodii G. (2001) The shuffling function of resolvases. *Gene* 269: 121-130.

Khush, G.S. (1999) Green revolution: preparing for the 21st century. *Genome* 42: 646-655.

Kim, A.I., Ghosh, P., Aarron, M.A., Bibb, L.A., Jain, S., Hatfull, G.F. (2003) Mycobacteriophage Bxb1 integrates into the *Mycobacterium smegmatis groELI* gene. *Mol Microbiol* 50: 463-473.

Kim, Y.G., Cha, J., Chandrasegaran, S. (1996) Hybrid restriction enzymes: zinc finger fusions to *Fok* I cleavage domain. *Proc Natl Acad Sci USA* 93: 1156-1160.

Kopertekh, L., Jűttner, G., Schiemann, J. (2004) PVX-Cre-mediated marker gene elimination from transgenic plants. *Plant Mol Biol* 55: 491-500.

Kopertekh, L., Schiemann, J. (2005) Agroinfiltration as a tool for transient expression of *Cre* recombinase *in vivo. Transgenic Res* 14: 793-798.

Koprowski, H., Yusibov, V. (2001) The green revolution: plants as heterologous expression vectors. *Vaccine* 19: 2735-2741.

Li, Z., Xing, A., Moon, B.P., Burgoyne, S.A., Guida, A.D., Liang, H., et al. (2007) A Cre/loxP-mediated self-activating gene excision system to produce marker gene free transgenic soybean plants. *Plant Mol Biol* 65: 329-341.

Liu, H.-K., Yang, C., Wei, Z.-M. (2005) Heat shock-regulated site-specific excision of extraneous DNA in transgenic plants. *Plant Sci* 168: 997-1003.

Liu, J., Skjorringe, T., Gjetting, T., Jensen, T.G. (2009) PhiC31 integrase induces a DNA damage response and chromosomal rearrangements in human adult fibroblasts. *BMC Biotechnol* 9: 31.

Loonstra, A., Vooijs, M., Beverloo, H.B., Allak, B.A., van Drunen, E., Kanaar, R., Berns, A., Jonkers, J. (2001) Growth inhibition and DNA damage induced by Cre recombinase in mammalian cells. *Natl Acad Sci USA* 98: 9209-9214.

Luo, H., Lyznik, A., Gidoni, D., Hodges, T.K. (2000) FLP-mediated recombination for use in hybrid plant production. *Plant J* 23: 423-430.

Luo, K., Duan, H., Zhao, D., Zheng, X., Deng, W., Chen, Y., Stewart Jr, C.N., McAvoy, R., Jiang, X., Wu, Y., He, A., Pei, Y., Li, Y. (2007) 'GM-gene-deletor': fused *loxP-FRT* recognition sequences dramatically improve the efficiency of FLP or CRE recombinase on transgene excision from pollen and seed to tobacco plants. *Plant Biotechnol J* 5: 263-274.

Matsunaga, E., Sugita, K., Ebinuma, H. (2002). Asexual production of selectable marker-free transgenic woody plants, vegetatively propagated species. *Mol Breed* 10: 95–106.

Mayo, H. (1980) *The Theory of Plant Breeding. Oxford*: Clarendon Press.

Meyer, P., Saedler, H. (1996) Homology dependent gene silencing in plants. *Ann Rev Plant Physiol Plant Mol Biol* 47: 23-48.

Miki, B., McHugh, S. (2004) Selectable marker genes in transgenic plants: applications, alternatives and biosafety. *J Biotechnol* 107: 193-232.

Mlynárová, L., Conner, A.J., Nap, J.-P. (2006) Directed microspore-specific recombination of transgenic alleles to prevent pollen-mediated transmission of transgenes. *Plant Biotechnol J* 4: 445-452.

Naiche, L.A., Papaioannou, V.E. (2007) Cre activity causes widespread apoptosis and lethal anemia during embryonic development. *Genesis* 45: 768-775.

Olivares, E.C., Hollis, R.P., Calos, M.P. (2001) Phage R4 integrase mediates site-specific integration in human cells. *Gene* 278: 167-176.

Onouchi, H., Yokoi, K., Machida, C., Matsuzaki, H., Oshima, Y., Matsuoka, K., Nakamura, K., Machida, Y. (1991) Operation of an efficient site-specific recombination system of *Zygosaccharomyces rouxii* in tobacco cells. *Nucleic Acids Res* 19: 6373-6378.

Ow, D.W. (2002) Recombinase-directed plant transformation for the post-genomic era. *Plant Mol Biol* 48: 183-200.

Petolino, J.F., Worden, A., Curlee, K., Connell, J., Moynahan, T.L.S., Larsen, C., Russell, S. (2010) Zinc finger nuclease-mediated transgene deletion. *Plant Mol Biol* 73: 617-628.

Poehlman, J.M., Sleper, D.A. (1995) The genetic basis of plant breeding. In: Breeding field crops, 4th edition. Iowa State University Press, Ames, Iowa, pp 19-82.

Porteus, M.H., Carroll, D. (2005) Gene targeting using zinc finger nucleases. *Nat Biotechnol* 23: 967-973.

Prakash, N.S., Bhojaraja, R., Shivbachan, S.K., Priya, G.G.H., Nagraj, T.K., Prasad, V., Babu, V.S., Jayaprakash, T.L., Dasgupta, S., Spencer, T.M., Boddupalli, R.S. (2009) Marker-free transgenic corn plant production through co-bombardment. *Plant Cell Rep* 28: 1655-1668.

Puchta, H. (2000) Removing selectable marker genes: taking the shortcut. *Trends Plant Sci* 5: 273-274.

Puchta, H., Hohn, B. (1996) From centiMorgans to basepairs: homologous recombination in plants. *Trends Plant Sci* 1: 340-348.

Sadowski, P.D.E. (1993) Site-specific genetic recombination: hops, flips, and flops. *FASEB J* 7: 760-767.

Sasaki, A., Ashikari, M., Ueguchi-Tanaka, M., Itoh, H., Nishimura, A., Swapan, D., Ishiyama, K., Saito, T., Kobayashi, M., Khush, G.S., Kitano, H., Matsuoka, M. (2002) A mutant gibberellin-synthesis gene in rice. *Nature* 416: 701-702.

Sauer, B. (1987) Functional expression of the *cre-lox* site-specific recombination system in the yeast *Saccharomyces cerevisiae. Mol Cell Biol* 7: 2087-2096.

Sauer, B., Henderson, N. (1988) Site-specific DNA recombination in mammalian cells by the Cre recombinase of bacteriophage P1. *Proc Natl Acad Sci USA* 93: 1156-1160.

Sauer, B., Henderson, N. (1989) Cre-stimulated recombination at *loxP*-containing DNA sequences placed into the mammalian genome. *Nucleic Acids Res* 17: 147-161.

Schwikardi, M., Dröge, P. (2000) Site-specific recombination in mammalian cells catalyzed by γδ resolvase mutants: implications for the topology of episomal DNA. *FEBS Lett* 471: 147-150.

Smith, M.C., Thorpe, H.M. (2002) Diversity in the serine recombinases. *Mol Microbiol* 44: 299-307.

Sreekala, C., Wu, L., Gu, K., Wang, D., Tian, D., Yin, Z. (2005) Excision of a selectable marker in transgenic rice (*Oryza sativa* L.) using a chemically regulated Cre/*loxP* system. *Plant Cell Rep* 24: 86-94.

Srivastava, V., Anderson, O.D., Ow, D.W. (1999) Single-copy transgenic wheat generated through the resolution of complex integration patterns. *Proc Natl Acad Sci USA* 96: 11117-11121.

Srivastava, V., Ow, D.W. (2001) Single-copy primary transformants of maize obtained through the co-introduction of a recombinase-expressing construct. *Plant Mol Biol* 46: 561–566.

Srivastava, V., Ow, D.W. (2004) Marker-free site-specific gene integration in plants. *Trends Biotechnol* 22: 627-629.

Stark, W.M., Boocock, M.R., Sherratt, D.J. (1992) Catalysis by site-sepcific recombinases. *Trends Genet* 8: 432-439.

Sternberg, N., Hamilton, D. (1981) Bacteriophage P1 site-specific recombination- I. Recombination between *lox*P sites. *J Mol Biol* 150: 467-486.

Stoll, S., Ginsburg, D.S., Calos, M.P. (2002) Phage TP901-1 site-specific integrase functions in human cells. *J Bacteriol* 184: 3657–3663.

Takahashi, T., Naito, S., Komeda, Y. (1992) Isolation and analysis of the expression of two genes for the 81-Kilodalton heat-shock proteins from *Arabidopsis*. *Plant Physiol* 99: 383-390.

Thomason, L.C., Calendar, R., Ow, D.W. (2001) Gene insertion and replacement in *Schizosaccharomyces pombe* mediated by the *Streptomyces* bacteriophage phiC31 site-specific recombination system. *Mol Gen Genet* 265: 1031-1038.

Thompson, C.J., Movva, N.R., Tizard, R., Crameri, R., Davies, J.E., Lauwereys, M., Botterman, J. (1987) Characterization of the herbicide-resistance gene *bar* from *Streptomyces hygroscopicus*. *EMBO J* 6: 2519-2523.

Thomson, J.G., Chan, R., Thilmony, R., Yau, Y.-Y., Ow, D.W. (2010a) PhiC31 recombination system demonstrates heritable germinal transmission of site-specific excision from the *Arabidopsis* genome. *BMC Biotechnology* 10:17 (on-line).

Thomson, J.G., Ow, D.W. (2006) Site-specific recombination systems for the genetic manipulation of eukaryotic genomes. *Genesis* 44: 465-476.

Thomson, J.G., Yau, Y.-Y., Blanvillain, R., Chiniquy, D., Thilmony, R. Ow, D.W. (2009) ParA resolvase catalyzes site-specific excision of DNA from the *Arabidopsis* genome. *Transgenic Res* 18: 237–248.

Thomson, J.G., Yau, Y.-Y, Chan, R., Thilmony, R. (2010b) Development of site-specific recombinase technology for precise crop plant genome modification. (*IAPB conference abstract, 2010*)

Thorpe, H.M., Wilson, S.E., Smith, M.C. (2000) Control of directionality in the site-specific recombination system of the *Streptomyces* phage phiC31. *Mol Microbiol* 38: 232-41.

Tobias, D.J., Manoharan, M., Pritsch, C., Dahleen, L.S. (2007) Co-bombardment, integration and expression of rice chitinase and thaumatin-like protein genes in barley (*Hordeum vulgare* cv. Conlon). *Plant Cell Rep* 26: 631-639.

Verweire, D., Verleyen, K., De Buck, S., Claeys, M., Angenon, G. (2007) Marker-free transgenic plants through genetically programmed auto-excision. *Plant Physiol* 145: 1220–1231.

Wang, M.-B., Waterhouse, P.M. (2000) High-efficiency silencing of a β-glucuronidase gene in rice is correlated with repetitive transgene structure but is independent of DNA methylation. *Plant Mol Biol* 43: 67–82.

Wang, Y., Chen, B., Hu, Y., Li, J., Lin, Z. (2005) Inducible excision of selectable marker gene from transgenic plants by the Cre/*lox* site-specific recombination system. *Transgenic Res* 14: 605-614.

Warwick, S.I., Simard, M.J., Légère, A., Beckie, H.J., Braun, L., Zhu, B., Mason, P., Séguin-Swartz, G., Stewart, C.N. (2003) Hybridization between transgenic *Brassica napus* L. and its wild relatives: *Brassica rapa* L., *Raphanus raphanistrum* L., *Sinapis arvensis* L., and *Erucastrum gallicum* (Willd.) O.E, Schulz. *Theor Appl Genet* 107: 528-539.

Woo, H.J., Cho, H.S., Lim, S.H., Shin, K.S., Lee, S.M., Lee, K.J., Kim, D.H., Cho, Y.G. (2009) Auto-excision of selectable marker genes from transgenic tobacco via a stress inducible FLP/*FRT* site-specific recombination system. *Transgenic Res* 18: 455-465.

Wulff, B.B.H., Kruijt, M., Collins, P.L., Thomas, C.M., Ludwig, A.A., De Wit, P.J.G.M., Jones, J.D.G. (2004) Gene shuffling-generated and natural variants of the tomato resistance gene *Cf-9* exhibit different auto-necrosis-inducing activities in *Nicotiana* species. *Plant J* 40: 942-956.

Yau, Y.-Y., Wang, Y.J., Thomson, J.G., Ow, D.W. (2010) Method for Bxb1-mediated site-specific integration *in planta*. Ed. James A. Birchler. In: *Plant Chromosome Engineering: Method in Molecular Biology*. Humana Press, Clifton, New Jersey).

Zhang, W., Subbarao, S., Addae, P., Shen, A., Armstrong, C., Peschke, V., Gilbertson, L. (2003) Cre/*lox*-mediated marker gene excision in transgenic maize (*Zea mays* L.) plants. *Theor Appl Genet* 107: 1157-1168.

Zhang, Y., Li, H., Ouyang, B., Lu, Y., Ye, Z. (2006) Chemical-induced autoexcision of selectable markers in elite tomato plants transformed with a gene conferring resistance to lepidopteran insects. *Biotechnol Lett* 28: 1247-1253.

Zhang, Y., Liu, H., Zhang, J.-T., Li, Y., Zhang, H. (2009) Generation of selectable marker-free transgenic tomato resistant to drought, cold and oxidative stress using the Cre/*loxP* DNA excision system. *Transgenic Res* 18: 607-619.

Zubko, E., Scutt, C., Meyer, P. (2000) Intrachromosomal recombination between *attP* regions as a tool to remove selectable marker genes from tobacco transgenes. *Nat Biotechnol* 18: 442-445.

Zuo, J., Niu, Q.-W., Møller, S.G., Chua, N.-H. (2000) An estrogen receptor-based transactivator XVE mediates highly inducible gene expression in transgenic plants. *Plant J* 24: 265-273.

Zuo, J., Niu, Q.-W., Møller, S.G., Chua, N.-H. (2001) Chemical-regulated, site-specific DNA excision in transgenic plants. *Nat Biotechnol* 19: 157–161.

In: Genetically Engineered Crops
Editors: Lindsay M. Grover

ISBN: 978-1-61122-526-6

Chapter 3

RISK ASSESSMENT OF SUGARCANE BORER RESISTANCE TO TRANSGENIC MAIZE EXPRESSING *BACILLUS THURINGIENSIS* PROTEINS

Fangneng Huang, B. Rogers Leonard, Xiaoyi Wu and Mukti N. Ghimire

Department of Entomology, Louisiana State University Agricultural Center,
Baton Rouge, Louisiana, U.S.A

ABSTRACT

Sugarcane borer, *Diatraea saccharalis*, is a major maize stalk-boring pest in many areas of the mid-south region of the United States. In Louisiana, transgenic maize expressing *Bacillus thuringiensis* (Bt) Cry1Ab protein was first commercially planted in 1999 and now is the primary tool for managing stalk boring pests. Resistance development in target pest populations is an important issue in the sustainable use of Bt crops. Since 2004, a coordinated research program has been established to assess the risk of *D. saccharalis* resistance to Bt maize. This paper summarizes the major accomplishments and presents up-to-date results of this research program. Field surveys showed that a complex of *D. saccharalis*; southwestern corn borer, *Diatraea grandiosella*; and European corn borer, *Ostrinia nubilalis* could occur in Louisiana maize fields. *D. saccharalis* is the dominant species across the state, accounting for 88% of the total maize borers sampled during 2004-2009. Both *D. saccharalis* and *D. grandiosella* were able to overwinter in the major maize growing areas of the state. During 2004-2009, a total of 2,061 feral individuals of *D. saccharalis* were collected from four geographical locations in Louisiana and examined for Cry1Ab resistance alleles using F_1/F_2 screening methods. Cry1Ab resistance allele frequency was low (0.0011) for the populations collected during 2004-2008, but increased in 2009 (0.0136). A highly Bt resistant strain (Cry1Ab-RR) of *D. saccharalis* was isolated from a field population by using the F_2 screen. The Cry1Ab-RR strain was able to survive and complete larval development on commercial Cry1Ab maize plants. Relative to a susceptible strain of *D. saccharalis*, the resistant strain showed an approximately 100-fold resistance to Cry1Ab and also exhibited a significant resistance to Cry1Aa and Cry1Ac. Larvae of the resistant heterozygotes demonstrated relatively high survivorship rates on Cry1Ab maize hybrids which were commonly planted in Louisiana, suggesting most Cry1Ab maize hybrids planted in the state did not express a "high dose" against *D. saccharalis*. Cry1Ab resistance in *D. saccharalis* was likely controlled by one (or a few linked) autosomal gene. However, effective dominance level of resistance varied depending on bioassay methods and maize hybrids, ranging from incompletely recessive to incompletely dominant. In addition, the Cry1Ab resistance in *D. saccharalis* was not associated with any fitness costs and the resistance was stable after 24 generations without selection. The results of the 6-year study suggest that there appears to be a high risk for development of Cry1Ab resistance in *D. saccharalis* if Cry1Ab maize continues to be widely used in the U.S. mid-south region.

Key words: *Diatraea saccharalis*, field maize, transgenic Bt crops, resistance monitoring, IRM, maize stalk borers.

INTRODUCTION

Advances in biotechnology have allowed scientists to transfer *Bacillus thuringiensis* (Bt) genes into crop plants. Since first being commercialized in 1996, Bt maize has gained widespread acceptance among United States maize producers; approaching >22 million hectares (or 63% of the total maize area) in 2009 (NASS 2009). However, the widespread acceptance of Bt maize should place a significant selection pressure on maize stalk borer populations and accelerate development of resistance, raising concerns about the sustainability of Bt maize as an effective pest management tool (Ostlie et al. 1997, Gould 1998). To enable the sustainable use of Bt maize technology, a mandatory insect resistance management (IRM) plan has been implemented for planting Bt maize in the United States. The current IRM strategy for Bt maize is named the "high dose/refuge" strategy.

Management of Bt resistance in sugarcane borer, *Diatraea saccharalis* (F.), is important in the mid-southern region of the United States (Huang et al. 2006). In recent years, *D. saccharalis* has expanded its geographic range and has become the dominant maize stalk borer species in many areas in the mid-southern states of the United States, especially in Louisiana and Texas (Castro et al. 2004a, Huang et al. 2006, Porter et al. 2005). Two previous studies from Louisiana also showed that field performance of Bt maize against *D. saccharalis* varied among individual Bt maize varieties (Castro et al. 2004b, McAllister et al. 2004). Because of the recent emerging problem of *D. saccharalis* on maize in the U.S. mid-southern region, this species has been listed as a target pest of Bt maize in the United States since 2005 (United States Environmental Protection Agency 2005a, b).

The success of the "high dose/refuge" IRM strategy requires that (1) Bt maize plants produce a high dose that can kill heterozygotes for Bt resistance; (2) resistance is controlled by a recessive gene; (3) resistance frequencies in field insect populations are initially very rare (e.g. 0.001); and (4) sufficient Bt susceptible insects are available to mate with resistant individuals survived from Bt plants (Bourguet et al. 2003). Since 2004, a cooperated research program with multiple approaches has been performed to assess the resistance risk of *D. saccharalis* to Bt maize. The major goals of the research program were 1) to validate the current high dose-refuge IRM strategy for managing resistance of *D. saccharalis* to Bt maize and 2) to provide information necessary to support/improve IRM strategies for managing *D.*

saccharalis resistance to Bt maize. This chapter summarizes the major accomplishments and reports up-to-date results of this research program.

Population Structures and Overwintering of Maize Stalk Borers in Louisiana

To determine the distribution and population structures of maize stalk borers in Louisiana, field surveys were conducted across the major maize growing areas from 2004-2009. During the six-year survey, a total of >10,000 maize stalk borers were collected and identified for species (Table 1). The survey showed that a complex of European corn borer, *Ostrinia nubilalis* Hübner; southwestern corn borer, *Diatraea grandiosella* Dyar, and *D. saccharalis* could occur in maize fields in Louisiana, while *D. saccharalis* was the most abundant species, accounting for 55-100% of the total borer populations sampled during the six years. *Diatraea grandiosella* and *O. nubilalis* occurred only in the northern region of the state, accounting for 11% and <1% of the total population, respectively. The field surveys also showed that *D. saccharalis* has virtually completely replaced *D. grandiosella* to become the pre-dominant maize stalk borer species in recent years in Louisiana (Table 1).

Table 1. Population structures of maize borers in Louisiana.

Year	% of total borer populations sampled in each year		
	Ostrinia nubilalis	*Diatraea grandiosella*	*Diatraea saccharalis*
2004	1.5	13	85
2005	1.0	44	55
2006	0.5	11	89
2007	0	2	98
2008	0	0	100
2009	0	0	100
Overall	<1	11	88

Field surveys were also conducted for overwintering of maize stalk borers in northeast and central regions of Louisiana from 2004-2007. During the late fall 2004 and winter months in 2005, >600 plant stubbles of maize and sorghum were sampled for overwintering borer larvae in Franklin and Tensas parishes in Louisiana. Approximately 5-10% of these stubbles contained at least one overwintering larva of *D. saccharalis* or *D. grandiosella* in both locations. Overwintering larvae of *D. saccharalis* also were observed in rice stubble in Franklin Parish. No overwintering larvae of *O. nubilalis* were found in the surveys. This was the first documented report that *D. saccharalis* was able to overwinter in northern Louisiana. The follow-up surveys during 2005-2007 also showed that the annual overwintering survival rate of *D. saccharalis* was closely related to the overall field occurrences in maize growing seasons. The clarification of the dominance of *D. saccharalis* in maize stalk borer populations and its ability to overwinter in northern Louisiana suggest that these two important factors should be incorporated in resistance management for the sustainable use of Bt maize for managing stalk borers in Louisiana.

Relative Susceptibility of *D. Saccharalis* to Cry1Ab Protein Expressed in Bt Maize Plants

Studies have shown that field resistance to Bt crops could develop more rapidly if the target insect pest is more tolerant to the Bt toxins expressed in the plants (van Rensburg, 2007, Huang and Andow 2010). Laboratory bioassays were conducted to compare the Cry1Ab susceptibility of *D. saccharalis* with those of *O. nubilalis* and *D. grandiosella*, the two most important maize borer species across the North Central and Mid-Western Regions of the US Maize Belt (Ostlie et al. 1997, Huang et al. 2006). For each borer species, Cry1Ab susceptibility was determined for a field-collected population and a laboratory strain. Larval mortality and insect growth were measured by exposing neonates to Bt leaf tissues and on meridic diet treated with Cry1Ab protein extracted from Bt maize leaves.

Laboratory bioassays showed that *D. saccharalis* was the least susceptible species to Cry1Ab protein among the three maize borer species. The LC_{50}s of *D. saccharalis* to Cry1Ab were 3- and 11-fold greater than those of *D. grandiosella* and *O. nubilalis*, respectively, and the LC_{90} values were 8- and 32-fold higher than those of *D. grandiosella* and *O. nubilalis.* The less susceptibility of *D. saccharalis* to Cry1Ab protein suggests that a high dose Bt maize for *O. nubilalis* and *D. grandiosella* might not be a high dose for *D. saccharalis* (Huang et al. 2006).

Resistance Allele Frequency to Cry1Ab Maize in Field Populations of *D. Saccharalis*

Resistance monitoring is considered one of the essential components in resistance management for Bt crops (United States Environmental Protection Agency 2001). Results of Bt resistance monitoring can be used to evaluate the success of IRM programs and more importantly, to provide a basis for the timely implementation of new management tactics to control resistant populations (Huang 2006). Since 2004, a coordinated monitoring program for detecting resistance to Cry1Ab maize in field populations of *D. saccharalis* has been implemented in Louisiana (Huang et al. 2007a, Huang and Leonard 2008, Yue et al. 2008). This monitoring program includes: 1) planting field maize after the optimum planting dates at multi-locations in the primary maize production areas of the state to attract and maintain late-season pest populations; 2) collecting larvae and pupae from these areas on multiple dates to ensure sufficient sample sizes for developing F_2/F_1 iso-family-lines; 3) rearing field-collected larvae to pupal stage and sexing all pupae within a location; 4) establishing two-parent family-lines for a F_2 screen or crossing field-derived moths with those from a Bt resistant strain that were used in a F_1 screen; 5) detecting Bt resistance alleles using F_2/F_1 screens on Bt maize leaf tissues; 6) confirming Bt resistance for potential positive family-lines identified in the F_2/F_1 screens; and 7) estimating Bt resistance allele frequencies. The Bt resistant strain (Cry1Ab-RR) of *D. saccharalis* used for developing F_1 populations in the F_1 screen was isolated from a field population by using the F_2 screen during 2004 (Huang et al. 2007a). The Cry1Ab-RR strain was able to survive and complete larval development on commercial Cry1Ab maize plants. The major goals of the monitoring program were to determine (i) if Bt resistance alleles in field populations of *D. saccharalis* was sufficiently low to meet the rare resistance requirement of the "high dose/refuge" IRM strategy and (ii) if there were early shifts in Cry1Ab resistance allele frequencies so that proactive measures for managing resistance could be deployed before field control failures.

Each year, larvae and pupae of *D. saccharalis* were collected from maize fields. Field collected larvae were reared to pupal stage. Field-derived moths were single-paired to establish two-parent family-lines for screening Bt resistance in F_2 generations or were crossed with those from a resistant strain for examining Bt resistance in F_1 generations. Insect samples collected during 2004, 2005, 2008, and 2009 were examined for Bt resistance alleles using the F_2 screening method (Huang et al. 2007a, Huang et al. 2008, Huang and Leonard 2008). If one of the two parents for a F_2 family-line carried a major Bt resistance allele, 6.25% of F_2 larvae were expected to be homozygous for Bt resistance and should survive in the F_2 screen (Andow and Alstad 1998). Maize borers sampled during 2006 and 2007 were examined for Bt resistance using the F_1 screening method (Yue et al. 2008). If the feral parent in a F_1 line possesses the same resistance allele as the resistant parent, 50% of the F_1 progeny were expected to be homozygous for Bt resistance and should survive in the F_1 screen. In the F_2 screen, a feral individual of *D. saccharalis* was defined to possess a major resistance allele if its F_2 larvae could survive for ≥12 days in the laboratory screens and could complete larval development on intact Bt maize plants (Huang et al. 2008). For the F_1 screen, the field collected parent was considered to carry a resistance allele if the larval survivorship rate of its F_1 progeny was not significantly different from the expected values (e.g. 50% survival rate) at the $\alpha = 0.05$ level (Yue et al. 2008). The Bt resistance allele frequencies were estimated using a Bayesian analysis as described in Andow and Alstad (1999).

Resistance allele frequency in a field population of *D. saccharalis* sampled during 2004. A total of 426 feral individuals of *D. saccharalis* collected from maize fields in Franklin Parish, Louisiana were screened for Cry1Ab resistance using the F_2 screening method (Huang et al. 2007a). One individual was identified to possess a major resistance allele to Cry1Ab maize. The estimated Cry1Ab resistance allele frequency in this population was 0.0023 with a 95% confidence interval of 0.0003-0.0064 (Table 2).

Table 2. Estimated frequency of Cry1Ab resistance alleles in Louisiana populations of *Diatraea saccharalis*.

Year	F_2/F_1 screen	No. feral individuals screened	No. individuals possessing major resistance alleles	Major resistance allele frequency and/or 95% CI
2004	F_2	426	1	0.0023 (0.0003-0.0064)
2005	F_2	560	0	(0 – 0.0027)
2006	F_1	331	1	0.0030 (0.0004-0.0084)
2007	F_1	251	1	0.0039(0.0005-0.0110)
2008	F_2	236	0	(0 – 0.0063)
2009	F_2	257	8	0.018 (0.008, 0.031)

Resistance allele frequency in field populations of *D. saccharalis* sampled during 2005. A total of 560 feral individuals from four populations of *D. saccharalis* were examined for Cry1Ab resistance using the F_2 screening procedures (Huang et al. 2008). The four insect populations were collected from Franklin, East Carroll, Tensas, and Rapides parishes, respectively. None of these collections possessed major Bt resistance alleles. The estimated Cry1Ab resistance allele frequency with a 95% probability for these populations was < 0.0027 (Table 2).

Resistance allele frequency in field populations of *D. saccharalis* sampled during 2006. A total of 331 feral individuals of *D. saccharalis* collected from the same locations as in 2005 were examined for Bt resistance using the F_1 screening method (Yue et al. 2008). A Bt resistance allele was detected in one individual. Cry1Ab resistance frequency in these populations was estimated to be 0.003 with a 95% confidence interval of 0.0004-0.0084 (Table 2).

Resistance allele frequency in field populations of *D. saccharalis* sampled during 2007. A total of 251 feral individuals of *D. saccharalis* collected from Tensas, Franklin, and Rapides Parishes were screened for Bt resistance using the F_1 screening method (Huang and Leonard 2008). One feral individual was identified to possess a major Bt resistance allele. The Bt resistance allele frequency for these collections was estimated to be 0.0039 with a 95% confidence interval of 0.0005-0.011 (Table 2).

Resistance allele frequency I n field populations of *D. saccharalis* sampled during 2008. A total of 236 feral individuals of *D. saccharalis* collected from three locations (Tensas, Franklin, and Rapides parishes) in Louisiana were examined for Bt resistance using the F_2 screening method. None of these individuals possessed major Bt resistance alleles. The estimated major Bt resistance allele frequency for these collections with a 95% probability was < 0.0063 (Table 2).

Resistance allele frequency in field populations of *D. saccharalis* sampled during 2009. A total of 257 feral individuals of *D. saccharalis* collected from two locations (Franklin, and Rapides parishes) in Louisiana were examined for Bt resistance using the F_2 screening method. Among the F_2 family-lines screened, 29 lines (or 29 feral individuals) were established by crossing field-collected individuals with a laboratory Cry1Ab-susceptible strain of *D. saccharalis*. Eight feral individuals were identified to possess a major resistance allele to Cry1Ab maize. The estimated major Bt resistance allele frequency for these collections was 0.018 with a 95% probability of 0.008-0.031.

In summary, a total of 2,061 feral individuals of *D. saccharalis* collected from four locations in Louisiana were examined for resistance to Cry1Ab maize during 2004-2009. Resistance allele frequency to Cry1Ab maize was low for populations collected during 2004-2008, but increased significantly for the populations sampled during 2009. Results of the six-year resistance monitoring suggests that it may be necessary to switch planting of the single Bt gene-expressed maize (e.g. YieldGard® maize) to pyramided Bt maize which contains two or more dissimilar Bt genes.

In addition, a Cry1Ab-resistant strain (Cry1Ab-RR) of *D. saccharalis* was isolated from a single family-line (line 52) in the F_2 screen during 2004 (Huang et al. 2007a). The Cry1Ab-RR strain has demonstrated to survive and complete entire larval development (neonate to pupal stage) on commercial transgenic maize plants expressing the Cry1Ab protein (Huang et al. 2007a). Laboratory bioassay confirmed that the survival of this strain on commercial Bt maize plants was due to resistance to Cry1Ab protein (Huang et al. 2007b). This resistant strain has been used in evaluation of a "high dose" qualification of commercial Bt maize hybrids and characterization of inheritance, cross-resistance, fitness, and stability of Cry1Ab resistance in *D. saccharalis*.

EVALUATION OF "HIGH DOSE" QUALIFICATION OF CRY1AB MAIZE AGAINST *D. SACCHARALIS*

The success of the "high dose/refuge" IRM strategy requires that Bt maize plants must produce a high dose that can kill the heterozygous genotype for Bt resistance in the target species (Ostlie et al. 1997). A Bt maize hybrid can be judged as "high dose" if it kills ≥ 95% heterozygotes (RS) for Bt resistance (FIFRA Scientific Advisory Panel 1998, United States Environmental Protection Agency 2001). The high dose qualification has not been directly evaluated for *O. nubilalis* and *D. grandiosella* because major resistance traits were not found in the two species (FIFRA Scientific Advisory Panel 1998, Bourguet et al. 2003). The availability of the Cry1Ab-RR strain of *D. saccharalis* provided an opportunity to directly test the "high dose" assumption. To validate if Cry1Ab maize hybrids commonly planted in Louisiana qualified as a "high dose" for *D. saccharalis*, multiple independent trials were conducted to evaluate the survival of resistant Cry1Ab-RR, -susceptible (Cry1Ab-SS), and –heterozygous (Cry1Ab-RS) genotypes on Bt maize plants in the greenhouse (Wu et al. 2007, Ghimire et al. 2010).

During 2005 and 2006, four trials were conducted to evaluate the "high-dose" qualification for seven commonly planted Cry1Ab maize hybrids (YieldGard®) in the greenhouse (Wu et al. 2007). Two trials were designed for evaluating vegetative plant stages (V7-V10) and two for testing reproductive growth stages (R1-R3). In these trials, larval survival of *D. saccharalis* was recorded 21days after infestation of neonates on potted plants. Larvae of Cry1Ab-RR genotype survived well on both non-Bt and Bt maize plants. Results of the greenhouse tests showed that all seven Cry1Ab maize hybrids were highly efficacious against Cry1Ab-SS and -RS genotypes during the vegetative plant stage, with a larval survival rate of < 2% for the Cry1Ab-SS and ≤ 5% for the heterozygotes. However, 8 -18% of the heterozygous genotype survived on reproductive stage plants in four of the seven Cry1Ab maize hybrids. The varied performance of Cry1Ab maize hybrids at different plant growth stages against the Cry1Ab-RS larvae of *D. saccharalis* suggests a differential seasonal expression of the Cry1Ab protein (Wu et al. 2007). The greenhouse tests also indicated that Bt maize hybrids expressing a "high dose" for *O. nubilalis* and *D. grandiosella* (United States Environmental Protection Agency 2001) may not produce a sufficient high dose for *D. saccharalis*.

During 2008-2009, two additional trials were conducted to evaluate six other commercial Bt maize hybrids by using the similar method as described in Wu et al. (2007). The six Bt maize hybrids included four Cry1Ab (YieldGard®) and two Cry1F (Herculex®) maize hybrids. Larval survivorship on plants of Cry1Ab or Cry1F maize after 21-25 days was 0-8.1% for Cry1Ab-SS, 1.3-34% for Cry1Ab-RS, and 19-51% for Cry1Ab-RR (Ghimire et al. 2010). Larvae of Cry1Ab-RR and -RS also caused significant plant injury to most of the Bt maize hybrids, especially for Cry1Ab maize. Compared to Cry1Ab maize, the two Cry1F maize hybrids appeared to be more effective against both Cry1Ab-RS and -RR genotypes. Larval survivorship of Cry1Ab-RR and –RS averaged 26.3 and 11.1% on Cry1F plants respectively, whereas on the four Cry1Ab hybrids it was 40.1% for Cry1Ab-RR larvae and 26.4% for Cry1Ab-RS genotype (Ghimire et al. 2010). Results from this study showed that all of the four Cry1Ab maize hybrids tested did not produce a sufficient dose to kill ≥95% of the

Cry1Ab-RS genotypes, and therefore these Bt maize apparently did not qualify as "high-dose" Bt maize as defined in the "high dose/refuge" IRM strategy for *D. saccharalis*.

INHERITANCE OF CRY1AB-RESISTANCE IN *D. SACCHARALIS*

Another key assumption of the "high dose/refuge" strategy is that resistance in the target pest species should be recessive or at least incompletely recessive so that resistant heterozygotes can be killed by high dose expressed Bt maize plants (Bourguet et al., 2003). Inheritance traits of the Cry1Ab-RR strain of *D. saccharalis* were analyzed using various genetic crosses. It included: 1) reciprocal parental crosses between Cry1Ab-SS and Cry1Ab-RR strains; 2) F_1 by F_1 crosses; and 3) backcrosses of F_1 with the Cry1Ab-RR strains (Wu et al. 2009a, Ghimire et al. 2010). Maternal effects and sex linkage were examined by comparing the larval mortality of the parental and the F_1 cross-populations on Cry1Ab diet. Number of genes associated with the resistance was evaluated by fitting the observed mortality of F_2 and backcross populations on Cry1Ab diet with a Mendelian monogenic inheritance model. Effective dominance levels (D_{ML}) of resistance were measured by comparing the larval mortality of the Cry1Ab-resistant, -susceptible, and -heterozygous populations on Bt maize leaf tissue and intact Bt maize plants. D_{ML} was estimated using the method described in Bourguet et al. (2000). D_{ML} ranges between 0 and 1; $D_{ML} = 0$ means that the resistance is completely recessive, while $D_{ML} = 1$ refers to a completely dominant resistance.

Table 3. Effective dominance (D_{ML}) of Cry1Ab resistance in *Diatraea saccharalis*.

Testing material	Corn growth stages	Testing materials	Year tested	D_{ML}
YieldGard®[a]	Vegetative	Leaf tissue	2007	0.26
	Vegetative	Intact plant	2005-2006	0.10-0.24
	Reproductive	Intact plant	2005-2006	0.08-0.26
YieldGard® (Bt-11)[b]	Vegetative	Leaf tissue	2009	0.19
	Vegetative	Intact plant	2009	0.32-0.69
Herculex® I[b]	Vegetative	Leaf tissue	2009	0.41
	Vegetative	Intact plant	2009	0.15-0.63
YieldGard® VT3[b,c]	Vegetative	Leaf tissue	2009	0.34
	Vegetative	Intact plant	2009	0.58-0.66

[a] Data from Wu et al. (2009).
[b] Data from Ghimire et al. (2010).
[c] YieldGard VT3 maize expresses the Cry1Ab protein (e.g. event MON 810 for controlling above ground lepidopteran pests) and Cry3Bb1 protein (e.g. event MON 863) for controlling underground coleopteran rootworm pests).

Diet incorporated bioassays indicated that Cry1Ab resistance in *D. saccharalis* was likely inherited as a single or a few tightly linked autosomal genes (Wu et al. 2009a). Wu et al. (2007) showed that Cry1Ab resistance in *D. saccharalis* was incompletely recessive on Bt maize leaf tissue and intact Bt plants (Table 3). D_{ML} estimated based on larval mortality on leaf tissue was 0.26, while it ranged from 0.08-0.26 based on the testing on intact plant tests

(Wu et al. 2009a). Comparing the results reported in Wu et al. (2009a), a more recent study showed that the Cry1Ab resistance in *D. saccharalis* was more dominant on six other Bt maize hybrids. Ghimire et al. (2010) reported that D_{ML} measured based on larval mortality on leaf tissue of the Bt (Cry1Ab or Cry1F) maize plants of the six hybrids ranged from 0.19 to 0.41, while it was 0.15 to 0.69 based on the intact plant tests in the greenhouse (Table 3). Results of the more recent studies reported in Ghimire et al. (2010) suggested that the Cry1Ab resistance in *D. saccharalis* was incompletely dominant to most of these Bt maize hybrids rather than recessive as required for the "high dose/refuge" IRM strategy.

CROSS-RESISTANCE OF CRY1AB-RESISTANT *D. SACCHARALIS* STRAIN TO OTHER BT PROTEINS

To delay resistance development, a gene-pyramiding strategy has been used to develop transgenic plants that express two or more Bt proteins for targeting a same group of insect pests. Such kind of Bt maize technology has recently become commercially available for managing lepidopteran maize pests including *D. saccharalis* (Monsanto 2010). Information on cross-resistance of insect pests among Bt proteins is essential in selection of proper Bt genes for the pyramiding strategy. Cross-resistance among Bt proteins has been evaluated in several target pests of Bt maize and Bt cotton. Most previous studies showed that selection for resistance to one Bt toxin can lead to resistance to others (Tabashnik, 1992; Ferré and Van Rie, 2002; Li et al., 2005), but the patterns of cross-resistance can vary among insect species and Bt toxins.

Susceptibility of Cry1Ab-susceptible and -resistant strains of *D. saccharalis* has been evaluated on a meridic diet treated with Cry1Aa, Cry1Ac, Cry1A.105, or Cry2Ab2 (Wu et al. 2009b) in the laboratory and on intact plants of Cry1F maize in the greenhouse (Ghimire et al. 2010). In the laboratory bioassays, neonates of Cry1Ab-SS and –RR strains of *D. saccharalis* were assayed on a meridic diet containing one of the four Cry proteins (Wu et al. 2009b). Larval mortality and body weight were recorded after 7 days. The bioassays showed that the Cry1Ab-RR strain was highly resistant to Cry1Aa (>80-fold) and Cry1Ac (45-fold), slightly cross-resistant to Cry1A.105 (4.1-fold), but more susceptible to Cry2Ab2 (-0.5-fold) (Wu et al. 2009b). A recent greenhouse study also demonstrated that the Cry1Ab-RR larvae of *D. saccharalis* could survive and cause considerable plant damage on two commercial Cry1F maize hybrids (Herculex®), indicating that these insects are also cross-resistant to Cry1F maize plants. This was the first demonstration of cross-resistance within commercial Bt maize events. The results suggest that resistance to one Bt toxin in *D. saccharalis* can lead to resistance to other Bt toxins and therefore careful selection of different Bt proteins is essential for the gene pyramiding strategy.

FITNESS OF CRY1AB-RESISTANCE IN *D. SACCHARALIS*

Understanding the fitness of resistant insects is important in development of effective IRM strategies. Relative fitness of Bt resistance has been examined for several target pests of Bt maize and Bt cotton (see details in Gassmann et al., 2009). Most published studies showed that Bt resistance with high resistance levels is associated with fitness costs.

To examine the relative fitness of Cry1Ab resistance in *D. saccharalis*, larval growth and development were examined on non-toxic diet and on non-Bt maize plants for five insect genotypes: a Cry1Ab-RR, a Cry1Ab-SS, a backcrossed and re-selected resistant strain (Cry1Ab-R'R'), and two F_1 progeny of the Cry1Ab-SS and Cry1Ab-R'R'(Wu et al. 2009c). Fitness of the five genotypes was examined by infesting neonates on a meridic diet in the laboratory and on intact non-Bt maize plants in the greenhouse. Biological parameters measured included neonate-to-pupa development time and pupation rate, larval survival, larval and pupal weight, and sex ratio. Unlike Bt resistance in most other lepidopteran species, in which Bt resistance are often associated with significant fitness costs (Gassmann et al. 2009), larvae of five insect genotypes of *D. saccharalis* grew normally on non-toxic diet and non-Bt maize plants. There were generally no significant differences among the five insect genotypes for all the parameters measured, suggesting a lack-of-fitness cost of the Cry1Ab resistance in *D. saccharalis*. The lack-of-fitness costs of Cry1Ab resistance in *D. saccharalis* also suggest a greater challenge in managing Bt resistance for this maize borer species.

STABILITY OF CRY1AB-RESISTANCE IN *D. SACCHARALIS*

A strain (Cry1Ab-RR-UNS) of *D. saccharalis* derived from the Cry1Ab-RR strain was maintained without exposure to any Bt toxins to determine if the Cry1Ab resistance was stable in absence of selection pressures. To examine resistance stability, larval (practical) mortality of the Cry1Ab-RR-UNS strain and -susceptible insects (Cry1Ab-SS) on diet treated with a trypsin-activated Cry1Ab toxin was assayed at 1st, 2nd, 9th, and 24th generations after selection pressures were removed by using a standard diet incorporated method (Huang et al. 2007b). Our previous studies showed that Cry1Ab at the concentration of 1-2 μg/g could be used as good discrimination doses for separating resistant and susceptible larvae of *D. saccharalis* (Huang et al. 2007b). Therefore, larval mortality at 7th day after exposure of neonates was compared among insect strains and generations at 0.125, 0.5, 2, and 8 μg/g. Larval mortality was measured as 'practical' mortality, which considered both the actual dead larvae and the surviving larvae that did not show a significant gain in body weight (< 0.1 mg per larva) as morbid or non-feeding insects (Huang et al. 2006). Larval mortality data were subjected to a two-way ANOVA with insect-generation and Cry1Ab concentration as the two main factors, followed by a LSMEANS test at the a = 0.05 level to determine treatment differences.

Larval mortality of both susceptible and resistant strains of *D. saccharalis* at each Cry1Ab concentration was generally consistent across generations (Table 4). Compared to the original resistant insects, mortality of Cry1Ab-RR-UNS strain did not increase in absence of selection pressures for 24 generations, suggesting Cry1Ab resistance in *D. saccharalis* is very stable.

Table 4. Larval mortality (% mean ± sem) of Cry1Ab-susceptible (Cry1Ab-SS) and –resistant (Cry1Ab-RR) *Diatraea saccharalis* feeding on diet containing Cry1Ab protein [a,b].

No. generations without selections	Cry1Ab concentration (μg/g)							
	0.125		0.5		2		8	
	Cry1Ab-SS	Cry1Ab-RR	Cry1Ab-SS	Cry1Ab-RR	Cry1Ab-SS	Cry1Ab-RR	Cry1Ab-SS	Cry1Ab-RR
0	53.1 ± 8.1 hi	1.4 ± 1.0 a	96.5 ± 2.0 l	9.8 ± 4.5 abc	100 ± 0.0 l	30.4 ± 3.0 ef	100 ± 0.0 l	46.2 ± 2.3 gh
1	40.4 ± 3.1 fg	1.9 ± 1.3 ab	78.5 ± 3.4 k	2.9 ± 1.7 abc	93.4 ± 2.3 l	17.0 ± 3.2 cd	100 ± 0.0 l	43.0 ± 1.4 gh
2	71.8 ± 6.1 jk	4.0 ± 1.4 abc	91.1 ± 1.5 l	12.1 ± 2.2 bc	100 ± 0.0 l	12.2 ± 1.6 bc	100 ± 0.0 l	62.4 ± 6.4 ij
9	37.0 ± 8.6 fg	8.2 ± 6.3 abc	69.7 ± 12.0 jk	3.3 ± 1.9 abc	92.0 ± 4.0 l	3.3 ± 3.3 abc	100 ± 0.0 l	23.4 ± 1.6 e
24	12.7 ± 4.6 c	0.0 ± 0.0 a	91.2 ± 1.4 l	1.1 ± 0.8 a	100 ± 0.0 l	4.3 ± 2.5 abc	100 ± 0.0 l	4.6 ± 3.0 abc

[a] Mean values in the table across all combinations of insect strains, generations, and Cry1Ab concentrations followed by a same letter are not significantly different ($P < 0.05$; LSMEANS test).

[b] A measure of larval mortality was defined as the number of dead larvae plus surviving larvae that did not demonstrate significant weight gains (<0.1 mg/larva) in a 7-d bioassay divided by the total number of larvae in the test.

Resistance stability in absence of selection pressures is often associated with fitness of the resistance. If a significant fitness cost is associated with resistance, the resistance allele frequency will decline in pest populations once the selection pressure is removed. Resistance development in field populations therefore can be delayed or reversed in the presence of fitness costs, especially with non-recessive costs that are more efficient than recessive costs for delaying resistance (Tabashnik et al., 2005; Gassmann et al., 2009). The stable resistance in absence of selection pressures observed for the Cry1Ab-resistant strain of *D. saccharalis* could be a result of several factors, including 1) resistance was controlled by a single gene (Wu et al. 2009a); 2) the Cry1Ab-RR strain had reached a plateau of homozygous alleles for Cry1Ab resistance before selection pressures were removed; and/or 3) Cry1Ab resistance was not associated with a fitness cost (Wu et al. 2009c). Stability of resistance to Bt toxins has also been examined in several other insect pests. Results vary among species and even among populations within same species. In most cases, resistance declined after the selection pressure was removed, yet the decline was incomplete, seldom returning to the pre-selection levels (Tabashnik et al. 1995, Tang et al. 1997, Huang et al. 1999, Sayyed et al. 2001).

CONCLUSION

Since first being commercialized in 1996, transgenic Bt maize has gained widespread acceptance for managing maize stalk borer pests in the United States. IRM is an important issue in the sustainable use of Bt crop technologies. Since 2004, a cooperative research program has been established to evaluate resistance risk of *D. saccharalis* to Bt maize in Louisiana. Field surveys have documented the overwintering, distribution, and population structures of maize stalk borer species in the state. Unlike any other regions of the United States where *O. nubilalis*/*D. grandiosella* are the major maize stalk boring pests, our survey showed that *D. saccharalis* is the dominant species in many areas of the mid-south region, especially in Louisiana.

One of the key factors for a successful resistance management plan is to have a cost-effective resistance monitoring system that can detect rare resistance alleles in field populations. Novel F_2/F_1 screening methods has been developed to detect rare Bt resistance alleles in *D. saccharalis*. With these novel screening methods, Bt resistance frequency in field populations has been monitored across the major maize growing areas in Louisiana since 2004. Bt resistance allele frequency was low (averaged 0.0011) for populations collected during 2004-2008, but increased for populations sampled during 2009. A landmark accomplishment of this research program was the detection of a major Bt resistance allele in a 2004 field population of *D. saccharalis*. This was the first major resistance allele to commercial Bt maize discovered in any maize borer species worldwide. A *D. saccharalis* strain containing this major resistance allele has been established in the laboratory for future studies.

Additional studies showed that Bt resistance in *D. saccharalis* was likely controlled by one autosomal gene. Most Cry1Ab-expressing maize hybrids (e.g. YieldGard® maize) planted in Louisiana did not express a sufficiently high Bt dose against *D. saccharalis*. Cry1Ab resistance in *D. saccharalis* was incompletely dominant to most of commercial Bt maize hybrids rather than recessive as required for the "high dose/refuge" IRM strategy. Cry1Ab-resistant larvae of *D. saccharalis* also demonstrated a high level of cross-resistance to several other Bt toxins commonly used as transgenes (e.g. Cry1Aa and Cry1Ac). The resistance was not associated with any fitness costs and was stable in absence of selection pressure. Results of the six-year study suggest that there appears to be a high risk for development of Bt resistance in *D. saccharalis* if Cry1Ab maize continues to be widely used in the U.S. mid-south region. Bt resistance surveys in *D. saccharalis* in the mid-south region should continue to help monitor this situation and insure the continued success of the Bt maize technologies.

In addition, our recent studies have demonstrated that Cry1Ab-resistance in *D. saccharalis* could be overcome by a pyramided Bt maize technology containing two different Bt genes (Ghimire et al. 2010). Bt maize hybrids containing such pyramided Bt genes (eg. Genuity™ VT Triple Pro and SmartStax™) were commercially available in the United States in 2010 crop seasons (Monsanto 2010). With the availability of these new Bt maize technologies, maize varieties expressing only a single Bt protein (e.g. YieldGard®, Herculex® I) will be phased out in the near future. Results from our research suggest that the new Bt maize with pyramided genes should provide a means for Bt resistance management in *D. saccharalis*.

ACKNOWLEDGMENT

This paper was reviewed by Drs. Carlos Blanco (USDA-APHIS) and Brenda Oppert (USDA-ARS). This article is published with the approval of the Director of the Louisiana Agricultural Experiment Station as manuscript No. 2010-234-5137.

REFERENCES

Andow, D.A., & Alstad, D.N. (1998) F_2 screen for rare resistance alleles. *Journal of Economic Entomology, 91*, 572–578.

Andow, D.A., & Alstad, D.N. (1999) Credibility interval for rare resistance allele frequencies. *Journal of Economic Entomology, 92*, 755–758.

Bourguet, D., Chaufaux, J., Séguin, M., Buisson, C., Hinton, J.L., Stodola, T.J., Porter, P., Cronholm, G., Buschman, L.L., & Andow, D.A. (2003) Frequency of alleles conferring resistance to Bt maize in French and US corn belt populations of the European corn borer, *Ostrinia nubilalis. Theoretical and Applied Genetics, 106*, 1225–1233.

Bourguet, D., Genissel, A., & Raymond, M. (2000) Insecticide resistance and dominance levels. *Journal of Economic Entomology, 93*, 1588–1595.

Castro, B. A., Riley, T.J., Leonard, B.R., & Baldwin, J. (2004a) Borers galore: Emerging pests in Louisiana corn, grain sorghum and rice. *Louisiana Agriculture, 47*, 4–6.

Castro, B.A., Leonard, B.R., & Riley, T.J. (2004b) Evaluation of *Bacillus thuringiensis* transgenic field corn for management of southwestern corn borer and sugarcane borer (Lepidoptera: Crambidae). *Journal of Economic Entomology, 97*, 2106–2116.

Ferré, J., & Van Rie, J. (2002) Biochemistry and genetics of insect resistance to *Bacillus thuringiensis. Annual Review of Entomology, 47,* 501–533.

FIFRA Scientific Advisory Panel (1998) Report of Subpanel on *Bacillus thuringiensis* (Bt) plant-pesticides and resistance management. *EPA SAP Report. Feb. 9, 1998*, http://www.mindfully.org/GE/FIFRA-SAP-Bt.htm.

Ghimire, M.N., Huang, F., Leonard, B.R., Head, G.P. & Yang, Y. (2010) Susceptibility of Cry1Ab-susceptible and -resistant sugarcane borer to transgenic corn plants containing single or pyramided *Bacillus thuringiensis* genes. *Crop Protection*. In press.

Gould, F. (1998) Sustainability of transgenic insecticidal cultivars: integrating pest genetics and ecology. *Annual Review of Entomology*, *43,* 701–726.

Gassmann, A.J., Carrière, Y., & Tabashnik, B.E. (2009) Fitness costs of insect resistance to *Bacillus thuringiensis. Annual Review of Entomology, 54,* 147–163.

Huang, F., Higgins, R.A., & Buschman, L.L. (1999) Heritability and stability of resistance to *Bacillus thuringiensis* in European corn borer (Lepidoptera: Pyralidae). *Bulletin of Entomological Research, 89,* 449–454.

Huang, F. (2006) Detection and monitoring of insect resistance to Bt crops. *Insect Science 13*, 73–90.

Huang, F., B. R. Leonard, B.R., & Gable, R.H. (2006) Comparative susceptibility of European corn borer, southwestern corn borer, and sugarcane borer (Lepidoptera: Crambidae) to Cry1ab protein in a commercial Bt-corn hybrid. *Journal of Economic Entomology, 99*,194–202.

Huang F., Leonard B.R., & Andow D.A. (2007a) Sugarcane borer resistance to transgenic *Bacillus thuringiensis*-maize. *Journal of Economic Entomology, 100*,164–171.

Huang, F., Leonard, B.R., & Wu, X. (2007b) Resistance of sugarcane borer to *Bacillus thuringiensis* Cry1Ab toxin. *Entomologia Experimentalis et Applicata 124*, 117–123.

Huang, F., and B.R. Leonard. 2008. Detection and monitoring of *Bacillus thuringiensis* resistance alleles in the sugarcane borer, *Diatraea saccharalis* (F.). pp 1–17. In

"*Genetically Modified Plants: New Research Trends*", (eds) by T.V. Wolf and J. P. Koch. Nova Science Publishers, Hauppauge, NY.

Huang, F., Leonard, B.R., Moore, S.H., Cook, D.R., Baldwin, J., Tindall, K.V., & Lee, D. (2008) Allele frequency of resistance to *Bacillus thuringiensis* Cry1Ab corn in Louisiana populations of sugarcane borer (Lepidoptera: Crambidae). *Journal of Economic Entomology, 101,* 492–498.

Huang, F., & Andow, D. (2010) Global status of transgenic *Bacillus thuringiensis*-crops and resistance management. *Entomologia Experimentalis et Applicata.* Submitted.

Li, H, Oppert, B., Higgins, R.A., Huang, F., Buschman, L.L., & Zhu, K.Y. (2005) Susceptibility of Dipel-resistant and -susceptible *Ostrinia nubilalis* (Lepidoptera: Crambidae) to individual *Bacillus thuringiensis* protoxins. *Journal of Economic Entomology, 98,* 1333–1340.

McAllister, C.D., Gravois, K., Schexnayder, K.P., & Reagan, T.E. (2004) Transgenic Bt-corn affects sugarcane borer in Louisiana. *Southwestern Entomologist, 29*, 263–269.

Monsanto (2010) Insect resistance management corn and cotton areas growers guide. http://www.monsanto.com/monsanto/ag_products/pdf/stewardship/2010_irm_guide.pdf.

National Agricultural Statistics Service. (2009) Acreage. USDA, Washington, DC. *http://usda.mannlib.cornell.edu/usda/current/Acre/Acre-06-30-2009.pdf.*

Ostlie, K.R., Hutchison, W.D. & Hellmich, R.L. (1997) Bt-corn & European corn borer, Long-term success through resistance management. *North Central Region Extension Publication 602.* University of Minnesota, St. Paul, MN, USA.

Porter, P., Troxclair, N., Schuster, G., Porter, D.O., Cronholm, G., Bynum, E. Jr., Patrick, C., & Davis, S.G. (2005) Texas corn production: Emphasizing pest management and irrigation. Texas Cooperative Extension, *The Texas A&M University System B-6177.* pp 71.

Sayyed, A.H., & Wright, D.J. (2001) Fitness costs and stability of resistance to *Bacillus thuringiensis* in a field population of the diamondback moth *Plutella xylostella* L. *Ecological Entomology, 26,* 502–508.

Tabashnik, B.E. (1992) Evolution of resistance to *Bacillus thuringiensis. Annual Review Entomology, 39,* 47–79.

Tabashnik, B.E., Finson, N., Johnson, M.W., & Heckel, D.G. (1995) Prolonged selection affects stability of resistance to *Bacillus thuringiensis* in diamondback moth (Lepidoptera: Plutellidae). *Journal of Economic Entomology, 88,* 219–224.

Tabashnik, B.E., Dennehy, T.J., & Carrière, Y. (2005) Delayed resistance to transgenic cotton bollworm. *Proceedings of the National Academy of Science of the United States, 102,* 15389–15393.

Tang, J.D., Gilboa, S., Roush, R.T., & Shelton, A.M. (1997) Inheritance, stability, and lack-of-fitness costs of field-selected resistance to *Bacillus thuringiensis* in diamondback moth (Lepidoptera: Plutellidae) from Florida. *Journal of Economic Entomology, 90,* 732–741.

United States Environmental Protection Agency. (2001) Biopesticides registration action document: *Bacillus thuringiensis* plant-incorporated protectants. *http://www.epa.gov/opp bppd1/biopesticides/ pips/Bt_brad2/ 5-benefits.pdf.*

United States Environmental Protection Agency. (2005a) Bacillus thuringiensis Cry3Bb1 protein and the genetic material necessary for its production (Vector ZMIR13L) in event MON 863 corn & Bacillus thuringiensis Cry1Ab delta-endotoxin and the genetic material necessary for its production in corn (006430, 006484) fact sheet. *http://www.epa.gov/pesticides*.

United States Environmental Protection Agency. (2005b) Bacillus thuringiensis Cry34Ab1 and Cry35Ab1 proteins and the genetic material necessary for their production (plasmid insert PHP 17662) in Event DAS-59122-7 corn & Bacillus thuringiensis Cry1F protein and the genetic material necessary for its production (plasmid insert PHI8999) in Event TC1507 corn Fact sheet. *http://www.epa.gov/pesticides*.

van Rensburg, J.B.J. (2007) First report of field resistance by the stem borer, *Busseola fusca* (Fuller) to Bt-resistance maize. South African Journal of Plant and Soil, *24,* 147−151.

Wu, X., Huang, F., Leonard, B.R., & Moore, S.H. (2007) Evaluation of transgenic *Bacillus thuringiensis* corn hybrids against Cry1Ab-susceptible and -resistant sugarcane borer (Lepidoptera: Crambidae). *Journal of Economic Entomology, 100,* 1880−1886.

Wu, X., Huang, F., Leonard, B.R., & Ottea, J. (2009a) Inheritance of resistance to *Bacillus thuringiensis* Cry1Ab protein in the sugarcane borer (Lepidoptera: Crambidae). *Journal of Invertebrate Pathology*, 102, 44–49.

Wu, X., Leonard, B.R., Zhu, Y.C., Abel, C.A., Head, G.P., & Huang, F. (2009b). Susceptibility of Cry1Ab-resistant and -susceptible sugarcane borer (Lepidoptera: Crambidae) to four *Bacillus thuringiensis* toxins. *Journal of Invertebrate Pathology, 100,* 29−34.

Wu, X., Huang, F., Leonard, B.R., & Ghimire, M. (2009c) Growth and development of *Bacillus thuringiensis* Cry1Ab-susceptible and -resistant sugarcane borer on diet and conventional corn plants. *Entomologia Experimentalis et Applicata, 133,* 199−207.

Yue, B., Huang, F., Leonard, B.R., Moore, S.H., Parker, R., Andow, D.A., Cook, D.R., Emfinger, K., & Lee, D.R. (2008) Verifying an F_1 screen for identification and quantification of rare *Bacillus thuringiensis* resistance alleles in field populations of sugarcane borer (Lepidoptera: Crambidae). *Entomologia Experimentalis et Applicata, 129,* 172−180.

In: Genetically Engineered Crops
Editors: Lindsay M. Grover

ISBN: 978-1-61122-526-6

Chapter 4

NPR1: A MULTIFUNCTIONAL REGULATORY PROTEIN IN PLANTS

S Sundaresha[1], Manjunath Prasad,[2]
S.C Chandrashekar[1] and Rohini Sreevathsa[*2]
[1]Department of Plant Pathology,
University of Agricultural Sciences, GKVK, Bangalore, India
[2]Department of Crop Physiology,
University of Agricultural Sciences, GKVK, Bangalore, India

ABSTRACT

Agricultural production of the world is severely affected by various stresses such as biotic and abiotic stress, contributes majorly by decreasing the average yields over 50% and biotic stress reduces the crop yield by 20 to 30%. Plants innate immunity is based on a surprisingly complex response that is highly flexible in its capacity to recognize and counteract different invaders. To combat invasion by microbial pathogens and herbivorous insects effectively, plants make use of pre-existing physical and chemical barriers as well as inducible defense mechanism that become activated upon attack. The plant hormone salicylic acid (SA), Jasmonic acid (JA) and ethylene are major players in the regulation of signaling networks that are involved in induced defense responses against pathogens, insects and abiotic stress as well. During fast few years, significant progress has been made in understanding the function of Nonexpressor of pathogenesis related genes 1 (NPR1), a key regulator of systemic acquired resistance (SAR), that is essential for transducing the SA signals to activate pathogenesis related (PR) gene expression. Therefore it has been a well known fact that NPR1 induce PR genes and thereby SAR. In the recent years, the involvement of NPR1 in abiotic stress is also being elucidated and it found to be induced by salinity, drought and low temperature by different regulatory pathways. Overall the signaling molecules mediated protein-protein (NPR1-PR) interaction induces the enhanced tolerance to biotic and abiotic stress in crop plants. This review would emphasize on understanding the multifunctional role of NPR1 in plants.

* Corresponding Author :rohinisreevathsa@rediffmail.com, Telephone: 91-080-2363713 Extn: 47

INTRODUCTION

Under any stress condition plant tries to defend itself, the perception of the stress and signal transduction to switch on adaptive/defense response are critical in determining the survival and reproduction of plants exposed to adverse environments. The adaptive or defence responses are controlled majorly by two pathways i.e. jasmonic acid and salicylic acid pathway and are characterized by the induction of pathogenesis related (PR) genes (Pieters *et. al.,* 2001). Depending on the stimulus, generally one of the pathways gets activated and tries to safeguard the plant. It is also reported that a cross-talk exists among these pathways and this is also determined by the external stimulus.

It earlier reported that PR genes are only induced by pathogens (Van Loon and Van Kammen, 1970), but later it was reported that abiotic stress conditions also influence their induction (Yamada *et al.,* 1997; Koike *et al.,* 2002; Dombrowski, 2003) and thus it was understood that these proteins have important roles in defending the plant under biotic and abiotic stresses. The PR1, PR2 and PR5 are influenced by salicylic acid pathway and are said to inhibit the growth of microorganism (Cao *et al.,* 1998; Metraux, 2001). Later studies have shown that PR1 was also induced by salinity, drought and low temperature (Hong *et al.,* 2005).

Plants, like animals, have evolved both innate and acquired immunity to counter attacks by microbial pathogens. Each plant genome encodes hundreds of so-called resistance (R) proteins that allow the plant to recognize specific pathogen-derived molecules known as avirulence (avr) factors. This R-avr recognition between plants and pathogens often triggers the hypersensitive response (HR), which involves programmed cell death, production of reactive oxygen species (ROS), synthesis of antimicrobial compounds at the site of infection, and ultimately leads to pathogen resistance (Dangl and Jones, 2001). Thus, these R proteins determine the innate immunity of each plant species and individuals within a species. Concomitant with the appearance of secondary resistance response known as systemic acquired resistance (SAR) is induced in the uninfected tissues (Sticher *et al.,* 1997). SAR provides long lasting resistance throughout the plant to subsequent infections by a broad range of pathogens.

Systemic acquired resistance (SAR) is an inducible form of plant defense conferring broad-spectrum immunity to secondary infection of plant tissues above the initial infection site. SAR is triggered by systemic increase in salicylic acid (SA) levels following local infection by certain phytopathogens (Durrant and Dong, 2004) and results in the transcriptional activation of ~10% of the genes in the *Arabidopsis* genome. NPR1 (nonexpressor of pathogenesis-related genes 1) is a key SAR regulator. NPR1 contains a BTB/POZ (broad-complex, tramtrac, bric-a-brac/poxvirus, zinc finger) domain and an ankyrin-repeat domain. In the absence of infection, NPR1 is predominantly oligomeric and sequestered in the cytoplasm. Upon pathogen challenge, NPR1 is reduced to a monomeric state and translocates to the nucleus (Mou *et al.,* 2003). Within the nucleus, NPR1 physically interacts with TGA-bZIP transcription factors, inducing expression of defense response genes via a largely unknown mechanism to activate SAR.

Despite substantial advances in plant disease control strategies, our global food supply is still threatened by a multitude of pathogens and pests. Plant diseases can dramatically reduce crop yield and impact of disease outbreaks is particularly acute in developing nations.

Pesticides provide effective protection but their application can be compromised by adverse environmental effects and by the emergence of resistant pathogen strains. Chemical controls are often beyond the means of farmers in developing nations. For these reasons, much effort has been invested towards understanding innate responses, compromised or genetically programmed suicide of infected cells (the hypersensitive response, HR) as well as tissue reinforcement and antibiotic production at the site of infection. These local responses can, in turn, trigger a long lasting systemic response (systemic acquired resistance, SAR) that primes the plant for resistance against a broad spectrum of the pathogens (Metraux, 2001).

Our focus will be on various regulatory roles played by NPR1 against biotic and abiotic stress.

The Fungal Cell Wall

The knowledge of fungal cell wall provides the basis for the control of the fungi as it proves to be the target for the action of pathogenesis related proteins (De Lucca and Walsh, 1999). Theis and Stahl (2004) have given the structures from the outer most part of the fungal cell, which is defined by the cell wall to plasma membrane and finally to several intracellular targets. These data should provide the basis for a greatest insight into the selectivity of antifungal proteins, as this specificity is most often due to fungal-specific features of the target structures.

Structure of the Cell Wall

The fungal cell wall accounts for approximately 20–30% of the dry weight of fungal cells (Kills, 1994). It protects the cell from physical damage and is responsible for its shape. Furthermore, there are several proteins associated with the cell wall which are involved in mating, adherence to substrates, vegetative (in) compatibility and hydrophobic properties of the cell. The three best-studied *Ascomycotina* species in terms of cell wall composition are currently *Saccharomyces cerevisiae*, *Candida albicans* and *Aspergillus fumigatus*. Smits *et al.* (1999) presented a comprehensive molecular model of the cell wall of *S. cerevisiae*, which represents Ascomycetes as shown in Figure 1. The cell wall of *S. cerevisiae* has three classes of carbohydrate polymers: chitin, β1,3-glucan and β1,6-glucan. Furthermore, it contains two classes of Glycosylated Cell Wall Proteins (CWPs) which are linked via either a Glycosylphosphatidylinositol (GPI) anchor or a Pir (protein with internal repeats) sequence to the cell wall (Kapteyn *et al.*, 1999). The β 1,3-glucan molecules are held together by hydrogen bridges and form a three-dimensional network which covers the whole cell. Chitin is attached at the interior side of the cell wall, whereas β1,6-glucan and Pir-CWPs are attached on the exterior side. Infrequent linkage of chitin to β1,6-glucan has also been observed.

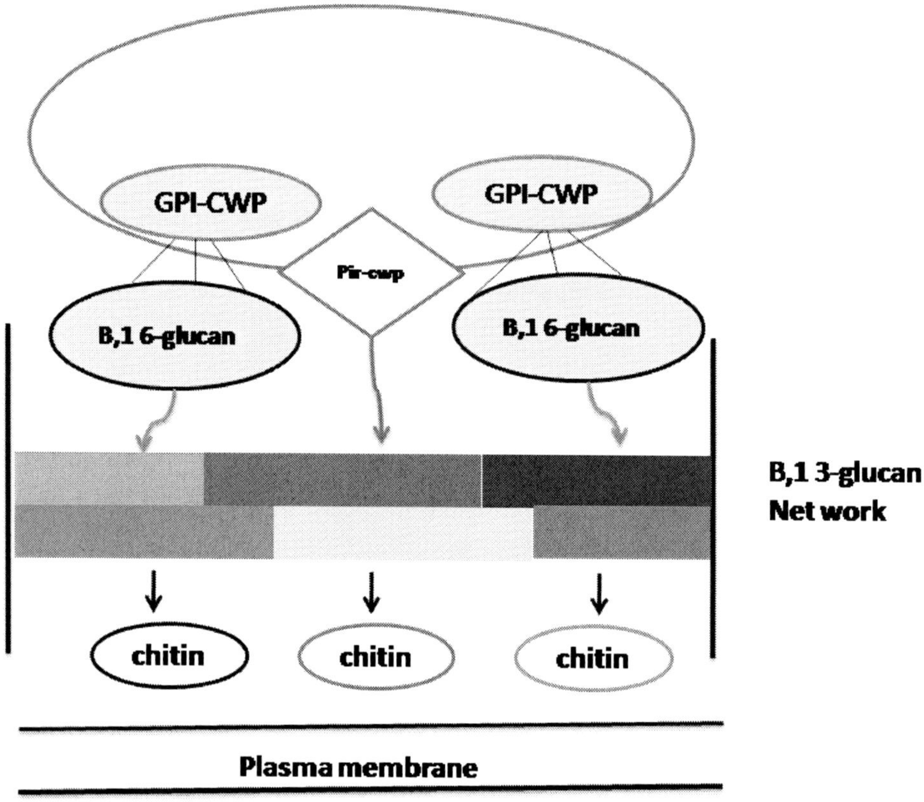

Figure 1. Molecular model of the cell wall of *Saccharomyces cerevisiae*. GPI, Glycosylphosphatidylinositol; CWP, Cell wall protein; Pir, Protein with internal repeats

There is no complete model of the cell wall of filamentous fungi to date. Apart from *Oomycetes*, which lack chitin in their cell walls, all three classes of carbohydrate polymers described in yeast are also present in cell walls of most filamentous fungi. This indicates that the molecular organisation of the hyphal cell wall is similar to that of *S. cerevisiae*. The concentration of chitin in the cell wall of filamentous fungi (approximately10%) is significantly higher than in yeast (2%). The surface properties of most species of the *Ascomycotina*, *Basidiomycotina* and the *Zygomycotina* are largely determined by an outer layer of glycoproteins (Barkai-Golan *et al.,* 1978), which are either anchored by GPI or bound in a Pir-like manner. Apart from their functions in anastomosis, hyphal aggregation and vegetative (in) compatibility, these proteins are involved in the adherence of fungal pathogens to their hosts (Gauer and Klotz 1997; Cormack *et al.,* 1999) and determine the antigenic properties of the cell. In addition, glycoproteins limit the permeability of the cell wall and therefore play a protective role (Nobel *et al.,* 1990). The cell wall of aerial hyphae is covered by a special class of proteins, designated as hydrophobins (Wosten *et al.,* 1999).

Carbohydrate polymers of fungal cell walls are mainly synthesised at the plasma membrane (Montijn *et al.,* 1999). The incorporation of chitin, β1,3-glucan and α1,3-glucan into the cell walls of filamentous fungi seems to be limited to the hyphal tips (Gooday, 1971); thus the composition and organisation of cell walls of apical cells differ from mature cell walls. In contrast, the incorporation of chitin into the lateral walls is delayed until after cytokinesis in *S. cerevisiae* (Shaw *et al.,* 1991). The unique fungal cell wall with its skeletal layer composed of chitin and β1,3-glucan is a target for a wide range of antifungal proteins. Fungal growth is affected by chitin binding proteins, chitinases and glucanases. Furthermore, several proteins besides interfering with fungal chitin and glucan-synthesis also cause growth inhibitory effect.

PLANT AND PATHOGEN INTERACTION

Since the onset of civilization, plant diseases had catastrophic effects on crops and the well being of human populations. For example, the fungus *Phytophthora infestans* caused the epidemic that triggered the Irish potato famine. These diseases resulted from agricultural practices that rely on monoculture of crops and planting genetically closely related crop species over wide geographical areas providing for the evolution of increasingly virulent pathogen forms. Protection of crops from disease can substantially improve agricultural production. The harmful effects of the pesticides on health and environment have limited their use. A striking feature of the response of plants to pathogen attack is the induction of a battery of defenses, suggesting that different protective mechanisms may have complementary roles in overall expression of resistance. Mycelial cell walls have complex architectures incorporating number of characteristic polymers including chitin and various glucans. Hence, combinations of hydrolytic enzymes with different substrate specificities may be required for efficient lysis of fungal pathogens.

With the beginning of the molecular era of plant biology in the early 1980's, a major area of research has been to identify, clone, and characterize various genes involved in disease resistance. As a result, many intriguing mechanisms, which plants have evolved to respond to pathogen infection, have been identified over the past 10 years, and remarkable progress has been made towards elucidating the multitude of genes that are involved in these responses. The identification of these genes has made it possible to subsequently evaluate their specific roles and importance in disease response pathways using transgenic plants developed with genetic engineering techniques (Figure 2).

PLANT DEFENSE SYSTEM

All organisms have evolved several defence systems in order to protect themselves against bacteria, fungi and viruses. Higher organisms have developed a complex network of humoral and cellular responses, called adaptive immunity. A second defence system, innate immunity, was discovered in the early 1980s, consisting of small cationic peptides with a broad antimicrobial spectrum. These proteins act immediately at sites of infection or inflammation. The production of proteins with antimicrobial activity was not limited to higher organisms but was also found in insects, plants and microorganisms. During the last 2 decades a broad range of proteins with very different structural features have been isolated and characterized from differing organisms ranging from bacteria to human beings. Over 500 cationic membrane-acting proteins with antimicrobial and antifungal activities have been identified to date.

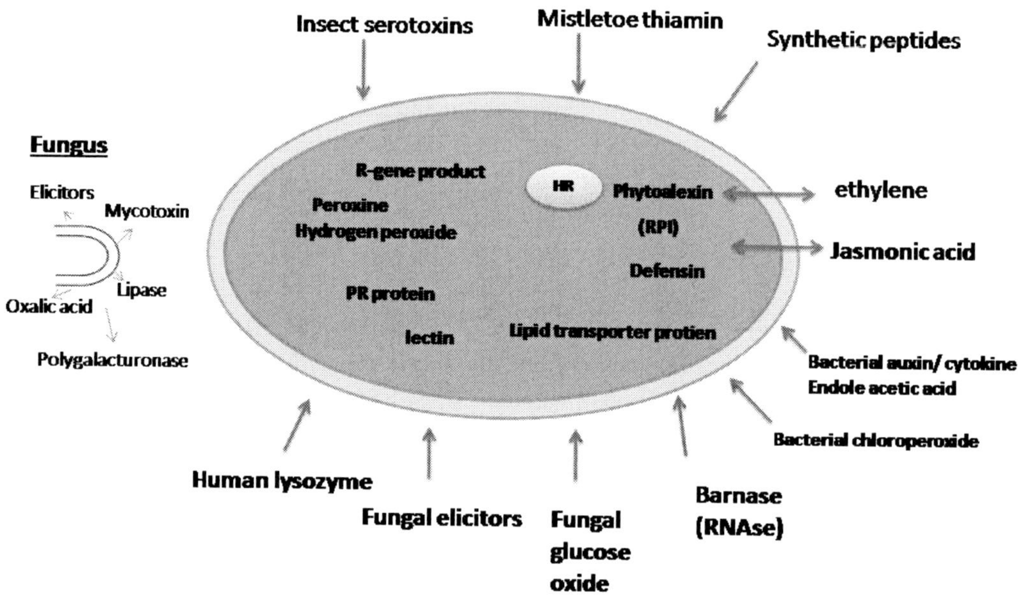

Figure 2. Components of plant pathogen interaction (from hypha on left), enhanced expression of plant-derived gene products (inside of cell) or gene products from nonplant sources (outside of cell)

Induced Resistance

To combat invasion by microbial pathogens and herbivour insects effectively, plants make use of pre-existing physical and chemical barriers, as well as inducible defense mechanism that become activated upon attack. Following primary infection by a microbial pathogen or attack by an insect pest, susceptible plants can develop an enhanced resistance to further attack (Kuc, 1982). This phenomenon is known as induced resistance. Induced resistance may be expressed locally at the site of infection as well as systemically. The potential of plants to react to an invader by triggering local and systemic response was first described by phytopathologists like Carbone and Arnaudi, (1930); Chester, (1933); and Gaumann (1946).

This systemically induced response protects the plant against subsequent invaders. Several biologically induced systemic defense responses have been characterised in detail. This include systemic acquired resistance (SAR), which is triggered by necrotizing pathogens (Sticher et al.,1997); induced systemic resistance (ISR) which is activated upon colonisation of roots by selected strains of nonpathognic rhizobacteria (Pieterse *et al.,* 2002); and wound induced defense, which is typically elicited upon tissue damage such as that caused by feeding insects (Kesselvr and Baldwin, 2002). Induced defense responses are regulated by a network of interconnecting signal transduction pathways in which the hormonal signals salicylic acid (SA), jasmonic acid (JA) and ethylene play a major role (Pieterse and Van Loon, 1999; Thomma *et al.,* 2001; Glazeberook, 2001) and other harmones such as Brasinosteroids and abscisic acid are also involved (Nakashita *et al.,* 2003; Audenaert *et al.,* 2002; Ton and Mauch-Mani, 2004). SA, JA and ethylene accumulate in response to pathogen infection or herbivore damage leading to activation of distinct and partly overlapping sets of defense related genes.

Systemic Acquired Resistance (SAR)

SAR is a mechanism of induced defense that confers long lasting protection against a broad spectrum of microorganisms. It is by far the best studied induced resistance response. The term was first coined by Ross to describe induced resistance in the upper leaves of tobacco plants, which had developed necrotic lesion on the lower leaves after inoculation with tobacco mosaic virus (TMV)(Ross, 1966). SAR requires the signal molecule salicylic acid (SA) and is associated with the accumulation of pathogenesis related proteins, which are thought to contribute to resistance. Activation of PR gene expression and SAR are dependent on transduction of SA signals and this in turn requires the function of the regulatory protein NPR1 (Shah *et al.,* 1997). Salicylic acid and NPR1 are the two main components in the induction of SAR.

NPR1-A Multifunctional Regulatory Protein

NPR1 is a regulatory protein that was identified in *Arabidopsis* through several genetic screens for SAR compromised mutants. Upon induction of SAR, NPR1 is translocated into the nucleus. NPR1 acts as a modulator of PR gene expression but does not bind DNA directly. NPR1 is a positive regulator of SAR and is also known as NIM1 (non-inducible immunity) or SAI1 (salicylic acid-insensitivity) was identified using various genetic screens.

NPR1 is a key regulator protein crossed with multiple defense pathways or multiple defense proteins. NPR1 may interact with different partners in different tissues, with the redox-regulated oligomer-to-monomer exchange controlling its subcellular localization in the cytoplasm or the nucleus and it regulates the various transcription factors depending on the external signal stimuli, thereby it activates or regulates the many defense protein or PR genes. This protein-protein interaction stabilises the plants to defend themselves to various stresses. In this chapter we outline the various roles of NPR1 in both biotic and abiotic stresses.

I. Npr1 (Non Expressor of Pathogenesis Related Protein), a Sar Regulator

NPR1 is a key regulator in the signal transduction pathway that leads to SAR. Mutation in NPR1 results in a failure to induce PR genes in systemic tissues and a heightened susceptibility to pathogen infection, whereas overexpression of the NPR1 protein leads to increased induction of the PR genes and enhanced disease resistance. NPR1 acts through salicylic acid pathway and transcription factors which express the PR genes in preventing the proximal spread of the pathogens in addition to their suggested role in SAR (Systemic acquired resistance).

Successful disease resistance of plants to potential pathogens depends primarily on timely recognition of the pathogen and triggering of an immediate and effective defense response.

Initial exposure to pathogens can trigger defense signaling leading to a broad-spectrum long-lasting immunity against a wide range of potential pathogens, called systemic acquired resistance (SAR) (Sticher *et al.,* 1997). SAR is strongly associated with elevated levels of salicylic acid (SA), which is essential for triggering SAR within plants (Gaffney *et al.,* 1993). Exogenous application of SA or its biologically active analogs like isonicotinic acid (INA) and benzothiadiazole (BTH) can trigger activation of SAR (Ryals *et al.,* 1996), which leads to a concerted expression of a battery of pathogenesis-related (PR) genes, for example *PR1*, *β-1, 3 glucanase* (*PR2* or *BGL2*) and *PR5* (Ward *et al.,* 1991). Although the mechanism behind the activation of SAR is not completely understood, one of the key factors which connects SA signaling to *PR* gene expression is NPR1 (Durrant and Dong, 2004). The *Arabidopsis thaliana* NPR1 gene, also known as *NIM1* and *SAI1*, was identified by genetic screening for SAR deficient mutants (Cao *et al.,* 1994; Glazebrook *et al.,* 1996; Shah *et al.*, 1997; Delaney *et al.,* 1995).

The *Arabidopsis npr1* mutants failed to respond to SA and could not trigger *PR1* expression or mount SAR upon treatment with SA, and were highly susceptible to various pathogens. Overexpression of NPR1 in *Arabidopsis* on the other hand, conferred marked enhancement in inducible disease resistance (Cao *et al.,* 1998) without any noticeable decrease in field fitness (Heidel *et al.,* 2004).

Molecular characterization of *Arabidopsis* NPR1 revealed that it is a transcription factor with distinct protein-protein interaction motifs and bears a bipartite nuclear localization signal (NLS) (Cao *et al.,* 1997 and Ryals *et al.,* 1997). NPR1 acts indirectly to induce transcription of *PR* genes by interacting with members of the TGA family of b-ZIP transcription factors. Upon activation, these TGA factors bind to the SA-responsive elements present in *PR* gene promoters and induce their transcription (Dong, 2004).

In *Arabidopsis*, NPR1 protein is normally detected as inactive oligomers formed by intermolecular disulphide bonds created in the cytosol. During SAR induction, a biphasic change in cellular reduction potential leads to reduction of NPR1 oligomers to active monomers, which subsequently migrate to the nucleus to induce *PR* gene expression (Kinkema *et al.,* 2000; Mou *et al.,* 2003).

Jasmonic acid (JA) and ethylene (ET) are also involved in activating defense signaling pathways. However, activation here induces expression of a different set of *PR* genes including antifungal proteins such as defensins (Penninckx *et al.,* 1996), thionin (Epple *et al.,* 1995). Several reports suggest that SA induced SAR is effective against biotrophic pathogens, while JA/ET-mediated signal transduction confers resistance against necrotrophic fungal pathogens (Thomma *et al.,* 2001). For example, infection with *Alternaria brassicicola,* which is a necrotrophic fungus, was shown to cause rapid increase in JA in the infected plant and induction of defensins.

Furthermore, defensins were shown to be induced by exogenous application of JA but not by SA (Pennickx *et al.,* 1996). It has also been reported that the *coi1* mutation in *Arabidopsis*, which blocks JA signaling, were severely compromised in resistance to *A. brassicicola*, but had no effect on resistance to the biotrophic *Peronospora parasitica* (Thomma *et al.,* 2001). Conversely, the mutation *npr1* and transgenic expression of the bacterial gene *salicylate hydroxylase* (*NahG*), which blocks SA signaling resulted in loss of resistance to *P. parasitica*, but had no effect on resistance to *A. brassicicola* (Thomma *et al.,* 1998).

Apparently, effective defense against biotrophic pathogens generally involves programmed cell death associated with hypersensitive response (HR) and SA- dependent signaling pathway. In contrast, necrotrophic pathogens can actually benefit from host cell death and are not controlled by cell death. SA-responsive defenses and therefore effective control relies on JA/ET signaling (Glazebrook, 2005). Although, role of NPR1 in induction of *PR* genes during SAR is well established, it appears that functional NPR1 is not essential for JA/ET-responsive defense signaling (Thomma *et al.,* 1998). Interestingly, functional NPR1 is required for activation of induced systemic resistance (ISR), yet another JA/ET responsive defense signaling triggered by non-pathogenic rhizobacteria but can confer effective resistance to various pathogens including root pathogen *Fusarium oxysporum* (Pieterse *et al.,* 1998). It would be interesting to examine the induction of NPR1 in response to the activation of different defense signaling pathways following diverse pathogenic infections.

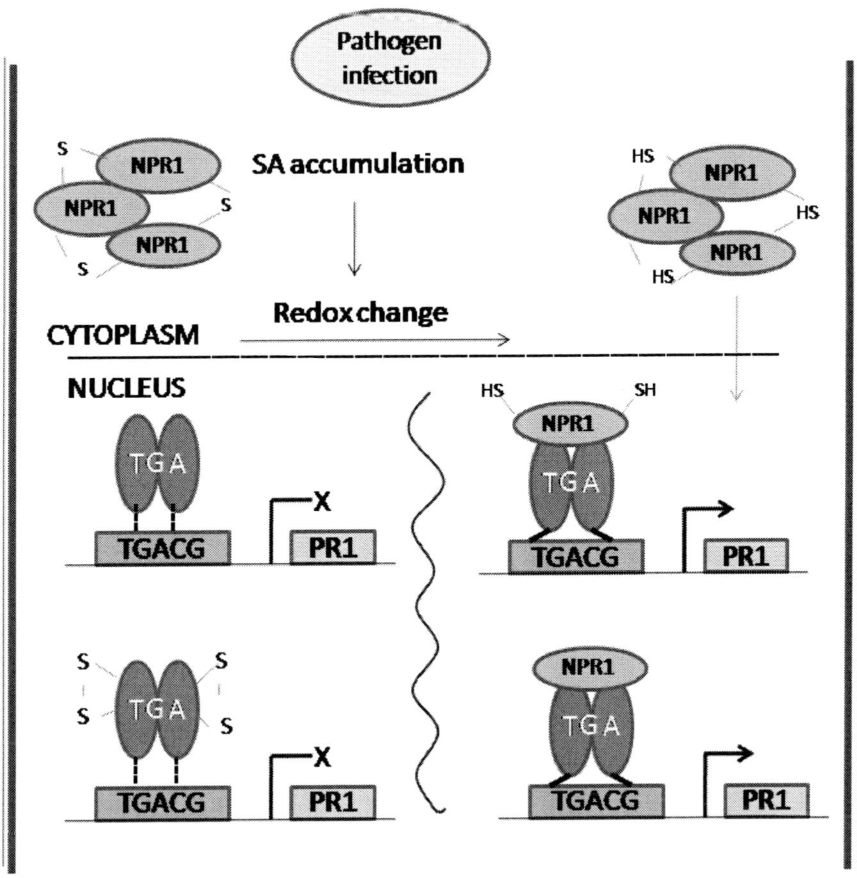

Figure 3. Model illustrating the interaction between SA-mediated redox changes, NPR1, and TGA transcription factors in SAR-related gene expression

In non-induced cells, oxidized NPR1 forms inactive oligomers that remain in the cytosol. Binding of TGAs to the cognate SA-responsive promoter elements (TGACG) (indicated by dotted lines in Figure 3) is not sufficient to activate the expression of PR-1 genes. Upon infection by a necrotizing pathogen, SA accumulates and plant cells attain a more reducing environment, possibly because of the accumulation of antioxidants. Under these conditions, NPR1 is reduced from an inactive oligomeric complex to an active monomeric state through the reduction of intermolecular disulfide bonds. Monomeric NPR1 is then translocated into the nucleus where it interacts with TGAs, such as TGA2. The binding of NPR1 to TGAs stimulates the DNA-binding activity of these transcription factors to the cognate *cis* element (represented by black boxes in figure 3), resulting in the activation of PR-1 gene expression. In non-induced cells, TGAs that do not interact with NPR1 in yeast two-hybrid assays, such as TGA1, are oxidized and form intramolecular disulfide bridges that prevent interaction with NPR1. Upon accumulation of SA *in planta*, the change in redox status reduces the disulfide bonds in these TGAs, resulting in a conformational change that allows interaction with NPR1 (Figure 3).

Redox Changes Connect the SA Signal with NPR1 Functioning

Clearly, NPR1 plays an important role in the SA mediated activation of defense-related genes by enhancing the DNA binding of TGAs to SA-responsive elements in their promoters. NPR1 proteins from different plant species contain conserved Cys residues that are capable of forming inter- or intra-molecular disulfide bonds, and that a mutation in one of these Cys residues resulted in a mutant *npr*1 phenotype, led (Mou *et al.,* 2003) to the hypothesis that NPR1 protein conformation may be sensitive to SA-induced changes in cellular redox status. In a series of elegant experiments, Mou *et al.* (2003) demonstrated that the induction of SAR is indeed associated with a change in redox state, possibly caused by the accumulation of antioxidants. Under these conditions, NPR1 was reduced from an inactive oligomeric complex to an active monomeric form. It seems that the monomeric form is required for PR-1 activation, as inhibition of NPR1 reduction prevented PR-1 gene expression. Mutation of two Cys residues that are crucial for NPR1 oligomer formation led to constitutive monomerization and nuclear localization of NPR1, and constitutive expression of the PR-1 gene. Thus, cellular redox changes that are induced as a result of SA accumulation connect the SA signal with NPR1 activity during SAR (Figure 3).

Transcription Factors and PR Gene Expression

The monomerized NPR1 after entering nucleus cannot directly interact with PRI promoters, as it does not have DNA binding domains. The presence of protein-protein interacting domains suggested that it might influence some transcription factors that are involved in inducing the expression of PR genes.

Using the yeast two-hybrid system with NPR1 as bait, two classes of interacting proteins were identified. They are NIMIN (NIM interactor) and TGA transcription factors. Three small structurally similar NIMIN proteins were identified and named as NIMIN1, NIMIN2, and NIMIN3. NIMIN1 and NIMIN2 interact with the C-terminus of NPR1, while NIMIN3 interacts with the N terminus (Weigel *et al.,* 2001). NIMINs contain stretches of acidic amino acids and are hypothesized to be transcription factors. NIMIN1 was studied in detail and it was found that it interacts with NPR1 by forming a complex. The so formed complex binds to LS7 region (as-l element found in the promoter of PR genes) and negatively regulates PR gene induction (Weigel *et al.,* 2005).

The second group of transcription factors with which NPR1 interacts belongs to TGA subclass of the basic leucine zipper (bZIP) family of transcription factors (TGAs) (Despres *et al.,* 2000). NPR1 interacts with the *Arabidopsis* TGA factors, TGA2, TGA3, TGA5, TGA6, and TGA7 but only weakly or not at all with TGAI and TGA4 (Despres *et al.,* 2000). NPR1 also interacts with TGA factors from tobacco and rice. Using truncated or mutant forms of

NPR1, the ankyrin-repeat domain in the middle of the protein was shown to be essential for binding TGA factors, while the N-terminal region appears to enhance the binding (Despres *et al.,* 2001).

II. NPR1 in Biotic Stress Tolerance

ISR Through Bacteria and NPR1-Regulator in Bacterial Disease Resistance

Nearly 25% of the bacteria in the rhizosphere are assumed to have beneficial effects on plants by improving their growth (known therefore as plant growth-promoting rhizobacteria, PGPR) or suppressing diseases. Such effects can result from direct supply of nutrients or plant hormones to the roots, or from suppression of harmful microorganisms via competition or the secretion of toxic compounds (Van Loon *et al.,* 1998; Bloemberg and Lugtenberg, 2001; Ramamoorthy *et al.,* 2001). In 1991, PGPR were reported to induce resistance to foliar or systemic pathogens, an effect that was observed in work on resistance of carnation to *Fusarium* wilt (Van peer *et al.,* 1991), cucumber to anthracnose (Wei *et al.,* 1991), and bean to halo blight (Alstrom, 1991). Systemic resistance induced by PGPR has been termed 'induced systemic resistance' (ISR) (Kloepper *et al.,* 1992; Pieterse *et al.,* 1996). For example, the non-pathogenic rhizobacterial strain *Pseudomonas fluorescens* WCS417r can elicit resistance in carnation, radish, tomato and *Arabidopsis* to pathogens, arid colonization of *Arabidopsis* roots with this strain improves resistance to bacterial leaf pathogens, fungal root pathogens and a leaf-infecting oomycete (Hoffland *et al.,* 1996; Pieterse *et al.,* 1998). Whether ISR is elicited, however, depends upon the particular plant/PGPR strain interaction. For example, the same bacterial strain may induce resistance in *Arabidopsis*, and not in radish, while radish might express ISR in response to a different strain of PGPR, to which *Arabidopsis* is unresponsive (Pieterse *et al.,* 2001). ISR can be associated with the accumulation of PR proteins and/or phytoalexins, changes in cell wall constitution, and traits that characterize SAR (Ramamoorthy *et al.,* 2001).

Different bacteria trigger ISR in different ways, and the efficacy of the resulting resistance also depends on the specific pathosystem. These elicitors act as external stimuli to the cell wall receptors that perceive the signal and induce the downstream signaling responses as JA, ET production, Oxidative burst *etc.* Further downstream signaling involves NPR1 and it was identified through mutant studies. NPR1 mutants' did not express ISR even in the presence of PGPR or by application of its elicitors. Some studies also suggested the role of SA in ISR specifically in systemic signaling and at times in cross talk (Cameron, 2000; Meatraux, 2001).

ISR Through Insects

Insect feeding has two different kinds of elicitors; they are insect saliva and cell wall products of damaged cells. These induce direct and indirect resistance in the plant. A physical or chemical trait that directly acts against further attack or reduce herbivore performance is known as direct resistance. The attraction of enemies of plant is related to indirect resistance. In response to wounding and or insect feeding, linolenic acid is released from the membrane lipids that are enzymatically converted into JA. JA is known to be central signaling molecule in inducing the ISR (Creelman and Mullet, 1997; Wasternack and Parthier, 1997) by transcriptionally activating the protease inhibitors and enzymes involved in the synthesis of volatile compounds or of the secondary metabolites like nicotine and other phenolics (Creelman and Mullet, 1997; Wasternack and Parthier, 1997; Karban and Baldwin, 1997; Boland *et al.*, 1999).

Oligosaccahrides (Bishop *et al.*, 1981), Oligogalacturonides (Doraes *et al.*, 1995, Norman *et al.*, 1999) and systemin released from the damaged cell walls acts as elicitors for general wound response (Pearce *et al.*, 1991). Systemin influences the release of linolenic acid, which activates the octadecanoid signaling cascade (Ryan, 2000). The elicitor activity of cell wall fragments was supported by the finding that cellulysin, a mixture of cell wall degrading enzymes from *Trichoderma viride* can induce several JA responsive volatiles in lima bean (Piel *et al.*, 1997). The action of cellulysin is followed by rapid induction of endogenous JA levels. JA and systemin might act as systemic signal. The JA induces some transcription factors through 'MAPK cascade, which interact with NPR1 and influence the expression of different defense genes (Ferry *et al.*, 2004). Meur *et al.* (2006) have shown that the resistance was possibly by a pathway similar to ISR via JA resulting from the converging of SAR and ISR pathway downstream of NPR1 by activation of different defense genes. Therefore, it is presumed that the role of NPR1 towards insect resistance is due to a synergistic activity of SAR and ISR.

NPR1- A Role in Fungal Disease Resistance

As explained earlier NPR1 induces PR protein expression via transcription factors. This role of NPR1 is exploited in the development of several transgenic plants for fungal disease resistance. NPR1 overexpressing *Arabidodpsis thaliana* plants display an enhanced resistance to bacterial pathogen *Pseudomonas syringae* pv. *maculicola* that causes leaf spot disease and are more resistant to oomycete pathogen *Peronospora parasitica* which causes downy mildew disease (Cao et al.,1998).

Over expressing BjNPR1 showed the broad spectrum resistance to the powdery mildew pathogen *Erisiphae cruciferarum* in mustard plants. The pathogen infection had activated a SA- dependent defense signaling, which in turn would induce PR1, but not defensin. Therefore activation of SA signaling pathway by *E. cruciferarum* induced NPR1 gene expression in mustard (Meur *et al.*, 2006).

Expression of NPR1/defensin (multigene in single cassette) in groundnut cv. TMV-2 showed the enhanced resistance to *Cercospora* spp. and also showed the less accumulation of aflatoxin in transgenic plants when compared with the wild type (unpublished data Sundaresha, *et al.*, 2010).

NPR1- Role in Viral Disease Resistance

NPR1 a multi key regulatory protein because of it's broad spectrum resistance to bacterial and fungal disease it also enhances viral replication inhibitor. Development of effective disease-resistance to broad range of pathogens (bacterial, fungal and viral disease) in crops usually requires tremendous resources and effort. Wan-Chi Lin et al. (2004) engineered the *Arabidopsis* NPR1 in tomato cultivar and the disease screens against eight important tropical diseases revealed that enhanced resistance to bacterial wilt, *Fusarium* wilt and moderate degree of enhanced resistance to gray leaf spot and bacterial leaf spot, in addition to innate tobacco mosaic virus resistance.

III. NPR1- Induced PR Genes in Abiotic Stress

PR genes were very well studied in their role of biotic stress resistance. These are antifungal, anti bacterial and anti- insect in nature. Different studies later have shown that PR genes have a role in abiotic stress and are found to be induced in these conditions. The PR1, 2 and 5, characterised in SAR, were induced under different abiotic stresses. PR1 was induced by salinity, drought and low temperature (Hong *et al.*, 2005). The PR2 (Glucanase) was found to be induced by drought, cold, manganese (Akiyama and Arumugam, 2001; Christoffers, 2003). The PR5 (Thaumatin like) gene was induced by drought, salt, cold, heat, UV, mercuric chloride and manganese (Frendo *et al.*, 1992; Christoffers, 2003; Hong *et al.*, 2004). The PR6 and 12, observed in ISR, was also induced under abiotic stress conditions like UV, salt, drought, electric current etc (Huang *et al.*, 2007). PR12 (defensin) was induced by cold, salinity and drought (Koike *et al.*, 2002; Do *et al.*, 2004).

The role of PR genes in abiotic stress was also confirmed through overexpression studies. The overexpression of the PI (*Protease inhibitor*) led to heavy metal and oxidative stress tolerance (Hong and Hwang, 2005; Sarowar *et al.*, 2005). The overexpression of thaumatins (PR5) has resulted in salt and drought tolerance (Rajam *et al.*, 2007).

Cross-Talk Between Salicylate and Jasmonate-Ethylene Signal Transduction Pathways

In nature, plants are under continuous biotic stress caused by different attackers, including bacteria, fungi, viruses, and insects. These are generally categorized into biotrophs and necrotrophs. The impact of biotrophs was curtailed by induction of systemic acquired resistance (SAR) and that of necrotrophs and insects by Induced Systemic Resistance (ISR) (Felton and Korth, 2000; Pieterse *et al.*, 2001). Both the induced resistances are signalled through Saliylic acid and jasmonic acid respectively and are characterized by the expression of PR genes. It is reported that single a PR protein accounts for 1 % of the total soluble protein in infected tobacco plants (Antoniw and Pierpoint, 1978) indicating that large quantities of resources can be allocated to resistance traits. Therefore, plant induces either of the pathways and inhibits the other. It is reported that SA- and JA-dependent signal transduction pathways mutually inhibit each other (Kunkel and Brooks, 2002), suggesting the necessity of shutting down certain defense responses to avoid exhaustion of resources. As a matter of fact, NPR1 is involved in the suppression of JA dependent genes in the presence of SA (Spoel *et al.*, 2003). In the absence of NIMIN proteins, hyper activated NPR1 may lead to enhanced repression of JA-mediated protection rendering plants even more susceptible to many necrotrophic pathogens. Some studies also testified that JA suppresses SA and its related gene expression (Niki *et al.*, 1998; Kunkel and Brooks, 2002; Glazebrook *et al.*, 2003).

Suppression of JA by SA

The activation of SAR has been shown to suppress JA signaling in plants, thereby prioritizing SA-dependent resistance to microbial pathogens over JA-dependent defense against insect herbivory (Felton and Korth, 2000; Pieterse *et al.*, 2001). Moreover, pharmacological and genetic experiments have shown that SA is a potent suppressor of JA-induced gene expression (Doherty *et al.*, 1988; Doares *et al.*, 1995; Harms *et al.*, 1998; Gupta *et al.*, 2000). The antagonistic effect of SA on JA signaling shows a striking resemblance to the effect of the nonsteroidal anti inflammatory drug acetylsalicylic acid (aspirin), a derivative of SA, on the formation of prostaglandins in animal cells. Prostaglandins are related structurally to JA and play a role in diverse biological processes such as inflammation at sites of infection or tissue injury (Straus and Glass, 2001). JA and prostaglandins originate biosynthetically from linolenic acid and arachidonic acid, respectively, which are released from cell membranes upon phospholipid hydrolysis. Linolenic acid and arachidonic acid are metabolized rapidly via the octadecaiod pathway, in which the enzymatic reactions leading to JA and prostaglandin formation are sithilar (Pan *et al.,* 1998). In a similar process in plants, aspirin has been shown to inhibit the activity of the counterpart of cyclooxygenase, allene oxide synthase, which catalyzes the same step in the octadecanoid pathway in plants, thereby affecting the formation of JA and the subsequent activation of stress-related gene expression (Pan *et al.,* 1998).

The suppression of JA by SA is well proven, but the mechanism of action is not clear until 2003, when Spoel et al (2009) reported the role of NPR1 in SA mediated suppression of JA and its dependent gene expression. Earlier, there was a speculation on the role of NPR1 in the immune system of animals, which is structurally similar to NPR1 (Cao *et al.,* 1997; Ryals *et al.,* 1997) and also helps in the induction of defense genes upon infection of microbes (Baldwin, 1996; Hatada *et al.,* 2000). This has inspired Spoel *et al.,* 2003 to work on the role of NPR1 in *Arabidopsis*. Using different mutants, they have analyzed that NPR1 present in the cytoplasm suppresses JA biosynthesis gene such as allene oxide synthase and thereby the JA dependent gene expression in the presence of salicylic acid. NPR1 if localized to nucleus cannot suppress the JA levels or its downstream gene expression. In rice, it was reported that NPR1 suppress JA dependent gene expression alone, but not the JA levels in the presence of SA and NPR1. The presence of NPR1 in cytoplasm for the above mentioned purpose was again proved here (Yuan *et al.,* 2007).

Suppression of SA by JA

There are speculations on JA suppression of SA. Some studies support the mutual antagonism of JA and SA stating that the SAR and ISR are indeed costly, as each PR protein is around 1 % of the total protein pool under these induced conditions (Antoniw and Pierpoint, 1978). *Arabidopsis* mutants or transformants that overexpress SAR are usually 'stunted' or yield fewer seeds than wild type plants (Bowling *et al.,* 1994; Greenberg *et al.,* 2000; Jirage *et al.,* 2001). Also, the application of chemical resistance elicitors such as SA, INA and BTH often leads to negative effects on further plant development. This phenomenon has been described as 'phytotoxicity' by many authors, yet in fact might result from allocation costs. Unfortunately, fewer studies have been conducted on SAR and the long-known 'limiting effect of disease resistance on yield' (Petersen and Tolstrup, 1985) and thus the phenomenon remains understudied. Early studies on energy costs caused by the resistance of barley to fungal attack dates back to the 1980s (Petersen and Stolen, 1981) and reported ~7% grain yield loss due to respiration increases during successful resistance expression.

The exact procedure for the suppression of SA by JA is not known, but studies on mutants of JA genes have shown increase of SA. These mutant studies state that the respective mutated genes are crucial in control of SA production. The results of Seo et al. (2007) suggest that WIPK and SIPK play an important role in JA production in response to wounding, and that they function co-ordinatively to control SA biosynthesis. SIPK has been reported to regulate SA biosynthesis (Samuel *et al.,* 2005). Other candidates mediating cross talk between JA and SA are NtMPK4 (Gomi *et al.,* 2005) and AtMPK4 (Petersen et al, 2000).

In the case of *P. syringae* infection, the appropriate plant defense response would be the activation of SA-mediated defenses and a decrease in jasmonate-dependent signaling that would otherwise increase susceptibility. Activation of NPR1- dependent defenses would accomplish both these goals. Berry et at, (2006) reported that *P. syringae* evolved the ability to produce COR (Coronatine) as a molecular mimic of JA-He to bypass this inhibition, thereby restoring *JINL-dependent* signaling to down-regulate the plant's SA dependent defense and increase its susceptibility. This explains that *JINI* functions in controlling the production of SA. *JINI* encodes an MYC family transcription factor involved in mediating a subset of jasmonate induced responses (Boter *et al.,* 2004; Lorenzo *et al.,* 2004).

Lee and Sano in 2007 stated that a particular ATPase from tobacco which functions in different cellular activities suppress SA. The Tobacco lines, in which *NtAAAl* (ATPase) was suppressed by the RNA-interference (RNAi), were found to exhibit an elevated resistance to *Pseudomonas syringae* infection. In RNAi transgenic plants, genes related to salicylic acid were up-regulated; whereas those related to jasmonic acid and ethylene were generally down-regulated. Simultaneous application of jasmonic acid with salicylic acid markedly cancelled the effect of salicylic acid in the control, but not much in *NtAAAI-RNAi* transgenic plants. The findings suggest that NtAAAl broadly functions in cellular metabolism, particularly responding to jasmonic acid and/or ethylene signals, it also interferes with salicylic acid signaling.

IV. NPR1 Degradation and Plant Defense Response

The regulation of NPR1 plays a very important role in effective SAR. The transcription of the target genes is influenced by turnover of nuclear NPR1. A proteosome mediated degradation of NPR1 in the nucleus controls the regulation of NPR1 and thereby the target genes (Spoel *et al.,* 2009). A model was proposed for the regulation of NPR1 under both uninduced (uninfected) and induced (infected) SAR situation to avoid constitutive defense response in the absence of infection.

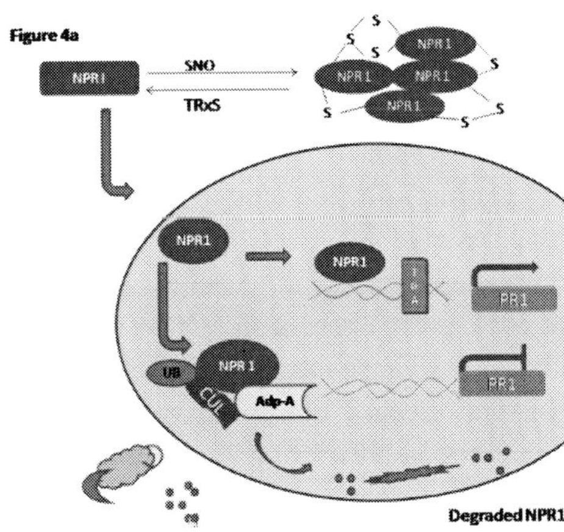

Figure 4 Continued

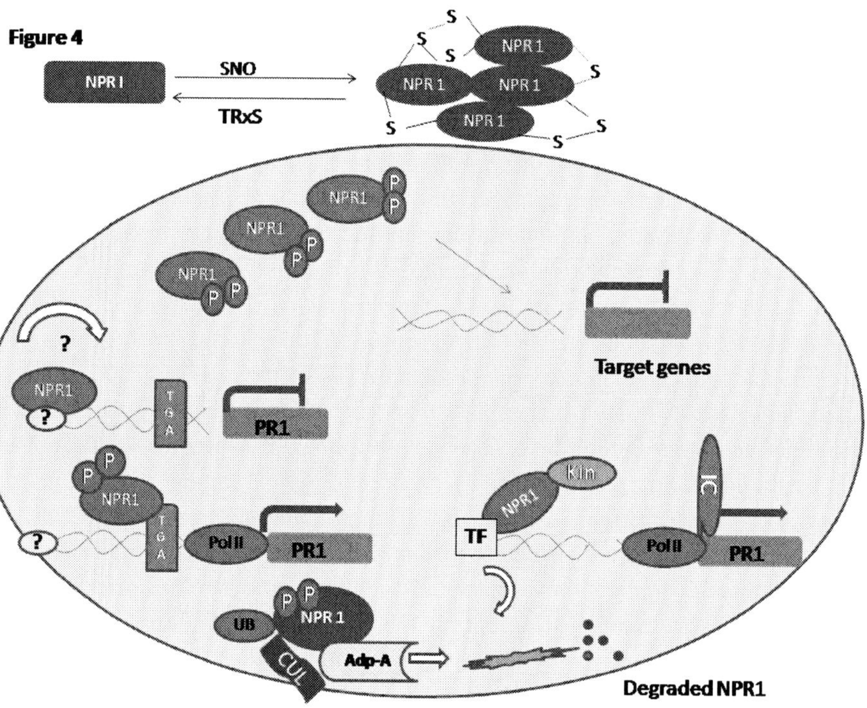

Figure 4. Proteosome-mediated degradation of NPR1. **a)** In uninduced cells, Ubiquitin (Ub)-dependent NPR1 degradation via a nuclear proteasome pathway prevents activation of NPR1 target genes. **b)** In induced cells, Degradation of NPR1 following target gene activation allows fresh NPR1 to be recruited for the next round of transcription initiation

A) Uninfected Cells

A ubiquitin dependent NPR1 degradation via nuclear proteosome pathway prevents activation of NPR1 target genes. Though a small amount of monomeric NPR1 translocates from cytoplasm to nucleus, the degradation of NPR1 by cullin based E3 ligase does not allow it to interact with TGA transcription factor. This does not allow activation of target genes (Figure 4a).

B) Infected Cells

A continuous monomerisation of NPR1, activation of target genes and degradation of the NPR1 is seen in the infected cells. The synthesis of NPR1 is controlled by WRKY transcription factor in the nucleus and it is oligomerised in the cytoplasm through intermolecular disulfide bonds. During infection, due to the presence of salicylic acid, a large amount of monomeric NPR1 enters the nucleus. These monomeric NPR1 as both phosphorylated and unphosphorylated forms bind to the TGA transcription factor (Rochon *et al.,* 2006) to further activate PR1 gene expression. Once the salicylic acid concentration comes down there is a CUL3 based E3 ligase mediated degradation of NPR1 (Figure 4b).

Given that NPR1 degradation occurs constantly, it remains unclear how the cell maintains a proper homeostasis between NPR1 oligomers and monomers. Tada *et al.,* (2008) recently showed that NPR1 is sequentially oxidized and reduced leading to NPR1 oligomerization and monomerization, respectively, following infection. Additionally, basal and salicylic acid-induced expression of NPR1 appears to be controlled by yet unidentified

WRKY transcription factors. This suggests the existence of a feedback loop that maintains the oligomeric form of NPR1 at a particular concentration in the cytoplasm. This may explain not only the onset of efficient SAR by transcriptional regulation coupled with proteolysis but also the inactivation of SAR once cellular salicylic acid concentrations decrease to basal levels.

CONCLUSION

Various external stimuli such as pathogenic elicitors and/or abiotic signals trigger NPR1 mediated activation of transcription factor to induce the protein-protein interaction and the respective signal transduction pathway (Figure 5). Therefore, NPR1 is a key regulatory component in the defense pathways interacting with various defense related proteins. It would be worthwhile to elucidate as to how a single protein can play multiple roles in different stress responses as outlined in the article. Further, knowledge of the proteins that form complexes with NPR1 and their characterization would be challenging.

Figure 5

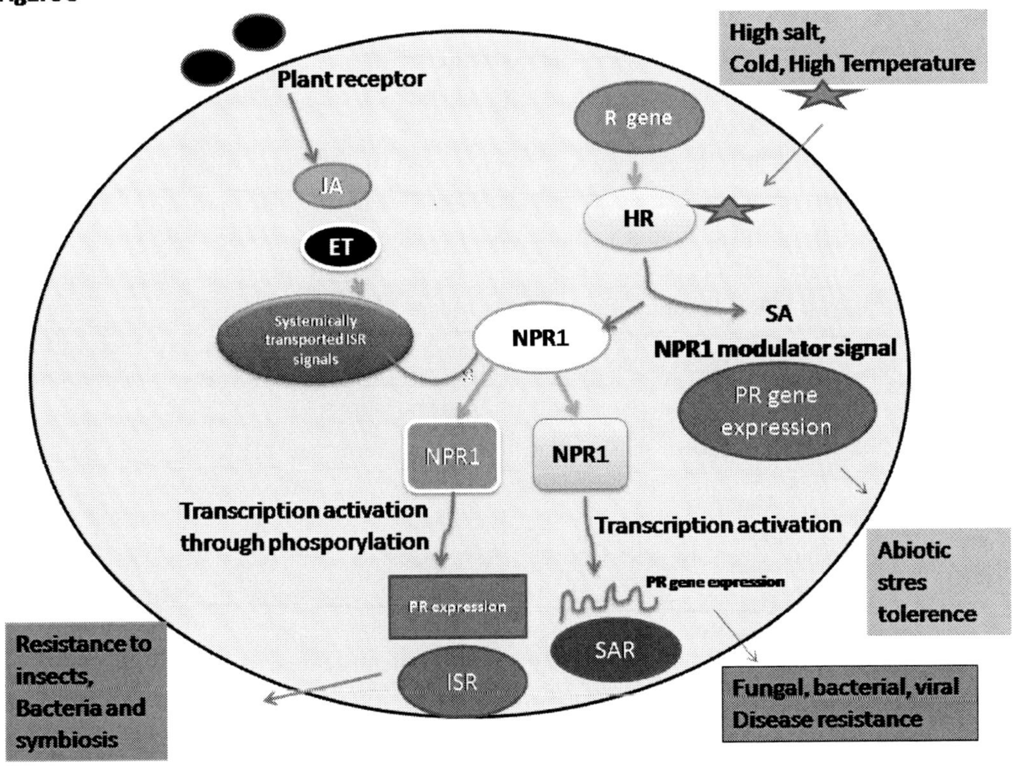

Figure 5. Overview of the role of NPR1 in various defense mechanisms.

REFERENCES

Akiyama, T. and Pillai Armugaum. (2001). Molecular C. Kming, characterization and in vitro expression of a novel endo-l, 3-glucanase up-regulated by ABA and drought stress in rice *(Oryza sativa* L). *Plant Science*, 161, 1089-98.

Alstrom, S. (1991). Induction of disease resistance in common bean susceptible to halo blight bacterial pathogen after seed bacterization with rhizosphere *Pseudomonas. Journal of General and Applied Microbiology*, 37, 495-501.

Antoniw, J.F. and Pierpoint, W.S. (1978). Purification of a tobacco leaf protein associated with resistance to virus infection. *Biochemical Society Transaction* (UK), 6, 248- 250.

Audenaert, K,. De Meyer, G. B., Hofte, M. M. (2002) Abscisic acid determeins basal susceptibilityof tomato to Botrytis cinerea and suppresses salicyclic acid –dependent signaling mechanisms. Plant Physiol 128: 491-501

Baldwin, A.S. (1996). The NF-kB and IkB proteins: New discoveries and insights. *Annual Review of Immunology*, 14, 649-683.

Barkai-Golan, R., Mirelman, D. and Sharon, N. (1978). Studies on growth inhibition by lectins of *Pencillia* and *Aspergilli. Archives of Microbiology,* 116: 119–124.

Berry, N. N., Joardhar, V., Sreet, I. H., Kunkel, B. N. (2006). The *Arabidopsis thaliana JASMONATE INSENSITIVE 1* gene is required for suppression of salycilic acid-dependent defenses during infection by *Psuedomonas syringae. Molecular Plant Microbe Interaction*, 19: 789-800.

Bishop, P.D., Makus, D.J., Pearce, G. and Ryan, C.A. (1981). *Proteinase inhibitor-inducing factor activity in tomato leaves residues in oligosaccharides enzymatically releasd from cell walls.* Procedings of National Academy of Sciences, USA, 78,3536-3540.

Bloemberg, G.V. and Lutenberg,B.J.J. (2001). Molecular basis of plant growth promotion and biocontrol by rhizobacteria. *Current opinion in plant biology*, 4, 343-350.

Boland, W., Koch, T., Krumm, T., Piel, J. and Jux, A. (1999). Induced biosynthesis of insect semiochemicals in plants. In: Chadwick DJ, Goode JA, eds. *Insect-plant interactions and induced plant defence.* Chichester, New York: Wiley, 110-131.

Boter. M, Ruiz-Rivero, Abdeen, A. and Prat, S. (2004). Conserved MYC transcription factors play a key role in jasmonate signaling both in tomato and *Arabidopsis. Genes and Development*, 18, 1577-1591.

Bowling, S.A., Guo, A., Cao, H., Gordon, A.S., Klessig, D.F. and Dong, X. (1994). A mutation in *Arabidopsis* that leads to constitutive expression of systemic acquired resistance. *Plant Cell*, 6, 1845-1857.

Cameron, R.K. (2000). Salicylic acid and its role in plant defense responses: What do we really know?. *Physiological and Molecular Plant Pathology*, 56, 91-93.

Cao, H., Bowling, S.A., Gordon, A.S. and Dong, X. (1994). Characterization of an *Arabidopsis* mutant that is nonresponsive to inducers of systemic acquired resistance. *Plant Cell,* 6, 583-1592.

Cao, H., Glazebrook, J., Clark, J.D., Volko, S. and Dong, X. (1997). The *Arabidopsis* NPRl gene that controls systemic acquired resistance encodes a novel protein containing ankyrin repeats. *Cell*, 88, 57-63.

Cao, H., Li, X. and Dong, X. (1998). *Generation of broad-spectrum disease resistance by overexpression of an essential regulatory gene in systemic acquired resistance.* Proceedings of Naional Academy of Sciences USA, 95, 6531–6536.

Carbone, D. and Arnaudi, C. (1930). *L'immunifa nelle piante.* Monografie dell'Istituto Sieroterapico Milanese, Milano v-cardenas MLO, Narvaez

Chester, K.S. (1933). The problem acquired physiological immunity in plants. *Quarterly Review of Biology,* 8, 275-324.

Christoffers, F.M.M., Braun,H.P., Guiller,L., Van Dorsselaer, A. And Horst, W.J. (2003). Effect of manganese toxicity on the proteome of the leaf apoplast in cowpea. *Plant Physiology*, 133,1935-1946.

Cormack, B. P., Ghori, N. and Falkow, S. (1999). An adhesin of the yeast pathogen *Candida glaberata* mediating adherence to human epithelial cells. *Science,* 285: 578–582.

Creelman, R.A. and Mullet, J.E. (1997). Biosynthesis and action of jasmonates in plants. *Annual review of plant physiology and plant molecular biolgy*, 48,355-381.

Dangl, J.L. and Jones, J.D.G. (2001). Plant pathogens and integrated defence responses to infection. *Nature, 411*, 826–833.

Delaney, T., Friedrich, L. and Ryals, J. (1995). *Arabidopsis signal transduction mutant defective in chemically and biologically induced disease resistance.* In: Proceedings of National Academy of Sciences, USA, 92, 6602–6606.

De Lucca, A. J. and Walsh, T. J. (1999). Antifungal peptides: novel therapeutic compounds against emerging pathogens. *Antimicrobial Agents and Chemotheraphy*, 43, 1–11.

Despres, C., DeLong, C., Glaze, S., Liu, E., and Fobert. (2000). The *Arabidopsis* NPR1/NIM1 protein enhances the DNA binding activity of a subgroup of the TGA family of bZIP transcription factors. *Plant Cell,* 12: 279-290.

Do, H.M., Lee, S.C., Jung, H.W., Sohn, K.M. and Hwang, B.K. (2004). Differential expression and in situ localisation of a pepper defensin (CADE1) gene in response to pathogen infection, abiotic elicitors and environmental stresses in *Capsicum annuum. Plant science*, 166, 1297-1305.

Doares, S.H., Syrovets, T., Weiler, E.W., Ryan, C.A. (1995). *Oligogalacturonides and chitosan activate plant defensive genes through the octadecanoid pathway.* Proceedings of National Academy of Sciences, USA, 92, 4095-4098.

Doherty, H.M., Selvvendran, R.R. and Bowles, D.J. (1988).The wound response in tomato plants can be inhibited by aspirin and related hydroxybenzoicacid. *Physiological and molecular plant pathology*, 33, 377-384.

Dombrowski, J. E. (2003). Sodium Chloride stress activation of wound-related genes in tomato plants. *Plant Physiology*, 132: 2098-2107.

Dong, X., (2004). NPR1, all things considered, Current Opinion in Plant Bioogy, 7, 547-552.

Durrant, W.E. and Dong, X. (2004). Systemic acquired resistance. *Annual Review of Phytopathology,* 42, 185–209.

Epple, P., Apel, K., and Bohlmann, H. (1995). An *Arabidopsis thaliana* thionin gene is inducible via a signal transduction pathway different from that for pathogenesis-related proteins. *Plant Physioog,* 109, 813-820.

Felton, G., and Korth, K. (2000). Trade-offs between pathogen and herbivore resistance. *Current opinion in Plant Biology*, 3: 309-314.

Ferry, N., Edwards, M. G., Gatehouse, J. A., Gatehouse, A. M. R. (2004). Plant-insect interactions: Molecular approaches to insect resistance. *Current opinion in Biotechnology*, 15: 155-161.

Frendo, P., Didierjean, L., Passelegue, E., Burkard, G. (1992). Abiotic stresses induce a thaumatin-like protein in maize; cDNA isolation and sequence analysis. *Plant Sicence*, 85: 61-69

Gaffney, T., Friedrich, L., Vernooij, B., Negrotto, D., Nye, G. and Uknes, S. (1993). Requirement of salicylic acid for the induction of systemic acquired resistance. *Science*, 261, 754-756.

Gauer, N. K. and Klotz, S. A. (1997). Expression, cloning, and characterization of a *Candida albicans* gene, *ALA1* that confers adherence properties upon *Saccharomyces cerevisiae* for extracellular matrix proteins. *Infectious Immunology*, 65, 5289–5294.

Gaumann, E. (1946). *Pflanzliche Infektionslehre*, Basel, Birkhauser, Basle.

Glazebrook, J. (2001) Genes controlling expression of defense responses in *Arabidopsis* - 2001 status. *Current Opinion in Plant Biology*, 4, 301-308.

Glazebrook, L. (2005). Contrasting mechanism of defense against biotrophic and necrotrophic pathogens. *Annual Review of Phytopathology*, 43: 205-227.

Glazebrook, J., Chen, W.J., Estes, B., Chang, H.S., Nawrath, C., Metraux, J.P., Zhu, T. and Katagiri, F. (2003.) Topology of the network integrating salicylate and jasmonate signal transduction derived from global expression phenotyping. *Plant Journal*, 34, 217-228.

Glazebrook, J., Rogers, E. E., Ausubel, F. M. (1996). Isolation of *Arabidopsis* mutants with enhanced disease susceptibility by direct screening. *Genetics*, 143: 973-982.

Gomi, K., Ogawa, D., Katou, S., Kamada, H., Nakajima, N., Saji, H., Soyano, T., Sasabe, M., Machida, Y., Mitsuhara, L, Ohashi, Y., Seo, S. (2005). A: mitogen-activated protein kinase NtMPK4 activated by SIPKK is required for jasmonic acid signaling and involved in ozone tolerance via stomatal movement in tobacco. *Plnat cell and Physiology*, 46, 1902-1914.

Gooday, G. W. (1971). An audioradiographic study of hyphal growth of some fungi. *Journal of General Microbiology*, 67, 125–133.

Greenberg, J. T., Silverman, F. P., Liang, H. (2000). Uncoupling salycilic acid-dependent cell death and defense-related responses from disease resistance in the *Arabidopsis* mutant acd5. *Genetics*, 156: 341-350.

Gupta, V., Willits, M. G., Glazebrook, J. (2000). *Arabidopsis thaliana EDS4* contributes to salycilic acid (SA)-dependent expression of defense responses: evidence for inhibition of jasmonic acid signaling by SA. *Molecular Plant Microbe Interaction*, 13: 503-511.

Harms, K., Ramirez, I. and Pella-Cortes, H. (1998). Inhibition of wound-induced accumulation of allene oxide synthase transcripts in flax' leaves by aspirin and salicylic acid. *Plant Physiology*,18, 1057-1065.

Hatada, E.N., Krappmann, D. and Scheidereit, C. (2000). NF-kB and the innate immune response. Current Opinion in Immunology,12, 52-58.

Heidel, A. J., Clarke, J.D., Antonovics, J. and Dong, X. (2004). Fitness costs of mutations affecting the systemic acquired resistance pathway in *Arabidopsis thaliana*. *Genetics*, 168, 2197-2206.

Hong, J.K. and Hwang, B.K. (2005). Induction of enhanced disease resistance and oxidative stress tolerance by overexpression of pepper basic PR-1 gene in *Arabidopsis*, *Physiological Plantarum*, 124, 267-277.

Hong, J.K., Jung, H.W., Lee, B.K., Dee, S.C., Lee, Y.K. and Hwang, B.K. (2004). An osmotin-like protein gene, CAOSM1, from pepper: differential expression and *in situ* localization of its mRNA during pathogen infection and abiotic stress. *Physiology and Molecular Plant Pathology*, 64, 301-310.

Hong, J.K., Lee, S.C. and Hwang, B.K. (2005). Activation of pepper basic PR-1 gene promoter during defense signaling to pathogen, abiotic and environmental stresses. Gene, 356, 169-180.

Huang, Y., Xiao, B. and Xiong, L. (2007). Characterisation of a stress responsive proteinase inhibitor gene with positive effect in improving drought resistance in rice. *Planta.* 226,73-85.

Jirage, D., Zhou, N., Cooper, Clarke, J.D., Dong, X. and Glazebrook, J. (2001). Constitutive salicylic acid-dependent signaling in cprl and cpr6 mutants requires P AD4. *Plant Journal,* 26, 395-407.

Kapteyn, J. C., Van den Ende, H. and Klis, F. M. (1999). The contribution of cell wall proteins to the organization of the yeast cell wall. *Biochemical and Biophysical Acta,* 6, 373–383.

Karban, R., and Baldwin, I.T. (1997). *Induced Responses to Herbivory.* The University of Chicago Press Ltd, Chicago, IL, USA.

Kessler, A., Baldwin, I.T. (2002). Plant response to insect herbivory: the emerging molecular analysis. *Annua Review Of Plant Biology,* 53, 299-328.

Kinkema, M., Fan, W. and Dong, X. (2000). Nuclear localization of NPRl is required for activation of PR gene expression. *Plant Cell*, 12, 2339-2350.

Klis, F. M. (1994). Review: cell wall assembly in yeast. Yeast, 10, 851–869.

Kloepper, J.W., Tuzun, S. and Kuc, J. (1992). Proposed definitions related to induced disease resistance. *Biocontrol Science and Technology*, 2, 349-351.

Koike, M., Okamoto, T., Tsuda, S. and Imai, R. (2002). A novel plant defensin-like gene of winter wheat is specifically induced during cold acclimation. Biochemical and Biophysical *Research Comrnunication*, 298, 46-53.

Kuc, J. (1982). Induced immunity toplant disease. *Bioscience*, 32,854-860.

Kunkel, B.N. and Brooks, D.M. (2002). Cross talk between signaling pathways in pathogen defense. *Current opinion in plant biology*, 5,325-331.

Lee, M.H. and Sano, H. (2007). Suppression of salicylic acid signaling pathwaysby an ATPase associated with various cellular activities (AAA) protein in tobacco plants. *Plant biotechnology*, 24, 209-215.

Lorenzo, O., Chico, J.M., Sanchez-Serrano, J. J. and Solono, R. (2004). JASMONATE-INSENSITIVE1 encodes a MYC transcription factor essential to discriminate between different jasmonate regulated dense responses in *Arabidopsis.* Plant cell, 16, 1938-1950.

Metraux, J.P. (2001). Systemic acquired resistance and salicylic acid: current state of knowledge. *European journal of plant pathology*, 107, 13-18.

Meur, S., Seebah, S., Suresh, A., Chuon, D., Beuerman, R. and Verma, C. (2006). Defensins: Family of antimicrobial peptides. *Journal of Biochemistry and Molecular Biology*, 39, 278-83.

Montijn, R. C., Vink, E., Muller, W. H., Verkley, A. J., Van den Ende, H. and Henrissat B. (1999), Localization of the synthesis of β1,6-glucan in *Saccharomyces cerevisiae. Journal of Bacteriology,* 181, 7414–7420.

Mou, Z., Fan, W. and Dong, X. (2003). Inducers of plant systemic acquired resistance regulate NPR1 function through redox changes. *Cell,* 113, 935-944.

Nakashita, H., Yasuda, M., Nitta, T., Asmi,T., Fujioka, S., Arai, Y., Sekinata, K., Takatsuto, S., Yamaguchi, I. and Yoshida, S. (2003). Brassinosteroid functions in a broad range of disease resistancein tobacco and rice. *Plant journal*, 33, 887-898.

Niki, T., Mitsuhara, I., Seo, S., Ohtsubo, N. and Ohashi, Y. (1998). Antagonistic effect of salicylic acid and jasmonic acid on the expression of pathogenesis-related *(PR)* protein genes in wounded mature tobacco leaves. *Plant Cell and Physiology*, 39, 500-507.

Nobel, J. G., Klis, F. M., Priem, J., Munnik, T. and Van den Ende, H. (1990). The glucanase-soluble mannoproteins limit cell wall porosity in *Saccharomyces cerevisiae. Yeast.* 6, 491–499.

Norman, C., Vidal, S. and Palva, E.T. (1999). Oligogalacturonide-mediated induction of a gene involved injasmonic acid synthesis in response to the cell-wall-degrading enzymes of the plant pathogen Erwinia caratovora. *Journal of Molecular Plant Microbe Interaction*, 12, 640-644.

Pan, Z.Q., Camara, B., Gardner, H.W. and Backhaus, R.A. (1998). Aspirin inhibition and acetylation of the plant cytochrome P450, allene oxide synthase, resembles that of animal prostaglandin endoperoxide H synthase. *Journal of Biological Chemistry*, 273, 18139-18145.

Pearce, G., Strydom, D., Johnson, S. and Ryan, C.A. (1991). A polypeptide from tomato leaves induces wound-inducible proteinase inhibitor proteins. *Science*, 253, 895-898.

Penninckx, I. A., Eggermont, K., Terras, F. R., Thomma, B. P., De Samblanx, G. W. and Buchala, A. (1996). Pathogen-induced systemic activation of a plant defensin gene in *Arabidopsis* follows a salicylic acid-independent pathway. *Plant Cell,* 8, 2309- 2323.

Petersen, M., Brodersen, P., Naested, H., Andreasson, E., Lindhart, U., Johansen, B., Nielsen, H., Lacy, M., Austin, M, and Parker, J. (2000). *Arabidopsis* MAP kinase 4 negatively regulates systemic acquired resistance. *Cell*, 103, 1111-1120.

Petersen, S.V. and Stolen, O. (1981). Effect of energy-requiring defense reactions on yield and grain quality in a powdery mildew-resistant barley cultivar. *Phytopathology*, 71,396-399.

Petersen, S.V. and Toistrup, K. (1985). The limiting effect of disease resistance on yield. *Annual Review of Phytopathology* , 23, 475-490.

Piel, J., Atzorn, R., Gabler, R., Kuhnemann, F. and Boland, W. (1997). Cellulysin from the plant parasitic fungus *Trichoderma viride* elicits volatile biosynthesis in higher plant via the octadecanoid signaling cascade. *FEBS Letter*, 416, 143-148.

Pieterse, C. M. J., Van Wees, S.C., Van Pelt, J. A., Knoester, M., Laan, R. and Gerrits, H. (1998). A novel signaling pathway controlling induced systemic resistance in *Arabidopsis*. *Plant Cell,* 10, 1571-1580.

Pieterse, C.M.J. and Van Loon, L.C (1999), Salicylic acid-independent plant defence pathways. Trends in Plant Sciences, 4, 52-58.

Pieterse, C.M.J., Ton, J. and Van Loon, L.C. (2001). Cross-talk between plant defence signaling pathways: Boost or burden? *Ag Biotech* Net 3, ABN 068.

Pieterse, C. M. J., Van Wees, S.C.M., Ton, J., Van Pelt, J.A. and Van Loon, L.C. (2002). Signaling in rhizobacteria-induced systemic resistance in *Arabidopsis thaliana. Plant Biology,* 4,535-544.

Rajam, M.V., Chandola, N., Goud, P.S., Singh, D., Kashyap, V., Choudhary, M. L. and Sihachakr, D. (2007). Thaumatin gene confers resistance to fungal pathogens as well as tolerance to abiotic stresses in transgenic tobacco plants. *Biology of Plantarum*, 51, 135-141.

Ramamoorthy, V., Viswanathan, R., Raguchander, T., Prakasam, V. and Samiyappan, R. (2001). Induction of systemic resistance by plant growth promoting rhizobacteria in crop plants against pests and diseases. *Crop protection*, 20,1-11.

Rochon, A., Boyle, P., Wignes, T., Fobert, P.R. and Despres, C. (2006). *Plant Cell*, 18, 3670–3685.

Ross, A.F. (1966). *Systemic effects of local lesion formation*. In viruses of plants, pp. 127-150. Edited by A.B.R. Beemster and J. Dijkstra. Amsterdam: North Holland Publishing compny.

Ryals, J. A., Neuenschwander, U.H., Willits, M.G., Molina, A., Steiner, H.Y., and Hunt, M.D. (1996). Systemic Acquired Resistance. *Plant Cell,* 8, 1809-1819.

Ryals, J., Weymann, K., Lawton, K., Friedrich, L., Ellis, D. and Steiner, H.Y. (1997), The *Arabidopsis* NIM1 protein shows homology to the mammalian transcription factor inhibitor I kappa B. *Plant Cell,* 9, 425-439.

Ryan, C.A. (2000). The systemin signaling pathway: differential activation of plant defensive genes. *Biochemical and Biophysical acta*, 1477, 112-121.

Samuel, M.A., Hall, H., Krzymowska, M., Drzewiecka, K., Hennig, J. and Ellis, B.E. (2005). SIPK signaling controls multipie components of harpin-induced cell death in tobacco. *Plant Journal,* 42, 406- 416.

Sarowar, S., Kim,Y.J., Kim,E.N., Kim, K.D., Hwang, B.K., Islam, R. and Shin, J.S. (2005). Overexpression of pepper basic pathogenesis-related protein 1 gene in tobacco plants enhances resistance to heavy metals and pathogen stress. *Plant cell Report*, 24, 216-224.

Seo, S., Katou, S., Seto, H., Gomi, K. and Ohash, Y. (2007). The mitogen activated protein kinases WIPK and SIPK regulate the levels of jasmonic and salicylic acids in wounded tobacco plants. *Plant Journal*, 49, 899-909.

Shah, J., Tsui F. and Klessig, D.F. (1997). Characterization of a salicylic acid-insensitive mutant (sai1) of *Arabidopsis thaliana,* identified in a selective screen utilizing the SA-inducible expression of the tms2 gene. *Molecular Plant Microbe Interaction,* 10, 69-78.

Shaw, J. A., Mol, P. C., Bowers, B., Silverman, S. J., Valdivieso, M. H. and Duran, A. (1991). The function of chitin synthase- 2 and synthase-3 in the *Saccharomyces cerevisiae* cell cycle. *Journal of Cell Biology*, 114, 111–123.

Smits, G., Kapteyn, J. C., Van den Ende, H., and Klis, F. M. (1999). Cell wall dynamics in yeast. *Current Opinion in Microbiology,* 2, 348–352.

Spoel,S.H., Annemart Koornneef, Susanne M. C. Claessens, Jerôme P. Korzelius, Johan A. Van Pelt, Martin J. Mueller, Antony J. Buchala, Jean-Pierre Métraux, Rebecca Brown, Kemal Kazan, L. C. Van Loon, Xinnian Dong and Corné M. J. Pieterse. (2003). NPR1 Modulates Cross-Talk between Salicylate- and Jasmonate-Dependent Defense Pathways through a Novel Function in the Cytosol. *The Plant Cell*,15, 760–770.

Spoel, S.H., Mou, Z., Tada, Y., Spivey, N.W., Genschik, P. and Dong, X. (2009). Proteasome-Mediated Turnover of the Transcription Coactivator NPR1 Plays Dual Roles in Regulating Plant Immunity. *Cell*, 137, 860-872.

Sticher, L., Mauch-Mani, B. and Metraux, J.P. (1997). Systemic acquired resistance, *Annual Review of Phytopathology,* 35, 235-270.

Straus, D.S. and Glass, C.K. (2001). Cyclopentenone prostaglandins: new insights on biological activities and cellular targets. *Medicinal Research Reviews*, 21, 185-210.

Tada, Y., Spoel, S.H., Pajerowska-Mukhtar, K., Mou, Z., Song, J., Wang, C., Zuo, J. and Dong, X. (2008). *Science*, 321, 952–956.

Theis, T. and Stahl, U. (2004). Antifungal proteins: targets, mechanisms and prospective applications. *Cellular and Molecular Life Sciences,* 61, 437–455.

Thomma, B., Eggermont, K., Penninckx, I., Mauch-Mani, B., Vogelsang, R. and Cammue, B. P. A. (1998). *Separate jasmonate-dependent and salicylate-dependent defense-response pathways in Arabidopsis are essential for resistance to distinct microbial pathogens.* In: Proceedings of National Academy of Siences USA, 95, 15107-15111.

Thomma, B.P., Penninckx, I.A., Broekaert W.F. and Cammue, B.P. (2001). The complexity of disease signaling in *Arabidopsis. Current Opinion in Immunology,* 13, 63-68.

Ton, J., and Mauch-Mani, B. (2004). Beta-amino-butyric acid-induced resistance against necrotrophic pathogens is based on ABA-dependent priming for callose. *Plant Journal*, 38: 119-130.

Van Loon, L.C., Bakker, P.A.H.M. and Pieterse, C.M.J. (1998). Systemic resistance induced by rhizosphere bacteria. *Annual Review of Phytopathology*, 36, 453-483.

Van Loon, L. C., and Van Kammen, A. (1970). Polyacrylamide disc electrophoresis of the soluble leaf proteins from *Nicotiana tabacum* var. 'Samsun' and 'Samsun NN' II. Changes in protein constitution after infection with tobacco mosaic virus. *Virology,* 40: 199-211.

Van Peer, R., Niemann, G. J., Schippers, B. (1991). Induced systemic resistance and phytoalexin accumulation in biological control of *Fusarium* wilt of carnation by sp. Strain WCS417r. *Phytopathology*, 81: 779-783.

Wan-Chi Lin, Ching-Fang Lu, Jia—Wei WU, Ming-Lung Cheng, Yu-Mei Lin, Ning-Sun Yang, Lowell Black, Sylvia, K. Green, Jaw-Fen Wang and Chiu-Ping Cheng. (2004). Transgenic tomato plants expressing the *Arabidopsis* NPR1 gene display enhanced resistance to a spectrum of fungal and bacterial diseases. *Transgenic Research,* 13, 567-581.

Ward, E.R., Uknes, S.J., Williams, S.C., Dincher, S.S., Wiederhold, and Alexander, D.C. (1991). Coordinate gene activity in response to agents that induce systemic acquired resistance. *Plant Cell,* 3: 1085-1094.

Wasternack, C. and Parthier, B. (1997). Jasmonic-signaled plant gene expression. *Trends in Plant Science.*, 2, 302-307.

Wei G, Cloepper, J. W., and Tuzun, S. (1991). Induced systemic resistance of Cucumber to *Colletotrichum orbiculare* by selected strains of plant growth promoting rhizobacteria. *Phytopathology*, 81: 1508-1512.

Weigel, R. R., Bauscher, C., Pfitzner, B., and Christiane, G. (2001). NIMIN-1, NIMIN-2 and NIMIN-3, members of a novel family of proteins from *Arabidopsis* that interact with NPR1/NIM1, a key regulator of systemic acquired resistance in plants. *Plant Molecular Biology*, 46: 143-160.

Weigel, R. R., Ursula, M., Pfitzner, B., and Christiane, G. (2005). Interaction of NIMIN1 with NPR1 modulates PR gene expression in *Arabidopsis. Planta Cell*, 17: 1279-1291.

Wosten, H. A. B., Richter, M. and Willey, J. M. (1999). Structural proteins involved in emergence of microbial aerial hyphae. *Fungal Genetics and Biology*, 27, 153–160.

Yamada, S., Komori, T., and Imaseki, H. (1997). Plant gene register. 97-131. *Plant Physiology*, 115: 314.

Yuan, Y., Zhong, S., Li, Q., Zhu, Z., Lou, Y., Wang, L., Wang, J., Wang, M., Li, Q., Yang, D. and He, Z. (2007). Functional analysis of rice NPR1-like genes eveals that OsNPR1/NH1 is the rice orthologue disease resistance with enhanced herbivore susceptibility. *Plant Biotechnology*, 5, 313-324.

In: Genetically Engineered Crops
Editors: Lindsay M. Grover

ISBN: 978-1-61122-526-6

Chapter 5

SAFETY ASSESSMENT OF GENETICALLY MODIFIED (GM) CROPS: CURRENT PROTOCOL, RESEARCH TREND AND FUTURE CHALLENGE

Chih-Hui Lin and Tzu-Ming Pan

Department of Biochemical Science and Technology, College of Life Science, National Taiwan University, Taipei, Taiwan, China

ABSTRACT

The rapid development of biotechnology and molecular biology has made the genetically modified (GM) crops became a part of agricultural production. From 1996 to 2010, a total of 144 GM crops events have been approved worldwide. Following the success in global market, safety issues of GM crops have become a public concern in the last decade. Therefore, GM crop safety assessment protocol has been developed following guidelines proposed by WHO/FAO and Codex. The current methodology of GM crops/foods safety assessment is relatively empirical and conservative. Thus, accurate hazard identification and risk assessment will continue to represent important part of safety assessment research. Today, many efforts have made in bioinformatics allergenicity prediction, animal models and the analysis of unintended effects. However, the knowledge gap between unintended effects and GM crop safety assessment will be the major challenge in the future.

Keywords: Genetically modified crop, Safety assessment, Unintended effects.

* Corresponding author: Department of Biochemical Science and Technology, College of Life Science, National Taiwan University, No. 1, Sec. 4, Roosevelt Road, Taipei, 10617 Taiwan. Tel: +886-2-33664519 ext. 10. Fax: +886-2-33663838. E-mail: tmpan@ntu.edu.tw

INTRODUCTION

Genetically modified (GM) crops are dominant products in current food production and world trading market due to its superior productivity. In 2009, 43% of global area of principal crops was GM (James, 2010). The global value of GM crop market was 10.5 billion USD in 2009, representing 20% of the global crop protection market. From 1996 to 2010, a total of 144 GM crops events have been approved worldwide (James, 2010). These statistics showed that GM crops safety is directly related to our daily life.

Following the success in global market, safety issues of GM crops have became a public concern, and the safety debates of GM crop are growing increasingly complex and emotional (Poppy, 2000). The safety of GM crops has become a complex of scientific, economical, emotional and even cultural issue. To avoid safety dispute and risk, scientific based GM crops safety assessment guidelines have been proposed by various international organizations such as Food and Agriculture Organization/World Health Organization (FAO/WHO) and Codex. The current GM crop safety assessment guidelines are based on proved methods including the evaluation of gene source, bioinformatics search, comparisons of key compositions, toxicological study, and simulated gastric fluid (SGF) assay. The advantages of current safety assessment protocol are simple, cost-effective and robust. It is important that there is no documented proof that any approved GM crop has caused allergic reactions owing to the transgenic protein, or that generation of significant endogenous allergenicity (Goodman et al., 2008).

One of the major debate of current safety assessment protocols is the "unintended effects", which is any unintended and unexpected changes of plant occurred during the process of genetic modification. The unintended effects of GM plant made the concept of "substantial equivalence", which was generally adopted by GM crops safety assessment guidelines, is not adequate to assess whether a transgenic food has to be considered as safe as its corresponding wild-type (Millstone et al., 1999; Corpillo et al., 2004). As the fast development of system biology and high throughput analysis, many studies have revealed the prevalence of unintended effects of GM crops (Cellini et al., 2004; Filipecki and Malepszy, 2006), which supports the concern of capability of "substantial equivalence".

Because of the indirect/unanticipated risk may raised from unintended effects, Codex guidelines for GM food has updated to include the analysis of unintended effect in 2003 (Haslberger, 2003). Besides the unintended effects, the dominance of stacked trait GM crops in the market has also make the concept of "substantial equivalence" obsolete, since there is no wild-type counterpart of stacked trait GM crop (Llorente et al., 2010).

In the other hand, most of recent strife about the safety of GM crops was raised from the debates that have no generally accepted assessment guideline, such as the environmental risk. For example, the controversy on the decision of Germany and France to cease the growing of MON810 GM maize in 2008 is still unsolved. The lack of consensus in the experiment design and the result interpretation makes the suspension of cultivation has no scientific justification (Ricroch et al., 2010).

In this report, we are focused on the safety assessment of GM crops for potential risk for human health. The current protocol, research trend and future challenges will be discussed in this article.

CURRENT GM CROP SAFETY ASSESSMENT GUIDELINES

The Development of GM Crop Safety Assessment

At an early stage in the introduction of biotechnology in modern plant breeding, efforts began to define internationally synchronized evaluation strategies for the safety of foods derived from GM organisms (GMOs). In 1990 the International Food Biotechnology Council (IFBC) published the first report on the issue of safety assessment of new varieties originated from biotechnology (IFBC, 1990). The comparative approach described in this report is the basis of later safety evaluation strategies. Other organizations, such as the Organization for Economic Cooperation and Development (OECD), FAO/WHO and the International Life Science Institute (ILSI) have developed further guidelines for safety assessment which have obtained broad international consensus among experts on food safety evaluation.

In 1993 the OECD formulated the concept of "substantial equivalence" as a main guideline for the assessment of GM foods, which has been further elaborated in the following years (OECD, 1993; OECD, 1996; OECD, 1998). In 1996 a consensus documents has been prepared by ILSI Europe on evaluation of the safety of novel foods (Jonas et al., 1996). Similar to the OECD's and FAO/WHO's consensus views, the ILSI document defines three scenarios based on the substantial equivalence to a reference food/ingredient. The three scenarios are envisioned in which the GM plant or food would be substantial equivalent, or substantially equivalent except for the inserted trait, or not equivalent at all. The compositional analysis of key components of GM plant, including key nutrients and natural toxicants, in addition to phenotypic and agronomic characteristics, is the basis of assessment of substantial equivalence.

A decision tree for testing GM foods for allergenicity has been also developed by ILSI in collaboration with the IFBC (Metacalfe et al., 1996). IgE binding tests are recommended with the new protein derived from known allergenic sources, using sera from individuals allergic to those sources, followed if necessary by skin-prick testing and double-blind food challenges. The decision-tree approach for new proteins derived from sources with no known history of allergenicity relies primarily upon sequences homology comparisons with known allergens, and on the stability of the protein in the human digest tract.

FAO and WHO have been organizing workshop and consultations on the safety of GMOs since 1990. At the Joint FAO/WHO Consultation in 1996 (FAO/WHO, 1996), it was recommended that the safety evaluation should be based on the concept of substantial equivalence. The following parameters should be considered to determine the substantial equivalence of a GM plant: molecular characterization; key nutrients; toxicants; and allergens. Following the early development of safety assessment by FAO/WHO, the Codex Alimentarius Commission of FAO/WHO is dedicated to develop international food standards. During the first session of Codex in Chiba (Japan) in March 2000 (FAO/WHO, 2000), definitions were agreed concerning the 'risk assessment' and 'risk analysis' of GM foods. Risk assessment includes food safety, substantial equivalence and long term health effects. Animal studies are deemed necessary to obtain information on the characteristics of newly expressed proteins, analogous to the conventional toxicity testing of food additives. At FAO/WHO Expert Consultation on Allergenicity held in Rome in January 2001, a new decision tree was developed (FAO/WHO, 2001). The new decision tree builds on the one developed by ILSI/IFBC (Metaclfe et al., 1996) and FAO/WHO (1996). Criteria for a positive outcome of sequence similarity include >35% identity in amino acid sequence between the expressed protein and a known allergen, or identity of six contiguous amino acids. Any positive outcome defines the product as allergenic, and further product development should be discontinued. A negative outcome of the specific *in vitro* serum screening will lead to further targeted serum screening, testing the expressed protein for pepsin resistance and immunogenicity in animal models. Targeted *in vitro* serum testing is done with sera from patients allergic to materials that are broadly related to the source of the original gene (FAO/WHO, 2001).

The Codex Alimentarius Commission adopted guidelines in 2003 to synchronize the premarket risk assessment process for plants derived from biotechnology (GM plants) in the global market. The Codex Alimentarius Commission guidelines abandoned the risk assessment based on a decision tree and adopted a weight-of-evidence approach. A decision tree was found to be too conservative in a situation where no single method is sufficiently predictive. A 35% identity over an 80 amino acid alignment was recommended as a sufficient prediction for potential cross-reactivity. Moreover, the analysis of unintended effects was included in the Codex guidelines. These recommendations have not been accepted by some regulators. Clearly, the existence of multiple documents with diverging recommendations coming from different organizations has resulted from regulatory authorities (Goodman et al., 2008). In summary, the major consensus of current safety assessment protocols including the detailed description of GM plant, the comparison of key components, the evaluation of potential toxicity and the evaluation of potential allergenicity.

The Description of GM Plant and Transgene

According to the Codex guidelines, the characteristic of following subjects should be described in order to evaluate the safety risk of GM plant: the GM plant and its use as food, the source of gene, the inserted DNA and flanking DNA at the insertion site and the expressed substance (Codex, 2003). The information is vital to the safety assessment, especially for the evaluation of potential allergenicity. The safety assessment guidelines developed by Taiwan Department of Health is based on Codex guidelines (DOH, 2010), in which detail descriptions of GM plant are required for assessing the safety risk. The description of GM plant should include following subjects:

1. The description of GM plant.
   A. The species of plant and the type of genetic modification.
   B. The purpose and the trait of genetically modification.
2. The description of host plant and the use for food.
   C. The common name and taxonomy of plant.
   D. History of cultivation and consumption, especially the safety risk to human.
   E. Genotype and phenotype related to the safety risk of host plant, which includes:
      i. Toxicity.
      ii. Allergenicity.
      iii. The safety risk of the close relatives of the host plant.

3. History of safe consumption, which includes:
   F. Conventional storage, transportation and breeding methods.
   G. Specific processing method to make the plant could be safely consumed.
   H. The role in the daily diet of people.
4. Transgene source organism and its close relatives.
   I. The common name and taxonomy of plant.
   J. History of safe consumption, which includes:
      iv. Information of endogenous toxin, anti-nutrient and allergen. Pathogenicity and the relation to known pathogens should be provided in the case of the origin of transgene is a microorganism.
      v. The history of food usage, and the possible non-food ingestion method.
5. The description of genetic modification.
   K. The transformation process.
      vi. The method used for transformation.
      vii. Information of transformation vector.
         [1] The source of vector (e. g. bacteria, virus, plant or synthetic)
         [2] Identification of vector
         [3] The anticipate function in plant.
      viii. Any intermediate vector used in the plant transformation process.
   L. Information of the DNA used to transform plant.
      ix. Characteristics of genetic materials, which includes selection marker, regulatory elements and any other functional sequence.
      x. Size and identification of transformation vector.
      xi. The anticipate position and orientation in the transgenic plant.
      xii. The anticipate function in the transgenic plant.
6. Characteristic of genetic modification.
   M. Information of transgenic DNA in the GM plant.
      xiii. Identification of transgenic DNA.
      xiv. Number of transgenic DNA in the GM plant.
      xv. The structure of each transgenic DNA.
         [4] The copy number of transgenic DNA.
         [5] The flanking sequence of transgenic DNA.
         [6] Other information that will help to evaluate safety risk (e.g. the expression level of transgene or unintended gene product).
      xvi. Any open reading frame that originated from transgenic DNA and the adjacent plant genomic DNA.
   N. Information of any new substance expressed in the GM plant.

If the gene source or other background description suggested that there is a potential risk of allergenicity (i.e. the source of gene is a commonly allergenic food), serum IgE tests using sera from patients allergic to the source are required to ensure that the protein encoded by the gene dose not bind IgE from those allergic to the source.

Comparison of Key Components

The method of comparison of key components between GM plant and its wild-type counterpart for safety assessment is based on the concept of substantial equivalence. In 1990, a joint consultation of the FAO and the WHO established that the comparison of a final product with one having an acceptable standard of safety provides an important element of safety assessment (WHO, 1991). In 1993 the OECD further elaborated this concept and considered the approach to safety assessment based on substantial equivalence as being the most practical approach to assessing the safety of foods and derived from modern biotechnology (OECD, 1993). A comparative approach focusing on the determination of similarities and differences between the GM food and its conventional counterpart aids in the identification of potential safety and is considered the most appropriate strategy for the safety assessment of GM foods.

The Joint FAO/WHO Expert Consultation on Biotechnology and Food Safety (1996) elaborated on compositional comparison as an important element in the determination of substantial equivalence. And the Joint FAO/WHO Expert Consultation (2000) concluded that the application of the concept of substantial equivalence contributes to a robust safety assessment framework. The comparison of key components should be seen as a key step in the safety assessment process although it is not a safety assessment in itself. A comparison of critical components can be carried out at the level of the food source (i.e. species) or the specific food product. Critical components are determined by identifying key nutrients, key toxicants and anti-nutrients for the food source. The comparison of key nutrients should be between the modified plant and non-modified counterpart with history of safe use. The data for the non-modified plant can be the natural ranges published in the literature or those measured levels in parental or other edible varieties of the species (FAO, 1996). Key components of a specific crop for comparison with a GM crop are first described by Consensus Documents compiled by the OECD's Task Force for the Safety of Novel Foods and Feeds (OECD, 2001). A series of consensus documents could be accessed through the OECD website (http://www.oecd.org/biotrack/), which provides useful guidance on the components should be minimally analyzed. For example, the suggested tomato constituents to be analyzed for food use includes proximate (protein, fat, total dietary fibre, ash and carbohydrates), minerals (magnesium and potassium), vitamins (vitamin C, K and folate), β-carotene, lycopene and tomatine (α-tomatine and dehydrotomatine) (OECD, 2008). The range of composition of various tomato fruit and products was also listed in the consensus document.

Allergenicity Assessment

Guidelines for allergenicity assessment of GM crops were published in three sequential documents that have been broadly recognized. The first comprehensive document was published in 1996 by the IFBC in collaboration of ILSI (Metacalfe et al., 1996). This was followed in 2001 by the FAO/WHO and in 2003 by the Codex Alimentarius Commission guidelines. Several elements are well established and have remained consistent throughout the three successive sets of recommendations. However, there is no single, definitive test for determine the allergenic potential of novel proteins. The Codex guidelines abandoned the risk assessment based on a decision tree (FAO/WHO, 2001) and adopted a weight-of-evidence approach (Figure 1.). In the Codex weight-of-evidence approach, the analysis of gene source, bioinformatics search and pepsin digestion is essential to evaluate the allergy risk of GM plant. Specific IgE test is needed to clarify allergy risk if any result of gene source, bioinformatics search and pepsin digestion test is positive.

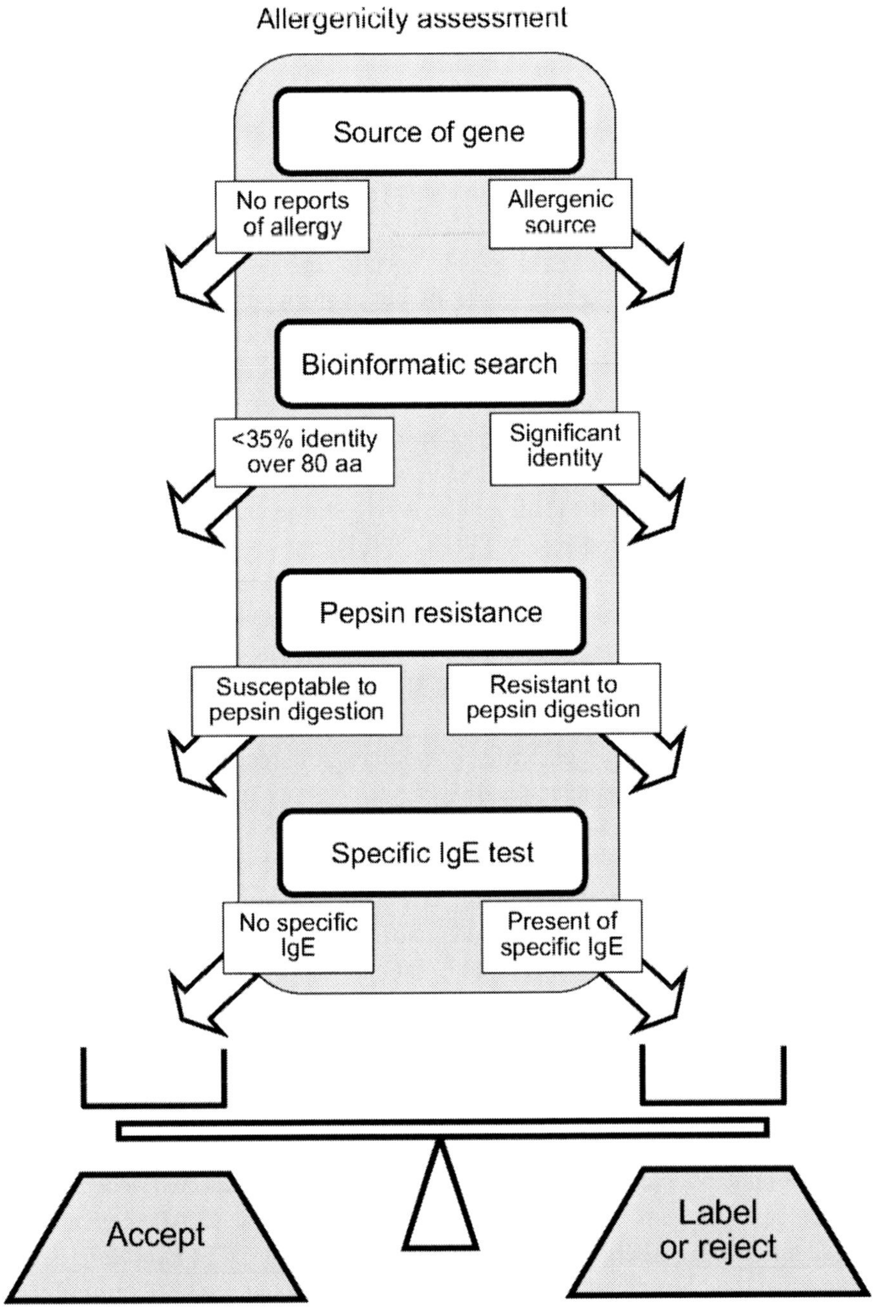

Figure 1. The diagram of weight-of-evidence approach suggested in the Codex guideline of GM crop safety assessment for allergenicity prediction.

Gene Source

The description of gene source is virtually included in the section of "description of GM plant and transgene". If the source of any transgene is a common allergenic organism (e.g. peanuts), specific IgE test using sera from patients allergic to the source are required to ensure that the protein encoded by the gene does not bind IgE from those allergic to the source.

Bioinformatics Analysis

The current approach of bioinformatics analysis is to assess the amino sequence similarity of transgenic protein to known allergen. The amino acid sequences of all transgenic proteins are needed to be analyzed. If any identity match is sufficiently high to suspect that the protein might cause allergic cross-reactions, specific IgE test should be carried out, using donors with specific allergies to the source of the sequence-matched allergen to evaluate specific IgE binding. As an integral part of weight-of-evidence approach (Figure 1.), this method is not meant to be a stand-alone test, but rather to identify proteins that would require serum testing.

Early guidelines proposed linear bioinformatics searches using the identity of 8-12 contiguous amino acids with a known allergen as an indicator of potential allergenic risk based on sequence similarity (Metacalfe et al., 1996). However, the use of a six contiguous amino acids was later recommended (FAO/WHO, 2001), but was subsequently found to yield an unacceptably high number of false positives (Hileman *et al.*, 2002; Silvanovich *et al.*, 2006). Stadler and Stadler (2003) reported that 6-mer match resulted in more than two-thirds of all proteins in Swiss-Prot being predicted to be allergens, and >40% of the human genome being predicted as such. Obviously, the use of short amino acid matching searches is not a useful approach for allergenicity assessment, but it has never been truly renounced. Consequently, a few regulatory authorities sometimes still require bioinformatics analyses based on 6-mer matches (Goodman et al., 2008).

Another method that generally adopted by allergenicity assessment guidelines is the sliding window of 80 amino acid searches, which is looking for identities greater than 35%. It has been concluded that using greater than 35% shared identity over any 80 amino acid section (sliding window) to estimate the potential of cross reactivity is highly conservative (i.e., many false positives) (Ladics et al., 2007; Thomas et al., 2005). Even so, a conservative threshold value of 35% identity over 80 amino acid section of the transgenic protein contained in both FAO/WHO and Codex documents was intended to identify conserved gene segments representing functional motifs, which might retain conformational epitope structure as well.

Choosing the appropriate allergen database to search is vital for reliable sequence comparison (Goodman et al., 2008). Database with up to date, comprehensive allergen records and reliable analysis tools are essential. Databases such as AllergenOnline (*http://www.allergenonline.org),* Allergen Database for Food Safety (ADFS, *http://allergen.nihs.go.jp/ADFS/index.jsp*) and Structural Database of Allergenic Proteins (SDAP,*http://Fermi.utmb.edu/SDAP/*) provide useful analysis tools based on the >35% identity in 80 amino acids alignment and 6 or 8 contiguous amino acids criteria.

The AllergenOnline database is the only database that is currently fully peer-reviewed regarding evaluation of published evidence of allergenicity. It provides access to a peer reviewed allergen list and sequence searchable database intended for identifying proteins that may present a potential risk of allergenic cross-reactivity. The predictive searches of the overall FASTA alignment, 80 amino acid alignments and 8 contiguous amino acid alignments are available. The allergenOnline database includes 13 (aero animal, aero fungi, aero insect, aero mite, aero plant, contact, food animal, food fungi, food plant, gliadin, protozoan, venom/salivary, and worm) categorized allergens.

ADFS is a web server system that was launched as a project of the Novel Foods and Immunochemistry of National Institute of Health Sciences (Japan), and this project was partly supported by a grant from the Ministry of Health, Labor and Welfare (Japan). The ADFS database has been constructed to include known allergens and B-cell epitope sequences, which also includes 13 categorized allergens (based on allergen type in AllergenOnline), with their accession numbers, epitope information, 3D-structure information, and sugar-containing information. The ADFS database provides allergenicity prediction tools of the 80 amino acids section and contiguous amino acids search tools (FAO/WHO method) and motif-based method. BLAST and epitope sequence search tools is also provided to analyze the sequence homology to known allergens.

SDAP is a database of allergenic proteins with various computational tools that can assist structural biology studies related to allergens. The predictive searches of the overall FASTA alignment, 80 amino acid alignments and contiguous amino acid alignments are available. Using the SDAP database, it is possible to retrieve information related to an allergen from the most common protein sequence and structure databases (SwissProt, PIR, NCBI, PDB), to find sequence and structural neighbors for an allergen, and to search for the presence of an epitope other the whole collection of allergens.

Other databases such as the National Center for Biotechnology Information (NCBI; *http://www.ncbi.nlm.nih.gov/*) non-redundant database and the Allergome *(http://www.allergome.org*) are available. The alternative of searching, with key-word limits can provide more updated sequences, but lack as accurate screening method for relevance.

It should be noted that the results of allergenicity tests with different databases may not completely consistent. Although based on the same criteria, the variation of allergen record registered in databases and the different algorithms used for allergenicity prediction may pose an inconsistency to the conclusion, if the identity of a transgenic protein to an allergen is approximate to the criteria. Table 1 is the result of bioinformatics search of Cucumber Mosaic Virus (CMV) coat protein, which is the transgenic protein expressed in the GM CMV-resistant tomato R8, showed that CMV coat protein has no significant identity to known allergen (Lin et al., 2010). The similarity of CMV coat protein sequence to known allergens is not significant, but the remarkable difference of the overall identity to allergens based on different database revealed a potential problem. Some allergen records are even only present in the result of specific databases. Table 2 is another example of inconsistent results of databases search (Lin et al., 2009). Sporamin, a Kunitz-type trypsin inhibitor with insect-resistance capability, is a major storage protein present in sweet potato (*Ipomoea batatas*), was used in the analysis. The results of bioinformatics search showed that the sequence of sporamin is significant similar (E score) to known allergens in overall sequence search, and part of 80 amino acid alignment results also showed that the sequence of sporamin has significant identity (>35%) to allergens. However, in the results of 80 amino acid alignment analysis, inconsistent conclusion was result from different database. The result of 80 amino acid alignment search of sporamin sequence in AllergenOnline database showed that the identity of sporamin to allergen Sola t 4 (accession no. P30941) was significant higher than 35% criteria (37.0%), while the result of same analysis in the SDAP database showed a negative result (<35% identity). The inconsistency suggests that controversy of allergenicity prediction may rise from the diversity of allergen records and algorithm of databases, if the analysis is solely based on one database. For these reason, we strongly recommend that multiple database should be used for the bioinformatics allergenicity test in order to avoid controversy caused by the diversity of databases.

Protein Digestibility in Simulated Gastric Fluid/Pepsin

The ability of the transgenic protein to resist digestion by simulated gastric fluid (SGF)/pepsin is recognized as a potential risk factor of allergenicity (Metacalfe et al., 1996). Several potent food allergens are known to be very stable in an *in vitro* pepsin digestion assay, whereas most dietary proteins are readily digestible (Asero et al., 2007.). Evaluation of the potential allergenicity of proteins derived from GM foods has involved a weight of evidence approach that incorporates an evaluation of protein digestibility in pepsin. For allergenicity evaluation, efforts should be taken to prevent the compromise of reproducibility and reliability of SGF test caused by the variation of assay parameters, and a validated SGF protocol is essential. Potential variations in assay parameters include: pH, pepsin purity, pepsin to target protein ratio, target protein purity, and method of detection may compromise the reliability of result.

A multi-laboratory evaluation of a common *in vitro* pepsin digestion assay protocol has been carried out by ILSI Health and Environmental Sciences Institute in 2004 (Thomas et al., 2004). A ratio of 10U of pepsin activity/microg test protein was selected for all tests (3:1 pepsin to protein, w:w), and digestions were performed at pH 1.2 and 2.0. SGF reaction was stopped immediately after sampling at 0.5, 2, 5, 10, 20, 30, and 60min. Protein digestibility was assessed from stained gels following SDS-PAGE of digestion samples and controls. It was showed that SGF reaction carried out with a ratio of 10 U of pepsin activity/μg test protein (3:1 pepsin to protein, w:w) at pH 1.2 is reproducible and yields consistent results when performed using the same proteins at different laboratories.

Other parameters of SGF assay includes an objective endpoint of digestion (e.g. 90%) (Ofori-Anti et al., 2004), the use of high quality pepsin (Sigma P6887 or equivalent), purified transgenic protein (purity >85%, concentration >5 mg/mL) and tricine-SDS-PAGE were recommended for the SGF test in the laboratory manual for allergenicity evaluation (Food Industry Research and Development Institute, Taiwan). The pepsin resistant protein control (e.g. soy trypsin inhibitor), pepsin susceptible protein control (e.g. bovine serum albumin) and a blank control (without transgenic protein) should be included to insure the reproducibility and reliability of SGF analysis.

The SGF test method, as well as the bioinformatics search used in the weight-of-evidence approach, however, is a preliminary prediction rather than a definitive test of protein allergenicity. There are some proteins not known to cause significant food allergies are also stable in SGF (Fu et al., 2002). And some proteins known to cause food allergy, especially those inducing only oral allergy syndrome-mild tingling or itch in the mouth, without substantial edema - are susceptible to pepsin digestion (Fu et al., 2002).

Some very stable proteins such as thaumatin-like proteins from apple and grape rarely cause allergy or possibly only mild reactions (Yagami et al., 2000), whereas others, like the lipid transfer proteins from a variety of sources, are very stable and may cause severe reactions (Vassilopoulou et al., 2007). Some of these stable proteins are inducible pathogenesis-related proteins and expression is variable in foods, which may complicate their recognition as allergens (Breiteneder and Radauer, 2004). There is also evidence that some important pepsin susceptible allergens become more stable with minor shift in pH (Untersmayr et al., 2005). Although the increased stability at moderate stomach pH values may help explain the allergenicity of some of these proteins, the use of standard pepsin stability testing at pH 1.2 or 2.0 still has good demonstrated predictive value.

An additional risk factor of a transgenic protein to cause food allergy is the abundance of the protein in food, as many major food allergens account for >1% of the protein in high-protein allergenic foods (Metacalfe et al., 1996). Others, such as lipid transfer proteins and parvalbumins are less abundant. Abundant, pepsin resistant proteins are more likely to survive digestion in sufficient quantities to become significant food allergens. The typical quantity consumed of specific foods would be expected to have an impact as well, so non-abundant, pepsin resistant proteins may be potent allergens, if a large amount of food is consumed. However, additional scientific data would be required to establish completely objective criteria for acceptance or concern based on stability and abundance. Currently the results are judged relative to common, potent food allergens.

Specific Serum IgE Test

Due to the methods of current allergenicity protocol including evaluation of gene source, bioinformatics search and pepsin digestion test are relatively indirect and empirical, specific serum IgE binding is deemed to be the most important evidence of allergenic potential for a protein. If any result of the evaluation of gene source, bioinformatics search and pepsin digestion test indicates potential allergenicity of a protein, specific serum IgE test should be performed according to the Codex guideline. Thus, specific serum IgE test is usually play a conclusive role in the weight-of-evidence approach. Despite specific serum IgE test is a relative straightforward method, the availability of specific sera, especially for rare allergic food, has limited the use of specific serum IgE test in the routine allergenicity evaluation.

Serum IgE testing to evaluate proteins from an allergenic source, or protein with sequence identity to a known allergen works very well if performed properly (Sten et al., 2004). A few relatively well-controlled studies have been used to evaluate GM crop safety, although the relevance of the donor selection has not always been clear (Sten et al., 2004). The design and interpretation of assays for specific IgE can be complex. Appropriate positive and negative control proteins or extracts of the allergenic source material are required to demonstrate assay validity. The positive test sera should typically allergic subjects who react to the gene source or sequence-matched allergen source. Negative control sera should typically include donors with allergies to other unrelated proteins as well as non-allergic subjects.

The test should be capable of detecting IgE binding to linear and conformational epitopes, sometimes requiring two separate assays. Test parameters include the molecular status of the protein and abundance of the protein in the source material should be confirmed. The test material must be representative of the form available in the GM food source. Demonstration of specificity of binding requires replicate samples with sera exposed to appropriate inhibitors (Goodman et al., 2008). Moreover, it must be recognized that there are no absolute thresholds of serum IgE binding that provide absolute measurement of safety or risk. Positive IgE tests without clinical relevance are common in clinical practice. Serum from individuals with strong carbohydrate-specific IgE antibodies should be avoided for GM assessment to ensure selection of appropriate donors who should have IgE directed against peptide epitopes rather than carbohydrate. Otherwise, carbohydrate-binding sera would lead to designating most glycoproteins as an allergenic risk, although it is widely accepted that the glycans are unlikely to cause clinical food allergy (van Ree, 2002). In the event the transgenic protein is glucosylated, alternative testing may be required to evaluate glycan structure or if IgE binding is demonstrated, the relevance should tested by basophil histamine release or *in vivo* allergen testing. Diagnosing allergic disease requires a holistic evaluation of diet, symptoms, SPT (skin prick test) and/or specific IgE and elimination diet or challenge test (Asero et al., 2007). Likewise, interpretation of IgE binding to GM proteins requires judgment. Strong, specific binding to the protein using appropriate donors should be taken as evidence of risk. However, low levels of binding that are not clearly specific and close in affinity to the suspected cross-reactive allergen may not indicate significant risk. If results are equivocal, SPT or challenges might be necessary to demonstrate the relevance of low levels of apparent specific IgE binding (Goodman et al., 2008).

Toxicology Study

Animal studies are recommended to obtain information on the characteristics of proteins which have no history of food use, analogous to the conventional toxicity testing of food addictives. Testing of whole foods may be considered if relevant changes in composition may take place. However, such studies should be considered on a case by case basis, taking the limitations of this type of study into account. The 2000 FAO/WHO Expert Consultation (FAO/WHO, 2000) recommended as minimum requirement a subchronic 90-day study, with possibly long-term studies needed if the results of a 90-day study indicate adverse effects, such as proliferative changes in tissues.

Testing of whole foods in laboratory animals has its specific problems, and considerable experience has been gained with toxicological testing of irradiated foods. The amounts of food to be administrated to animals are limited due to effects on satiety and possible negative interference with the nutritional balance of the animal diet (Kuiper et al., 2001). Feeding animals with whole foods at exaggerated dose level may induce a series of adverse effects that would mask potential adverse effects caused by the genetic modification. The highest test dosages should be the maximum amount of novel food product that can be included in a balanced animal diet, while the lowest dosage should be comparable to the expected amount in the human diet. Furthermore, the bio-availability of food constituents may be different when ingested as part of the food matrix (Kuiper et al., 2001).

Appropriate safety testing with whole foods should be carried out in parallel with toxicity studies on specific, isolated proteins. A combination of animal experiments and *in vitro* experiments with mammalian cell toxicity tests may carry out to focus on the identification of biomarkers for exposure and makers that are predictive in an early phase of exposure of chronic toxicity (Diplock et al., 1999).

TRENDS OF GM CROP SAFETY ASSESSMENT

Bioinformatics Analysis

The current bioinformatics approach should remarkably reduce the allergy risk of GM plant. However, methods such as the sliding window of 80 amino acid searches, which is looking for identities greater than 35%, has been concluded as highly conservative (i.e., many false positives) (Ladics et al., 2007; Thomas et al., 2005). Hence, although these sequence analogy approaches are useful, it is likely that many potentially beneficial protein products are unnecessarily eliminated in the early evaluation process. Additional approaches to sequence/structure analysis would be welcome.

One alternative of the >35% criteria of 80 amino acid search is to focus on overall sequence alignments, as suggested by Ladics et al. (Ladics et al., 2007). Fifty percent overall identity of amino acid sequence has been suggested as the criteria of the risk of potential cross-reactivity (Goodman et al., 2008). On the basis of literature searches, only a few examples of endogenous proteins from sources suspected of cross-reactivity for proteins showed 35% to 50% identity over entire sequence, and quantitative IgE binding and basophil histamine release demonstrate only partial reactivity (Goodman et al., 2005). Thus, another alternative would be to increase the criteria of identity for the 80-amino acid alignment closer to a level (possibly > 50% identity) where there are examples of at least weak *in vitro* cross-reactivity using sera from individuals having allergic symptoms to the proteins sources (Goodman et al., 2008).

The structure analysis of protein is also a potential tool for allergenicity prediction. It was reported that the allergenicity prediction based on similarities of protein motifs was superior to that based on amino acid sequence comparison (Stadler and Stadler, 2003). Although there are currently no known unique motifs that identify a protein as an allergen, a better understanding of structure attributes could prove valuable for allergenic potential.

The current sequence search algorithms are based on FASTA, which is a program for amino acid/nucleotide sequence comparisons and database searches, have very limit capability for protein structure analysis. Therefore, databases such as SDAP and ADFS provide several additional tools for structural analyses. In the ADFS, a motif-based method is provided to predict the allergenicity of specific protein by the similarity (by pftool) to 71 allergen motifs, which extracted from 1,053 allergen sequences. In the SDAP, all allergens are classified into their closet protein families as defined by the Pfam database of information about protein domain and families (Radauer et al., 2008). In a previous report, sequence motifs unique to the allergen in three families (storage proteins, Bet v 1, and tropomyosin) were found to overlap with known IgE epitopes (Ivanciuc et al., 2009a), suggesting that such motifs might be useful is separating non-allergenic from allergenic members within protein families. Another tool available in the SDAP is the computation of property distance (PD) values for measuring peptide similarity. The PD index is a statistically validated method to detect discrete regions of proteins that have a high probability of cross-reaction with IgE from allergic patients (Ivanciuc et al., 2009b). Finally, the SDAP provides tool to examine similarity to allergenic proteins with respect to three-dimensional (3D) structure. There are 433 reliable 3D models now available from the SDAP database.

The current bioinformatics tools for allergenicity prediction are proved to be effective. However, the highly conservative prediction is also unnecessarily eliminated safe proteins in the early evaluation process. Besides the refinements of current protocol, more sophisticated structure activity tools may prevent the unnecessarily rejection of potential useful proteins in the developmental process. Further discussion between scientists and the regulatory community are needed to determine the best use of these new *in silico* tools.

Animal Model for Allergenicity Assessment

There is considerable interest in the development of animal models for the identification and characterization of the allergic feature of proteins. The use of animal model to predict allergenicity has been recognized since the first discussion on safety assessment of transgenic protein (Metcalfe et al., 1996). The FAO/WHO recommendations called for evaluating each new GM crop with studies in two separate species of animals and/or using two routes of sensitization in one species. However, there are still no validated animal models for predicting allergenicity to food proteins, even though many models have been successfully applied.

As with any toxicity assessment, an appropriate animal model should be produce sensitization and/or elicitation of allergic symptoms are a physiologically relevant dose, via the relevant route of exposure (generally through ingestion) in a standard (readily available) mouse strain. However, factors such as oral tolerance, digestibility, pH, and/or solubility of the sensitizing food extract can change the results obtained in the oral animal models (Bowman and Selgrade, 2008). From some regulators point of view, it would be desirable to test purified, bacterial-expressed, transgenic protein because the transgenic protein, not the food crop, is regulated. However, this may not adequately mimic real world exposure. Whereas, transgenic crop extract is relatively available, for positive and negative controls extracts are more easily obtained than purified proteins.

There was general agreement that significant progress has been made toward developing animal models, but that none of the currently available animal models are ready for validation. More work is needed on consensus of appropriate end points, particularly those that reflect anaphylactic activity. It will also be important to resolve the role that matrix plays before designing a validation study. In addition, before validation can be considered, decisions are have to be made regarding which mouse strains and adjuvants to include, as well as the ideal dose of test materials. Appropriate test substance that represent a range from highly (commonly) allergic to poorly (rarely) allergenic need to be selected (Selgrade et al., 2009).

It is important to emphasize that currently none of these approaches has been validated for the purpose of allergenicity identification of a safety assessment. However, the available evidence suggests that an accurate and robust animal model, in combination with the other approaches used currently, could be of considerable utility in safety assessment. Therefore, as recommendation of Codex, animal models are excluded at the present safety assessment, but continuing research to evaluate potentially predictive animal models is encouraged. Caution should be taken on testing potential products at this time as there is no scientific validation of animal models used for risk evaluation.

Evaluation of Unintended Effects

Concerns have been raised that the current approach of using targeted analyses to compare the composition of GM crops to their traditional counterparts is biased (Millstone et al., 1999) and does not take into account the possibility of unintended effects that could result directly or indirectly from the genetic modification. The potential occurrence of such "unintended effects'' is currently one of the concerns being raised regarding the GM crops/foods.

To clarify the concept of unintended effects, the following definitions have been adopted (Cellini et al., 2004):

> ''Intended effects'' are those that are targeted to occur from the introduction of the gene(s) and that fulfill the original objectives of the genetic modification.
>
> ''Unintended effects'' represent a statistically significant difference in the phenotype, response, or composition of the GM plant compared with the parent plant, not including the expected effect of the transgenic gene. Such comparisons should be made when GM and non-GM counterparts are grown under the same regimes and environments.
>
> ''Predictable unintended effects'' are those unintended changes that may be explicable in terms of our current knowledge of plant biology by metabolic pathway integration and interconnections.
>
> ''Unpredictable unintended effects'' are those changes falling outside our present level of understanding. Predictable and unpredictable unintended effects may or may not prove to have relevance in terms of product safety, but must be taken into account when assessing risk.

The safety of conventionally bred crops is granted based on a history of safe use. However, some cases (extremely rare) have been reported where unintended effects have given rise to safety concerns. In current regulation of GM crops, characterization is a legal requirement, and is part of the safety assessment. GM crops are therefore better characterized than conventionally bred crops, including knowledge on the site and nature of the genetic modification. GM crop characterization currently includes compositional analysis of selected nutrients and toxins, which a relative comparison is made to the conventional crops.

Criticisms of this current strategy are that it is open to bias and may never pick up unintended effects. Profiling techniques may help address these potential limitations. Profiling techniques such as genomics, proteomics, and metabolomics provide a 'global' overview within the crop. These techniques aim to be unbiased with regard to the choice of genes, proteins, and metabolites profiled. Methodologies are still rapidly developing; they are not yet (and may never be) comprehensive. All current profiling approaches are based on comparison of GM materials with selected controls. The data generated have a great potential to increase our knowledge of plant physiology and metabolic networks, and will also improve targeted analyses. Vast quantities of data can be generated from these methods; however, subsequent interpretation of the data is limited by the present knowledge gap. To date, there is a lack of data on which to determine the useful contribution of these techniques to GM crop safety assessments (Cellini et al., 2004).

It is important that unintended effects do not necessarily relate to health hazards. Ideally, only those parameters that fall outside the range of natural variation would be considered further in safety assessment. However, there is a lack of information on the natural variation within and between given plant cultivars for all the parameters that may now be measured. Safety assessment will simplify if the identification and safety significance of differences is known. A major drawback would be the lack of adequate databases and analysis tools to interpret the safety significance of compounds with unknown identity and/or function. Numerous differences and/or differences of unknown function may lead to the consideration of more extensive safety testing. And further research is required to validate profiling methodologies. More computational analysis tools are needed to be developed to extract relevant information from raw data sets. The comprehensiveness of coverage given by all profiling techniques is also needed to be improved. The development of publicly available databases of crop composition and profiles is an absolute requirement to determine natural variation of compounds within and between given plant species.

The safety assessment of GM crops should focus primarily on the intended traits (target genes and products) at the present time. Profiling techniques should not at present be an official requirement. However, their further development should be actively encouraged (Cellini et al., 2004).

THE FUTURE CHALLENGES

The current methodology of GM crops/foods safety assessment is relatively empirical and conservative. Thus, accurate hazard identification and risk assessment will continue to represent important challenges for the future. Current methods of allergenicity prediction should be carefully reviewed and the works on the improvement and validation of bioinformatic tools could be helpful. However, the validation of bioinformatics methods may not be practical until the relevance of allergenicity to protein sequence/structure is clear.

There is considerable interest in the development of animal models for the identification and characterization of transgenic proteins to be allergenic. The major challenge to apply animal model for allergenicity is the lack of consensus and validation of experiment design and parameters. However, the available evidence suggests that an accurate and robust animal model, in combination with the other approaches used currently, could be of considerable utility in safety assessment.

Table 1. The bioinformatics search result of the Cucumber Mosaic Virus coat protein (Lin et al., 2010).

Allergen name	Organism	Accession no.	Database	Identity, %	E score[#] (full FASTA)	Description
Dol m 5	*Dolichovespula maculata* (Hornet)	P10736	AO SD	20.61 12.39	1.9 >0.01	Venom allergen 5.01
Cla h 4	*Davidiella tassiana*	P40918	AO SD	23.35 19.72	2.4 >0.01	Heat shock 70 kDa protein
Der p 8	*Dermatophagoide s pteronyssinus* (Home dust mite)	P46419	AO SD	40.47 10.55	3.5 >0.01	Glutathione transferase μ class
Der p 15	*Dermatophagoide s pteronyssinus* (Home dust mite)	AAX37326	AO	40.47	3.5	Glutathione transferase homologues
Ory s TAI	*Oryza sativa* (Rice)	BAA07710	AO SD	30.00 10.55	6.0 >0.01	α-amylase/trypsin inhibitor
Cyp 1.01	c*Cyprinus carpio* (common carp)	CAC83658	SD	8.72	>0.01	Parvalbumin
Ani s 4	*Anisakis simplex* (Herring worm)	P16347	AO	34.21	3.8	Allergen

*Database: "AO"=AllergenOnline, "SD"=SDAP.

[#] E score $< 1 \times 10^{-7}$ indicates significant homology for AllergenOnline database.

The aim of the safety assessment of GM crops/foods is to demonstrate that the GM crop/food is as safe as its traditional counterpart (where one exists) or does not introduce any additional or new risks to the health of the consumer. However, it is generally accepted (Millstone et al., 1999) that the current concept of substantial equivalence is not adequate to assess whether a transgenic food has to be considered as safe as its corresponding wild-type. The most recent Codex guideline has included the analysis of unintended effects to respond the demand of more in-depth GM crop safety assessment. And the rapid development of high-throughput analytical methodologies has made highly integrated, global comparison of GM plant to its wild type counterpart possible for unintended effects analysis. However, monitoring changes at the gene, transcript, protein and metabolite level in GM crops, as required by the safety assessment procedures, will result in enormous amounts of data. And the relevance of searching for unintended effects with respect to safety assessments has raised the question of whether the more information we have available would, in reality, reduce the uncertainties in the safety assessment. The knowledge gap between the context of systems biology and the safety of GM crops will be the main challenge to the study of unintended effects in the future. Efforts are need to integrate information and methodologies. Ultimately, the development of unintended study may lead to a comprehensively predictive GM crop safety assessment.

Table 2. The bioinformatics search result of the sporamin protein (Lin et al., 2009).

Allergen name	Accession no.	Database	Max similarity	80merE score (full FASTA)	Description
Sola t 4	P30941	AO	37.00%	2.5e-11	Serine protease inhibitor
		AD	NA	7.7e-9	
		SD	-	2.9e-11	
	CAA45723	AO	37.00%	1.1e-11	Aspartic protease inhibitor
		AD	NA	7.9 e-9	
		SD	37.50%	1.3e-11	
Gly m TI	P01070	AO	35.30%	1.7e-10	Kunitz type trypsin inhibitor
		AD	NA	1.4e-7	
		SD	38.75%	1.9e-10	
	Q39898	AO	35.30%	1.7e-10	Trypsin inhibitor A
		AD	NA	1.4e-7	
	Q39869	AO	-	1.4e-9	Kunitz type trypsin inhibitor precursor
		AD	NA	1.6e-7	
		SD	-	1.5e-9	
	Q39899	AO	36.50%	3.3e-10	Trypsin inhibitor subtype B
		AD	NA	1.6e-7	
	P25272	AO	-	1.6e-10	Kunitz type 1 trypsin inhibitor precursor
		AD	NA	2.0e-6	
		SD	-	1.9e-10	
Sola t 3	P20347	AO	-	7.2e-9	Serine protease inhibitor II
		AD	NA	6.2e-5	
		SD	-	7.6e-9	
Tri F073/F170	[a]P16347	AD	NA	1.1e-8	α-amylase and trypsin inhibitor

*Database: "AO"=AllergenOnline, "AD"=ADFS, "SD"=SDAP
Similarity: "NA"= not available, "-"= negative result

ACKNOWLEDGMENTS

Reviewed by: Professor Fang, T. J., Department of Food Science and Biotechnology, National Chung Hsing University

Professor Shyu, Y. T., Department of Horticulture, National Taiwan University.

REFERENCES

Asero, R., Ballmer-Weber, B. K., Beyer, K., Conti, A., Dubakiene, R., Fernandez-Rivas, M., Hoffmann-Sommergruber, K., Lidholm, J., Mustakov, T., Oude Elberink, J. N., Pumphrey, R. S., Stahl Skov, P., van Ree, R., Vlieg-Boerstra, B. J., Hiller, R., Hourihane, J. O., Kowalski, M., Papadopoulos, N. G., Wal, J. M., Mills, E. N. &Vieths, S. (2007). IgE-mediated food allergy diagnosis: Current status and new perspectives. *Mol. Nutr. Food Res.*, 51, 135-147.

Breiteneder, H. & Radauer, C. (2004). A classification of plant food allergens. *J. Allergy Clin. Immunol.*, 113, 821-830.

Bowman, C. C. & Selgrade, M. K. (2009). Utility of rodent models for evaluating protein allergenicity. *Regul. Toxicol. Pharmacol.*, 54, S58-S61.

Cellini, F., Chesson, A., Colquhoun, I., Constable, A., Davis, H. V., Engel, K. H., Gatehouse, A. M. R., Karenlampı, S., Kok, E. J., Leguay, J.-J., Lehesranta, S., Noteborn, H. P. J. M., Pedersen, J. & Smith, M. (2004). Unintended effects and their detection in genetically modified crops. *Food Chem. Toxicol.*, 42, 1089-1125.

Codex. (2003). *Principles for the Risk Analysis of Foods Derived from Modern Biotechnology* (CAC/GL 44-200). Codex Alimentarius Commission. Codex *ad hoc* intergovernmental task force on food derived from biotechnology, Yokohama, Japan.

Corpillo, D., Gardini, G., Varia, A. M., Basso, M., Aime, S., Accotto, G.. P. & Fasano, M. (2004). Proteomics as a tool to improve investigation of substaintial equivalence in genetically modified organisms: The case of virus-resistant tomato. *Proteomics*, 4, 193-200.

Diplock, A. T., aggett, P. J., Ashwell, M., Bornet, F., Fern, E. B. & Roberfroid, M. B. (1999). Scientific concepts of functional foods in Europe: consensus document. *Brit. J. Nutr.*, 81, S1-S27.

DOH. (2010). *Guideline for food safety assessment of genetically modified foods derived from recombinant-DNA organisms.* DOH 0991301116, Department of Health of Taiwan, Taipei, Taiwan.

FAO/WHO. (1996). *Biotechnology and Food Safety.* Report of a Joint FAO/WHO Expert consultation, Food and Agriculture Organization of the United Nations, Rime, Italy.

FAO/WHO. (1996). *Biotechnology and Food Safety.* Report of a Joint FAO / WHO Expert consultation, Food and Agriculture Organization of the United Nations, Rime, Italy.

FAO/WHO. (2000). *Report of the first session of the Codex ad hoc intergovernmental task force on food derived from biotechnology* (ALINORM 01/34). Food and Agriculture Organization of the United Nations, Rime, Italy.

FAO/WHO. (2001*). Allergenicity of genetically modified foods.* Report of a joint expert consultation on foods derived from biotechnology. Food and Agriculture Organization of the United Nations, Rime, Italy.

Filipecki, M. & Malepszy, S. (2006). Unintended consequences of plant transformation: a molecular insight. *J. Appl. Genet.*, 47, 277-286.

Fu, T. J., Abbott, U. R. & Hatzos, C. (2002). Digestibility of food allergens and nonallergenic proteins in simulated gastric fluid and simulated intestinal fluid-a comparative study. *J. Agric. Food Chem.*, 24, 7154-7160.

Goodman, R. E., Hefle, S. L., Taylor, S. L. & van Ree, R. (2005). Assessing genetically modified crops to minimize the risk of increased food allergy: a review. *Int. Arch Allergy Immunol.*, 137, 153-166.

Goodman, R. E., Vieths, S., Sampson, H. A., Hill, D., Ebisawa, M., Taulro, S. L. & Van Ree, R. (2008). Allergenicity assessment of genetically modified crops-what makes sense? *Nat. Biotechnol.*, 26, 73-81.

Haslberger, A. G. (2003). Codex guideline for GM foods include the analysis of unintended effects. *Nat. Biotechnol.*, 21, 739-740.

Hileman, R. E., silvanovich, A., Goodman, R. E., Rice, E. A., Holleschak, G., Astwood, J. D. & Hefle, S. L. (2002). Bioinformatic methods for allergenicity assessment using a comprehensive allergen database. *Int. Arch. Allergy Immunol.*, 128, 208-291.

IFBC. (1990). Biotechnologies and food: assuring the safety of foods produced by genetic modification. *Regul. Toxicol. Pharmacol.*, 12, S1-S196.

IFBC. (1996). Biotechnologies and food: assuring the safety of foods produced by genetic modification. *Regul. Toxicol. Pharmacol.*, 12, S1-S196.

Ivanciuc, O., Garcia, T., Torres, M., Schein, C. H. & Braun, W. (2009a). Characteristic motifs for families of allergenic proteins. *Mol. Immunol.* 46, 559-568.

Ivanciuc, O., Midoro-Horiuti, T., Schein, C. H., Xie, L., Hillman, G. R., Goldblum, R. M. & Braun, W. (2009b). The property distance index PD predicts peptides that cross-react with IgE antibodies. *Mol. Immunol.*, 46, 873-883.

James, C. (2010). *Global Status of Commercialized Biotech/GM crops: 2010. ISAAA Brief* No. 41. ISAAA, Ithaca, NY, USA.

Jonas, D. A., Antignac, E., Antoine, J. M., Classen, H. G., Huggett, A., Knudsen, I., Mahler, J., Ockhuizen, T., Smith, M., Teuber, M., Walker, R. & De Vogel, P. (1996). The safety assessment of novel foods. Guidelines prepared by ILSI Europe Novel Food Task Force. *Food Chem. Toxicol.*, 24, 931-940.

Kuiper, H. A., Kleter, G. A., Noteborn, H. P. & Kok, E. J. (2001). Assessment of the food safety issues related to genetically modified foods. *Plant J.*, 27, 503-528.

Ladics, G.. S., Bannon, G.. A., Silvanovich, A. & Cressman, R. F. (2007). Comparison of conventional FASTA identity searches with the 80 amino acid sliding window FASTA search for the elucidation of potential identities to known allergens. *Mol. Nutr. Food Res.*, 51, 985-998.

Lin, C. H., Lu, C. T., Lin, H. T. & Pan, T. M. (2009). Safety assessment and detection method of genetically modified Chinese Kale (*Brassica oleracea* cv. *alboglabra*). *J. Agric. Food Chem.*, 57, 1876-1881.

Lin, C. H., Sheu, F., Lin, H. T. & Pan, T. M. (2010). Allergenicity assessment of genetically modified Cucumber Mosaic virus (CMV) resistant tomato (*Solanum lycopersicon*). *J. Agric. Food Chem.*, 24, 2302-2306.

Llorente, B., Alonso, G.. D., Bravo-Almonacid, F., Rodriguez, V., Lopez, G.. M., Carrari, F., Torres, H. N. & Flawia, M. M. (2010). Safety assessment of nonbrowning potatoes: opening the discussion about the relevance of substantial equivalence on next generation biotech crops. *Plant Biotech. J.*, doi:10.1111/j.1467-7652.2010.00534.x.

Metacalfe, D. D., Astwood, J. D., Townsned, R., Sampson, H. A., Taylor, S. L. & Fuch, R. L. (1996). Assessment of the allergenic potential of food derived from genetically engineered crop plants. *Crit. Rev. Food Sci. Nutr.*, 36, S165-S186.

Millstone, E., Brunner, E. & Mayer, S. (1999). Beyond 'substantial equivalence'. *Nature*, 401, 525-526.

OECD. (1993). *Safety evaluation of food derived by modern biotechnology*. Organization for Economic Cooperation and Development, Paris, France.

OECD. (1996). *Food safety evaluation. Organization for Economic Cooperation and Development,* Paris, France.

OECD. (1998). *Report of OECD workshop on the toxicological and nutritional testing of novel foods*. Organization for Economic Cooperation and Development, Paris, France.

OECD. (1993). *Safety evaluation of food derived by modern biotechnology*. Organization for Economic Co-operation and Development, Paris, France.

OECD. (2001). *Consensus documents. Organization for Economic Co-operation and Development,* Paris, France.

OECD. (2003). *Considerations for the Safety Assessment of Animal Feed Stuffs Derived from Genetically Modified Plants*. Organization for Economic Co-operation and Development, Paris, France.

OECD. (2008). *Consensus document on compositional considerations for new varieties of tomato: key food and feed nutrients, toxicants and allergens*. Organization for Economic Co-operation and Development, Paris, France.

Ofori-Anti, A. O., Ariyarathna, H., Chen, L., Lee, H. L., Pramod, S. N. & Goodman, R. E. (2004). Establishing objective detection limits for the pepsin digestion assay used in the assessment of genetically modified foods. *Regul. Toxicol. Pharmacol.* 39, 87-98.

Poppy, G. (2000). GM crops: environmental risks and non-target effects. *Trends Plant Sci.*, 5, 4-6.

Radauer, C., Bublin, M., Wagner, S., Mari, A. & Breiteneder, H. (2008). Allergens are distributed into few protein families and possess a restricted number of biochemical functions. *J. Allergy Clin. Immunol.*, 121, 847-852.

Ricroch, A., Berge, J. B. & Kuntz, M. (2010). Is the German suspension of MON810 maize cultivation scientifically justified? *Transgenic Res.*, 19, 1-12.

Silvanovich, A., Nemeth, M. A., Song, P., Herman, R., Tagliani, L. & Bannon, G.. A. (2006). The value of short amino acid sequence matches for prediction of protein allergenicity. *Toxicol. Sci.,* 120, 491-503.

Selgrade, M. K., Bowman, C. C., Ladics, G. S., Privalle, L. & Laessig, S. A. (2009). Safety assessment of biotechnology products for potential risk of food allergy: implications of new research. *Toxicol. Sci.*, 110, 31-39.

Stadler, M. B. & Stadler, B. M. (2003). Allergenicity prediction by protein sequences. *FASEB J.*, 17, 1141-1143.

Sten, E., Skov, P. S., Andersen, S. B., Torp, A. M., Olesen, A., Bindslev-Jensen, U., Poulsen, L. K. & Bindslev-Jensen, C. (2004). A comparative study of the allergenic potency of wild-type and glyphosate-tolerant gene-modified soybean cultivars. *APMIS*, 112, 21-28.

Thomas, K., Bannon, G., Helfe, S., Herouet, C., Holsapple, M., Ladics. G., MacIntosh, S. & Privalle, L. (2005). In silico methods for evaluating human allergenicity to novel protein: International Bioinformatics Workshop Meeting Report. *Toxicol. Sci.,* 88, 307-310

Thomas, K., Aalbers, M., Bannon, G. A., Bartels, M., Dearman, R. J., Esdaile, D. J., Fu, T. J., Glatt, C. M., Hadfield, N., Hatzos, C., Hefle, S. L., Heylings, J. R., Goodman, R. E., Henry, B., Herouet, C., Holsapple, M., Ladics, G. S., Landry, T. D., MacIntosh, S. C., Rice, E. A., Privalle, L. S., Steiner, H. Y., Teshima, R., Van Ree, R., Woolhiser, M. & Zawodny, J. (2004). A multi-laboratory evaluation of a common *in vitro* pepsin digestion assay protocol used in assessing the safety of novel proteins. *Regul Toxicol Pharmacol.*, 39, 87-98.

Untersmayr, E., Poulsen, L. K., Platzer, M. H., Pedersen, M. H., Boltz-Nitulescu, G., Skov, P. S. & Jensen-Jarolim, E. (2005). The effects of gastric digestion on codfish allergenicity. *J. Allergy Clin. Immunol.*, 115, 377-382.

van Ree, R. (2002). Carbohydrate epitopes and their relevance for the diagnosis and treatment of allergic diseases. *Int. Arch. Allergy Immunol.* 129, 189-197.

Vassilopoulou, E., Zuidmeer, L., Akkerdaas, J., Tassios, I., Rigby, N. R., Mills, E. N., van Ree, R., Saxoni-Papageorgiou, P. & Papadopoulos, N. G. (2007). Severe immediate allergic reactions to grapes: part of a lipid transfer protein-associated clinical syndrome. *Int. Arch. Allergy Immunol.* 143, 92-102.

WHO. (1991). *Strategies for assessing the safety of foods produced by biotechnology*. World Health Organization, Geneva, Swiss.

Yagami, T., Haishima, Y., Nakamura, A., Osuna, H. & Ikezawa Z. (2000). Digestibility of allergens extracted from natural rubber latex and vegetable foods. *J. Allergy Clin. Immunol.* 106, 752-762.

In: Genetically Engineered Crops
Editors: Lindsay M. Grover

ISBN: 978-1-61122-526-6

Chapter 6

GENETIC MODIFICATION FOR HERBICIDE TOLERANCE IN CROP PLANTS: GLYPHOSATE AS A PARADIGM

T.C. Suma, Rohini Sreevathsa* and R. Devendra
Department of Crop Physiology, UAS, GKVK, Bangalore, India

ABSTRACT

Competition from weed plants causes great losses in agriculture production. It is considered that the yield loss due to competition from weeds far exceeds that of pests and diseases. It is important to remove competition by weeds at an early stage of crop growth. In this context, herbicides (post and pre emergent) play an important role in weed control. Different herbicides are known belonging to different chemical groups according to their mode of action. Glyphosate is one of the most used broad spectrum herbicides which act as a competitive inhibitor of the enzyme 5-Enol Pyruvyl Shikimate 3-Phosphate Synthase (EPSPS) hampering aromatic amino acid biosynthesis in plants. However, the effect of herbicides can be lethal to crop plants as well. To avoid this, herbicide tolerant crops would be an option. Both conventional breeding and transgenic technology can be used to produce herbicide tolerant crops. The herbicide tolerance therefore offers the opportunity to spray crops at the most effective time to kill weed species without damaging the crop plants. Various strategies conferring glyphosate tolerance in particular will be discussed in this chapter. A large number of glyphosate tolerant crops have been generated and commercialized. This chapter would provide an updated account of the glyphosate tolerant transgenic crops. The potential risks and benefits of the herbicide tolerant crops will also be emphasized. Further, the importance of the herbicide tolerant genes as selectable markers for stringent screening and evaluation of the transgenic crops will also be addressed.

* Corresponding author: Email: rohinisreevathsa@rediffmail.com

INTRODUCTION

Weeds are a major problem in agriculture. They compete with crops for moisture and nutrients. Loss of yield due to weed infestation is variable and is more pronounced in crops grown under rain fed conditions. Some parasitic weeds draw water and nutrients from crop plants and can inflict severe damage. Further, weeds serve as alternate hosts to pathogens and also harbor pests. Control of weeds during early stages of crop growth, when the young seedlings of crop plants are unable to compete with hardy weeds, is crucial for capturing yield potential. For this reason, labour demand for weeding operation is high during early phase of crop cycle and manual weed control over large areas is not feasible from the point of labour supply and monetary costs. Some weeds that are wild relatives of crop plants are difficult to distinguish from crop plants at early stages and pose challenge for manual weeding. Under these situations, chemical weed control is relevant for realizing higher productivity and production.

A plant is considered a weed if it is;

- A plant that is out of place and not intentionally sown
- A plant that grows where it is not wanted or welcomed
- A plant whose virtues have not yet been discovered
- A plant that is competitive, persistent, pernicious, and interferes negatively with human activity.

There are approximately 250,000 species of plants worldwide; of those, about 3 percent, or 8,000 species, behave as weeds. Of those 8,000, only 200 to 250 are major problems in worldwide cropping systems. A plant is considered a weed if it has certain characteristics that set it apart from other plant species. Weeds possess one or more of the following characteristics that allow them to survive and increase in nature:

- Abundant seed production
- Rapid population establishment
- Seed dormancy
- Long-term survival of buried seed
- Adaptation for spread
- Presence of vegetative reproductive structures
- Ability to occupy sites disturbed by humans

Since weeds are so prevalent in many areas of the landscape, management techniques are necessary to maintain order. Weed management is most successful when it involves an integrated approach using a variety of methods. The common methods used to manage weeds include prevention by cultural, mechanical, biological, and chemical means.

COMMON METHODS IN MANAGING WEEDS

Prevention

Preventative methods are used to stop the spread of weeds. Preventing the introduction of weeds is usually easier than controlling them after establishment. Preventative practices include cleaning tillage and harvesting equipment of weed seeds and vegetative structures; planting certified, weed-free crop seed; and controlling weeds in barnyards, around structures, and along fencerows, roadways, and ditch banks.

Cultural

Cultural and crop management techniques provide a healthy crop to best compete with weeds. Crop competition can be an inexpensive and effective aid to weed management if used to its fullest advantage. Examples of cultural techniques include following soil test recommendations for fertilizer and lime; selecting the best crop varieties; planting dense crop populations at the proper timing; scouting fields regularly for weeds, insects, and diseases and controlling them when necessary; and including crop rotations in the system.

Mechanical

Mechanical or physical techniques either destroy weeds or make the environment less favorable for seed germination and weed survival. These techniques include hand pulling, hoeing, mowing, plowing, disking, cultivating, and digging. Mulching (straw, wood chips, gravel, plastic, etc.) can also be considered a mechanical control means since it uses a physical barrier to block light and impede weed growth.

Biological

Biological weed control involves the use of other living organisms, such as insects, diseases, or livestock, for the management of certain weeds. There have been a few success stories of weed species being managed with insect or disease biocontrol agents.

Chemical

Herbicides can be defined as crop protecting chemicals used to kill weedy plants. Herbicides provide a convenient, economical, and effective way to help manage weeds. Herbicides can be classified based on weed control spectrum, labeled crop usage, chemical families, mode of action, application timing/method and others. For this publication,

Table 1. Important herbicide groups and examples.

Mode of Action (Class)	Site of Action	WSSA Group	Family	Active Ingredient	Trade Name(s)*
Plant growth regulators (PGR)	IAA like	4	phenoxy 2,4-DB MCPA MCPP (mecoprop)	2-4-D Butyrac various various	various
		4	benzoic acid	dicamba	Banvel, Clarity, Distinct, Vanquish
			carboxylic acid (pyridines)	aminopyralid clopyralid flufoxypyr picloram triclopyr	Milestone Stinger, Lontrel, Transline Starane Tordon Garlon
	auxin transport inhibitors	19	phthalamates semicarbazones	naptalam diflufenzopyr	Alanap component of Distinct, Status
Amino acid biosynthesis inhibitors	ALS enzyme	2	imidazolinone	imazapic imazamox imazamethabenz imazapyr imazaquin imazethapyr	Plateau Raptor Assert Arsenal Scepter Pursuit
			sulfonylurea	chlorimuron chlorsulfuron foramsulfuron halosulfuron iodosulfuron metsulfuron nicosulfuron primisulfuron prosulfuron rimsulfuron sulfometuron sulfosulfuron thifensulfuron triasulfuron tribenuron	Classic Glean, Telar Option Permit, Sandea, SedgeHammer Autumn, Equip Cimarron, Escort Accent Beacon Peak Matrix, Resolve Oust Maverick Harmony GT Amber Express
			sulfonylaminocarbonyl-triazolinones	flucarbazone propoxycarbazone	Everest Olympus
			triazolopyrimidine (sulfonamides)	cloransulam flumetsulam	FirstRate Python
	EPSP enzyme	9	amino acid derivative (glycines)	glyphosate	Roundup, Touchdown, Accord, Honcho, many others
Fatty acid (lipid) biosynthesis inhibitors	ACCase enzyme	1	aryloxyphenoxy propionates	clodinafop diclofop fenoxaprop fluazifop quizalofop	Discover Hoelon Acclaim, Puma Fusilade Assure II
			cyclohexanediones	clethodim sethoxydim tralkoxydim	Select, Prism Poast, Vantage Achieve
			phenylpyrazoline	pinoxaden	Axial

Table 1. (continued)

Mode of Action (Class)	Site of Action	WSSA Group	Family	Active Ingredient	Trade Name(s)*
Seedling growth inhibitors (root and shoot)	microtuble inhibitors	3	dinitroanilines	benefin	Balan
				ethalfluralin	Sonalan, Curbit
				oryzalin	Surflan
				pendimethalin	Prowl, Pre-M, Pendulum
				prodiamine	Barricade, Endurance
				trifluralin	Treflan, Tri-4
			pyridines	dithiopyr	Dimension
			benzamides	pronamide	Kerb
			benzoic acids	DCPA	Dacthal
			carbamates	asulam	Asulox
	cell wall biosynthesis inhibitors	20	nitriles	dichlobenil	Casoron
		21	benzamides	isoxaben	Gallery
Seedling growth inhibitors (shoot)	unknown	15	chloroacetamides	acetochlor	Harness, Surpass, Topnotch
				alachlor	Micro-Tech
				dimethenamid	Frontier, Outlook
				metolachlor	Dual, Pennant
				propachlor	Ramrod
			oxyacetamides	flufenacet	Define
	lipid synthesis inhibitors	8	thiocarbamates	butylate	Sutan
				EPTC	Eptam, Eradicane
				pebulate	Tillam
				vernolate	Vernam
	cell division inhibitors	8	phosphorodithioates	bensulide	Prefar
		15	acetamides	napropamide	Devrinol
Photosynthesis inhibitors (mobile 1)	photosystem II	5	triazines	ametryn	Evik
				atrazine	Atrazine
				propazine	Milo Pro
				prometon	Primito
				simazine	Princep
			triazinones	hexazinone	Velpar
				metribuzin	Lexone, Sencor
			uracils	bromacil	Hyvar
				terbacil	Sinbar
(mobile 2)	photosystem II	7	ureas	diuron	Karmex
				linuron	Lorox
				siduron	Tupersan
				tebuthiuron	Spike
Photosynthesis inhibitors (nonmobile; "rapid-acting")	photosystem II	6	nitriles	bromoxynil	Buctril
			benzothiadiazoles	bentazon	Basagran
			phenyl-pyridazines	pyridate	Tough

Table 1. (continued)

Mode of Action (Class)	Site of Action	WSSA Group	Family	Active Ingredient	Trade Name(s)*
Cell membrane disrupters	PPO enzyme	14	diphenyl ethers	acifluorfen fomesafen lactofen oxyfluorfen	Blazer Reflex, Flexstar Cobra Goal
			N-phenyl-phthalimides	flumioxazin flumiclorac	Valor Resource
			oxadiazoles	oxadiazon	Ronstar
			triazolinones	carfentrazone sulfentrazone	Aim Authority, Spartan
	photosystem I	22	bipyridyliums	diquat paraquat	Reward Gramoxone, Boa
Pigment inhibitors (bleaching)	diterpenes (carotenoid biosynthesis)	13	isoxazolidinones pyridazinones	clomazone norflurazon	Command Zorial
	4-HPPD enzyme	27	isoxazoles	isoxaflutole	Balance
			triketones	mesotrione tembotrione topramezone	Callisto Laudis Impact
Phosphorylated amino acid (N-metabolism disrupters)	GS enzyme	10	amino acid derivatives (phosphinic acids)	glufosinate	Liberty, Finale, Rely, Ignite
Unknown	?		?	dazomet endothall fosamine metam pelargonic acid cinnamon oil citric acid clove oil corn gluten mean thyme oil vinegar (acetic acid)	Basamid Aquathol Krenite Vapam Scythe Matran

herbicides will be grouped according to mode and site of action, which are also important in understanding herbicide resistance in weeds. Contact herbicides kill only the plant parts contacted by the chemical, whereas systemic herbicides are absorbed by the roots or foliage and translocated (moved) throughout the plant. Herbicide activity can be either selective or nonselective. Selective herbicides are used to kill weeds present if applied at an adequate rate.

Herbicide Mode and Site of Action

To be effective, herbicides must (1) adequately contact plants, (2) be absorbed by plants, (3) move within the plants to the site of action without being deactivated, and (4) reach toxic levels at the site of action. The term "mode of action" refers to the sequence of events from absorption into plants to plant death, or, in other words, how an herbicide works to injure or kill the plant. The specific site the herbicide affects is referred to as the "site or mechanism of action." Understanding herbicide mode of action is helpful in knowing what groups of weeds are killed, specifying application techniques, diagnosing herbicide injury problems, and preventing herbicide resistant weeds.

A common method of grouping herbicides is by their mode of action. Although a large number of herbicides are available in the marketplace, several have similar chemical properties and herbicidal activity. Herbicides with a common chemistry are grouped into "families." Also, two or more families may have the same mode of action, and thus can be grouped into "classes." Table 1 lists several groups of herbicides and information related to their mode of action.

HERBICIDES AS AMINO ACID BIOSYNTHESIS INHIBITORS

Most herbicides control weeds by targeting and inhibiting a protein or enzyme in plants (OECD, 1999; Cole *et al.*, 2000). Amino acid biosynthesis-inhibiting herbicides inhibit enzymes of the plant and consequently block the biosynthesis of the building blocks of proteins, amino acids. Because animals do not synthesize all needed amino acids but obtain some amino acids from plants or bacteria, amino acid biosynthesis inhibiting herbicides tend to have less impact on animals than herbicides with other modes of action (Reade and Cobb, 2002). Due to this effect, amino acid biosynthesis pathways are very desirable targets for new herbicide discovery. In fact, a significant portion of newly commercialized herbicides belong to the acetohydroxyacid synthase (AHAS) inhibitors (DeFelice, 1998; Cole *et al.*, 2000). The same trend is also evident with herbicide-tolerant crops. Crops that are tolerant to amino acid biosynthesis inhibitors have dominated the herbicide-tolerant crop market (Duke, 2005).

There are three major enzymes that can be inhibited by commercial amino acid biosynthesis-inhibiting herbicides (Fig 1) (LaRossa and Falco, 1984; Mousdale and Coggins, 1991; Vaughn and Duke, 1991; Singh and Shaner, 1995). They are AHAS (EC 4.1.3.18), also called acetolactate synthase (ALS), in the branched chain amino acid biosynthesis pathway, EPSP synthase (EC 2.5.1.19) in the shikimate pathway, and GS (EC 6.3.1.2) in ammonium assimilation (LaRossa and Falco, 1984; Kishore and Shah, 1988; Mousdale and Coggins, 1991; Sherman *et al.*, 1996; Reade and Cobb, 2002).

Herbicidal inhibitors of branched-chain amino acid biosynthesis; AHAS catalyzes two reactions in the branched-chain amino acid biosynthesis pathway: the condensation of two pyruvate molecules to acetolactate and the synthesis of 2-acetohydroxybutyrate from pyruvate and 2-ketobutyrate (Fig 1) (Singh and Shaner, 1995). Acetolactate leads to the final synthesis of leucine and valine, while 2-acetohydroxybutyrate is a precursor of isoleucine. On the other hand, the product amino acids function as feedback regulators of the activity of AHAS (Stidham and Singh, 1991; Singh and Shaner, 1995; Shaner *et al.*, 1996). AHAS is a nuclear-encoded protein, which is transported into plastids where the biosynthesis of the branched-chain amino acids primarily occurs (Vaughn and Duke, 1991; Singh and Shaner, 1995; Saari and Mauvais, 1996; Reade and Cobb, 2002).

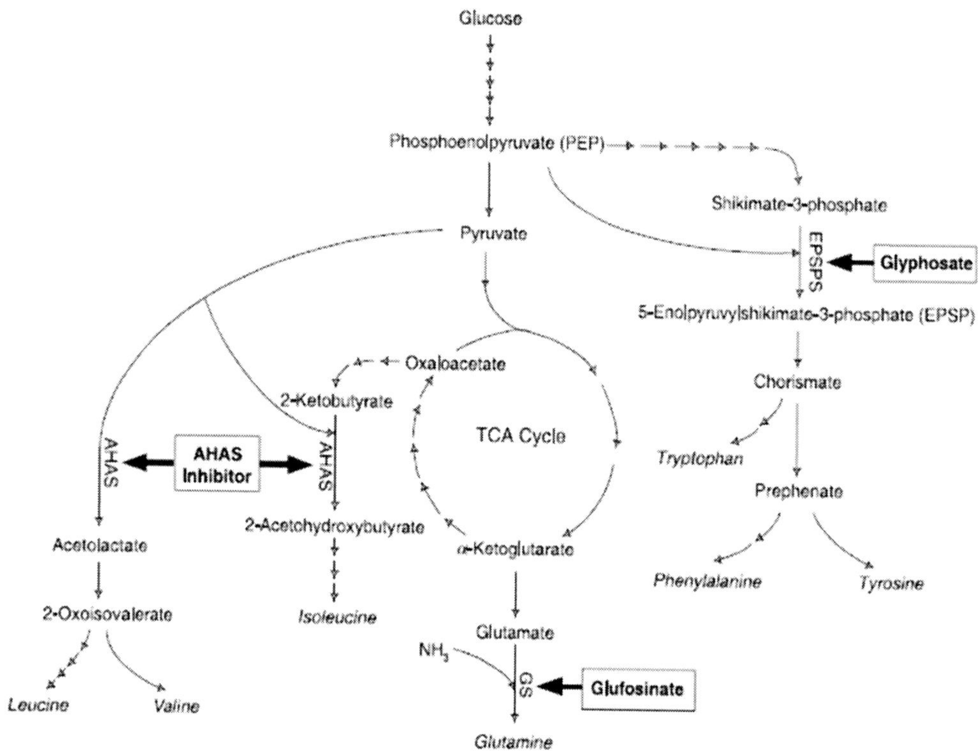

Figure 1. Biosynthesis pathways of branched-chain amino acids and glutamine in plants showing three enzymes that are inhibited by amino acid biosymthesis inhibitors. Acetohydroxyacid synthase (AHAS) inhibitors, glyphosate and glufosinate inhibit AHAS, 5-enolpyruvylshikimate-3-phosphate synthase (EPSPS) and glutamine synthetase (GS) respectively.

There are five chemical families of AHAS-inhibiting herbicides that have been commercialized (Fig 2) (Shaner and Singh, 1997; Vencill, 2002; Mallory-Smith and Retzinger, 2003). They are imidazolinones, sulfonylureas, triazolopyrimidines, pyrimidinylthio (or) oxy-benzoates, and sulfonylamino-carbonyltriazolinones. About 30 commercial herbicides are from the family of sulfonylureas, but only 2–6 are from each of the rest chemical families (Mallory-Smith and Retzinger, 2003; Herbicide Resistance Action Committee, 2005).

The AHAS inhibitors have been postulated to bind to the AHAS enzyme at the entry site for the AHAS substrate or the substrate access channel (Ott *et al.*, 1996; Pang *et al.*, 2003). It is generally believed that different families of AHAS inhibitors bind to AHAS at different but overlapping sites (Singh and Shaner, 1995; Ott *et al.*, 1996). The inhibition of AHAS by imidazolinones is uncompetitive with respect to pyruvate (Stidham and Singh, 1991; Shaner and Singh, 1997). Herbicide binding reduces or blocks AHAS catalytic activity and results in the deficiency of branched chain amino acids. Consequently, plants die from starvation for leucine, valine, and isoleucine (Shaner and Singh, 1992).

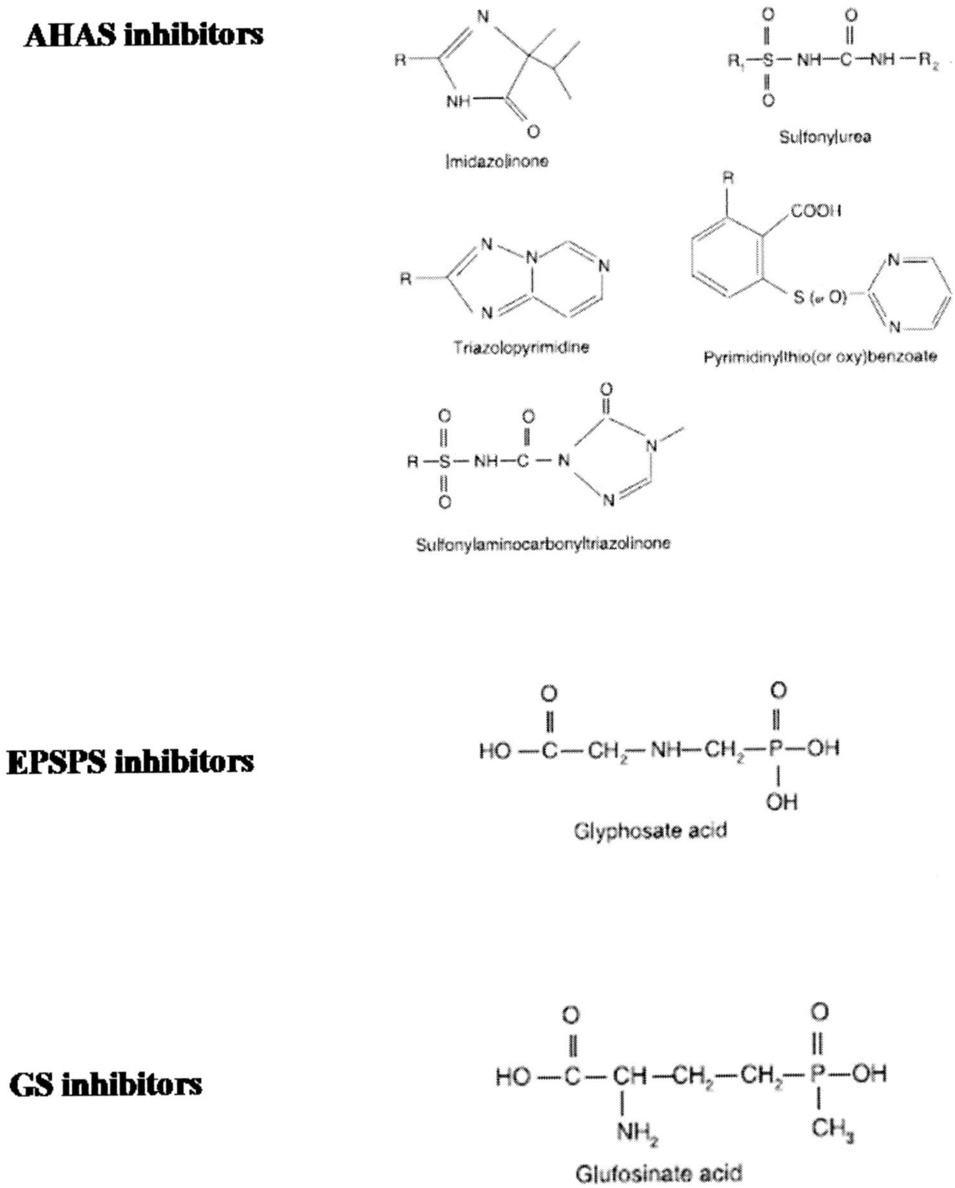

Figure 2. Basic molecular structures of herbicides that inhibit amino acid biosynthesis. AHAS inhibitors belong to five distinct families of chemicals. R, R1 and R2 are additional moieties besides the basic structures.

Glyphosate inhibition of aromatic amino acid biosynthesis: EPSP synthase catalyzes EPSP synthesis from phosphoenolpyruvate (PEP) and shikimate-3-phosphate in the shikimate pathway. EPSP leads to the final synthesis of tryptophan, phenylalanine, and tyrosine (Fig 1) (Dill, 2005). EPSP synthase is nuclear-encoded, and the biosynthesis of aromatic amino acids primarily occurs in plastids (Vaughn and Duke, 1991; Reade and Cobb, 2002).

Glyphosate interferes with aromatic amino acid biosynthesis by inhibiting EPSP synthase (Fig 1 and 2) (Cole, 1985; Kishore and Shah, 1988; OECD, 1999). It is believed that there is a close overlap of the binding sites of PEP and glyphosate on the EPSP synthase (Franz *et al.*,

1997; Dill, 2005). Glyphosate is competitive with respect to PEP in binding to EPSP synthase but uncompetitive with respect to shikimate-3-phosphate (Mousdale and Coggins, 1991; Vaughn and Duke, 1991; Franz *et al.*, 1997; Dill, 2005). Glyphosate binding reduces or blocks EPSP synthase catalytic activity and causes a deficiency of aromatic amino acid supply; consequently, plants die from starvation for tryptophan, phenylalanine, tyrosine, and secondary products from the shikimate pathway, and drainage of vital carbon into a useless pool of shikimate (Shaner and Singh, 1992).

Glufosinate inhibition of glutamine biosynthesis: GS is a nuclear-encoded protein that catalyzes conversion of L-glutamate to L-glutamine by assimilating ammonia in the cytoplasm and plastids but predominately in the chloroplast of green tissues (Vaughn and Duke, 1991; Reade and Cobb, 2002) (Fig 1). This reaction is a major mechanism for ammonia assimilation in plants (Mousdale and Coggins, 1991). Glufosinate is racemic, and its L-isomer is a GS inhibitor (Fig 2). Molecular structures of L-glufosinate and L-glutamate are very similar, and the inhibition is competitive with respect to substrate L-glutamate (Hess, 2000; Reade and Cobb, 2002). In contrast, D-glufosinate is not a GS inhibitor and has no herbicidal activity (OECD, 2002; Ruhland *et al.*, 2004). Glufosinate inhibition of GS can result in glutamine deficiency and accumulation of ammonia in the plant (Fig 1). However, ammonia accumulation is not directly responsible for the toxic effects of glufosinate. Glufosinate indirectly inhibits light reaction in photosynthesis, and inhibition of electron flow in photosynthesis under light causes membrane disruption (Hess, 2000).

GLYPHOSATE

Glyphosate is a very effective non-selective herbicide. Before introduction of GRCs, glyphosate was used in non-crop situations, before planting the crop, or with specialized application equipment to avoid contact with the crop. Glyphosate use increased more than six-fold between 1992 and 2002, to become the most used herbicide in the United States (Gianessi and Reigner, 2006).

Glyphosate is an aminophosphonic analogue of the natural amino acid glycine and the name is a contraction of glycine-, phospho-, and -ate. Glyphosate was first discovered to have herbicidal activity in 1970 by John Franz, while working for Monsanto (Murtaza *et al.*,2001).

Glyphosate was first reported as a herbicide in 1971. Three related products are now manufactured under the name glyphosate: glyphosate-isopropyl ammonium and glyphosate-sesquisodium patented by Monsanto, and glyphosate-trimesium patented by ICI (now Zeneca). In pure chemical terms, glyphosate is an organophosphate in that it contains carbon and phosphorous. However, it does not affect the nervous system in the same way as organophosphate insecticides, and is not a cholinesterase inhibitor.

Glyphosate is a broad spectrum, non-selective systemic herbicide. It is effective in killing all plant types including grasses, perennials and woody plants. As a herbicide, glyphosate works by being absorbed into the plant not only through its leaves but also through soft stalk tissues. It is then transported throughout the plant where it acts on various enzyme systems inhibiting amino acid metabolism in what is known as the shikimic acid pathway. This pathway exists in higher plants and microorganisms but not in animals. Plants treated with glyphosate slowly die over a period of days or weeks, and because the chemical is transported throughout the plant, no part survives.

The acute toxicity of glyphosate itself is very low. According to the World Health Organization, the oral LD_{50} in the rat for pure glyphosate is 4,230 mg/kg, or 5,600 mg/kg according to Monsanto (Monsanto Company, Toxicology of glyphosate and Roundup herbicide, 5 June 1985, Department of Medicine and Env. Health, Monsanto Co. Missouri, US.). The low acute toxicity of glyphosate can be attributed to its biochemical mode of action on a metabolic pathway in plants (called the shikimic acid pathway) that does not exist in animals (Carlisle and Trevors, 1988). However, glyphosate can also disrupt functions of enzymes in animals. In rats, it was found to decrease the activity of some detoxification enzymes when injected into the abdomen (Cox, 1995). In general, controlled toxicity tests report adverse symptoms from exposure to glyphosate only at extremely high doses, i.e. several grams per kg body weight.

Glyphosate dominates world herbicide usage due to its broad spectrum, ease of use and environmental attributes. The introduction of glyphosate resistant crops (GRCs) has increased the amount of the herbicide applied to soybeans, corn and canola over 10-fold since 1995. Why is glyphosate such a good herbicide? The three primary reasons are 1) the target site, 2) the ability of the herbicide to translocate in plants and 3) the inability of plants to rapidly detoxify the herbicide.

Glyphosate kills plants by inhibiting 5-enolpyruvylshikimate-3-phosphate synthase (EPSPS). EPSPS is a key enzyme in the shikimate biosynthetic pathway that is necessary for the production of the aromatic amino acids, auxin, phytoalexins, folic acid, lignin, plastoquinones and many other secondary products. Over 30% of the carbon fixed by plants passes through this pathway. Inhibition of EPSPS by glyphosate deregulates the pathway, leading to even more carbon flowing through the pathway with accumulation of shikimate and shikimate-3-phosphate. Up to 16% of the plant's dry matter can accumulate as shikimate. Glyphosate occupies the binding site on EPSPS for phosphoenol pyruvate, a substrate of EPSPS, by mimicking an intermediate state of the enzyme-substrates complex. There are two forms of EPSPS in nature, EPSPS I, which is found in plants, fungi, and most bacteria, and is sensitive to glyphosate, and EPSP II, which is found in glyphosate resistant bacteria and is not inhibited by glyphosate. It is the gene for an EPSPS II that has been used to genetically engineer resistance in crops.

The shikimate pathway is most active in meristematic tissue. Hence, glyphosate has to translocate to the meristematic tissue to be effective. Glyphosate translocates in the plant from a source to sink direction. Up to 70% of absorbed glyphosate can translocate out of the treated leaves to the root and shoot apices. However, glyphosate translocation is self-limiting and only occurs for the first 48-72 h after application. The reason for this self-limiting phenomenon is not clear, but is related to the site of action of the herbicide, since there is greater translocation in glyphosate resistant crops compared to susceptible plants. Glyphosate's ability to translocate readily in plants results in it controlling not only annual, but also perennial weeds .

Glyphosate is the only herbicide that acts by blocking the shikimate pathway through inhibition of 5-enolpyruvyl- shikimate-3-phosphate synthase (EPSPS). Inhibition of EPSPS results in reduced aromatic amino acids and deregulation of the shikimate pathway (Duke *et al.*, 2003). The latter effect causes massive flow of carbon into the pathway, with accumulation of high levels of shikimic acid and its derivatives. Glyphosate is particularly effective because most plants metabolically degrade it very slowly or not at all, and it translocates well to metabolically active tissues such as meristems. Its relatively slow mode of action allows movement of the herbicide throughout the plant before symptoms occur.

Herbicide Resistance

The herbicide resistance is the major trait that has been engineered into crops and herbicide-resistant crops (GRCs) occupy the largest area under transgenic crops in the world. Herbicide-resistant transgenic crops (GRCs) constitute nearly 72% of the 81 m ha under transgenic crops. Since the advent of large-scale cultivation of transgenic crops in 1996, GRCs have maintained the lead in terms of area coverage. GRCs were readily accepted in developed countries, where weed-control strategies almost always rely on the use of herbicides. The relevance of GRCs on small and fragmented farm holdings of India is, however, contentious. The proponents advocate that the new technology is scale-neutral and its benefits should be made available to Indian farmers.

Genetic Engineering of Herbicide Tolerance

Before the emergence of plant genetic engineering, options for selective crop protection against herbicides were limited. Specific herbicides could be used in the crops that were naturally resistant to the herbicide. In rare cases, resistance could be induced in crop varieties through mutations (Sebastian *et al.*, 1989).

For example, monocots are naturally resistant to triazine and hence triazine could be used as selective herbicide in monocot crops to control dicot weeds. Developments in plant genetic engineering and knowledge of biochemical action of herbicides on plants spurred innovative approaches to engineer crops to withstand herbicides. These strategies usually involve isolation and introduction of a gene from another organisms, mostly bacteria, which is able to overcome the herbicide-induced metabolic blockage. For example, tolerance to the herbicide glufosinate (Basta®) is conferred by the bacterial gene bar, which metabolizes the herbicide into a non-toxic compound (Thompson *et al.*, 1987). Glyphosate (another most popular herbicide) resistance is achieved by the introduction of either *Agrobacterium* gene CP4 that codes for a glyphosate-insensitive version of the plant enzyme, EPSP-synthase, or GOX gene from *Achromobacter*, which codes for glyphosate oxidoreductase in the breakdown of glyphosate (Barry *et al.*, 1992). A number of other genes have been identified that can alleviate the herbicide action through various ways (such as detoxification, sequestration, etc.) and thus confer resistance to the plants carrying them. Thus genetic engineering technology has made it possible to tailor crop varieties to resist specific herbicides by introducing relevant genes. Consequently, the range of selective herbicides has now greatly expanded, wherein specific genotypes and varieties can be conferred resistance rather than generic crops displaying resistance to specific herbicides. These developments have provided new opportunities to promote herbicides through development and marketing of genetically engineered GRCs (Bhat and Chopra, 2006).

Biotechnology and its use to modify the genetic makeup of living organisms has become a topic of heated discussion in recent years. Confusion is plentiful on the topic of biotechnology and genetically modified organisms (GMOs). Biotechnology includes a vast array of tools used in research and modification of biological systems. These include: genetic mapping, molecular based disease diagnosis, gene therapy, forensic science, and genetic transformation.

STRATEGIES FOR ENGINEERING HERBICIDE RESISTANCE

The engineering of herbicide resistance demonstrates the way in which quiet different strategies can be used to achieve the same objective. Mullineaux (1992) identified four different strategies for engineering herbicide resistance. These are,

Over-Expression of the Targeted Protein

This strategy effectively involves titrating the herbicide out by over production o the targeted protein. For example, if the herbicide is a specific inhibitor of one particular enzyme, production of sufficient excess enzyme will partially overcome the inhibition. Over expression can be achieved by the integration of multiple copies of the gene and or the use of strong promoter plus translation enhancer to drive expression of the gene.

Of the 4 strategies described above, 3 have been actually tested in the laboratory, 2 of which form the basis of the current commercial plantings of glyphosate resistant crop. One of the earliest approaches to engineering glyphosate resistance involved the over expression of a plant EPSPS gene. This was facilitated by the isolation of petunia c DNA from glyphosate resistant tissue culture. The step wise selection of petunia cells capable of growing in presence of increasing amounts of glyphosate led to the isolation of cultures in which the levels of EPSPS enzyme was much higher than normal. This was found not to be due to increased expression of the EPSPS gene, but was the result of a gene amplification, such that there were multiple (upto 20) copies of the EPSPS gene in an otherwise normal petunia genome. The EPSPS enzyme was not itself mutated i.e. the resistance was simply due to the increased amount of enzyme.

The result of this experiment was a 40 fold increase in EPSPS activity in the transgenic plants and a tolerance to glyphosate sprayed in the field at a dose 2-4 times higher than that required to kill wild type plants.

Mutation of the Targeted Protein

The logic behind this approach is to find a modified target protein that substitutes functionally for the native protein and which is resistant to inhibition by the herbicide, and to incorporate the resistant target protein gene in to the plant genome. Several sources of resistant proteins can be exploited. Note that both the over expression on mutated target protein strategies require knowledge of mode of action of the herbicide.

Mutated EPSPS gene has been isolated from a number of glyphosate resistant bacteria. In one, a mutated *aroA* gene from *Salmonella typhimurium* was inserted between the promoter and terminator sequence of the *ocs* gene of the *Agrobacterium tumefaciens* Ti-plasmid. Only a moderate increase of tolerance was obtained. Resistance was limited, as the prokaryote did not have a plastid transit peptide sequence, so the resistance enzyme was not transported to the chloroplast.

Although successful plant regeneration methods have been developed, the technology has not provided regeneration in other crops for using transformation protocols, which is a serious limitation to the exploitation of gene transfer to its full potential. Even if the crop of interest is amenable to *in vitro* regeneration, it is unfortunate that plant regeneration from single transformed cells often produces mutations ranging from single base changes or small rearrangements to the loss of the entire chromosomes. I addition significant epigenetic changes (like DNA methylation) can also occur (Phillips *et al.*, 1994). It is often necessary to generate and screen a large number of independent plant lines transformed with the same construct to find lines that have suffered minimum genetic damage and carry a simple insertion event (Birch, 1997; Hansen and Wright, 1999). In the light of these constraints, it becomes necessary to evolve transformation strategies that do not depend on tissue-culture regeneration or those that substantially eliminate the intervening tissue-culture steps.

Detoxification of the Herbicide, Using a Single Gene from a Foreign Source

Detoxification is a means of converting herbicide to a less toxic form and or removing it from the system. This strategy can be contrasted with the previous two because it does not require detailed knowledge of the site of action.

Enhanced Plant Detoxification

The aim here is to improve the natural plant defenses against complex compounds. This requires detailed information about endogenous plant detoxification pathways and the mechanisms by which compounds are recognized and targeted for detoxification by the plant.

Plastid Transformation

The wake-up call came when a gene conferring insect resistance that had proven notoriously difficult to express from the plant's nuclear genome was put into the chloroplast genome and its gene product (a *Bacillus thuringiensis* toxin protein) accumulated to extraordinarily high levels (McBride *et al.*, 1995). Since then, chloroplast transformation has been the subject of considerable excitement among biotechnologists. When genetically engineering herbicide resistance into plants, the level of resistance often positively correlates with the accumulation levels of the resistance conferring gene product (protein). The feasibility to use chloroplast genetic engineering for weed control has been explored in several studies that aimed at producing glyphosate-tolerant tobacco plants.

In plants, EPSPS is a nuclear-encoded chloroplast-targeted enzyme. Attempts to optimize glyphosate tolerance in transplastomic tobacco plants have involved three strategies: testing of Glyphosate tolerant EPSPS genes from different sources, adjusting the codon usage of the EPSPS gene to the codon usage in plastids where usually triplets with A or T in third codon position are strongly preferred, and testing different expression signals (promoters, 50 untranslated regions, N-terminal fusions) to maximize EPSPS expression (Daniell *et al.*, 1998; Ye G-N *et al.*, 2001). Of these strategies, the choice of the 5' untranslated region (containing the Shine–Dalgarno sequence and thus determining the rate of translation initiation) had by far the strongest effect on the attainable level of herbicide tolerance (Ye G-N *et al.*, 2001). EPSPS accumulation levels varied more than 10 000-fold and, with the best performing constructs, protein accumulation reached values of over 10% of the TSP of the plant. These transplastomic plants were tolerant to very high glyphosate doses, but revealed a so far unresolved paradox: chloroplast localization of the EPSPS gene required 250-fold higher enzyme accumulation to reach similar tolerance levels as nuclear expression of the identical EPSPS gene (Ye G-N *et al.*, 2001, Ralph Bock, 2007).

A Built-In Strategy for Containment of Transgenic Plants

Simple and reliable containment strategies for transgenes are highly desirable. In this method, the gene(s) of interest is tagged with a RNA interference cassette, which specifically suppresses the expression of the bentazon detoxification enzyme **CYP81A6** and thus renders transgenic rice to be sensitive to bentazon, a herbicide used for rice weed control. Transgenic rice plants generated by this method using a new glyphosate resistant 5-enolpyruvylshikimate-3-phosphate synthase (EPSPS) gene from **Pesudomonas putida** as the gene of interest (*G6* ; gb: EU169459), and demonstrated that these transgenic rice plants were highly sensitive to bentazon but tolerant to glyphosate, which is exactly the opposite of conventional rice. Field trial of these transgenic rice plants further confirmed that they can be selectively killed at 100% by one spray of bentazon at a regular dose used for conventional rice weed control (Chaoyang Lin, 2008).

Intein-Mediated Protein Trans-Splicing System

One of the approaches recently outlined is intein-mediated protein trans-splicing in which the gene to be inserted is split into two halves and the activity of the transgene product is reconstituted upon self excision of inteins and concomitant ligation of the truncated protein products – exteins (Chin *et al.*, 2003; Sun *et al.*, 2006). The potential for trans-splicing of transgenes using inteins in conjunction with plastid engineering to provide for a more effective transgene containment strategy is discussed here.

Inteins are intervening protein sequences that catalyze multi-step, post-translational protein splicing. In contrast to introns, they are transcribed and translated with their host proteins. More than 200 inteins have been identified from 34 different types of proteins (InBase; http://www.neb.com/neb/inteins.html; (Gogarten, 2002; Perler, 2002). Inteins are 134–1650 amino acids in size and have been identified in the genomes of eubacteria, eukaryota and archaea.

Most inteins consist of two domains – a self-splicing domain and an endonuclease domain. The endonuclease domain, present only in large inteins, does not appear to be necessary for protein splicing (Gogarten, 2002; Perler, 2002). The mini-inteins comprise a self-splicing domain and can be artificially split into two fragments (N- and C-terminal) with conserved motifs necessary for splicing. Hence, the N-terminal fragment contains the A and B motifs and the C-terminal fragment contains the F and G motifs (Perler, 1997). When the split intein fragments are fused with different exteins or peptide precursors they can conduct protein trans-splicing to assemble two distant proteins (exteins) or precursor molecules into a mature hybrid protein through four steps: two acyl rearrangements at the two splicing junctions, a trans-esterification between the two junctions, and a cyclization of the asparagine residue at the C-terminal junction (Perler, 1997).

The most commonly used, naturally occurring split intein is the Ssp DnaE intein of Synechocystis sp.PCC6803 with N-terminal and C-terminal regions of 134 and 36 amino acids of, respectively (Wu, 1998). In Synechocystis a split copy of the replicative DNA polymerase III catalytic alpha subunit (DnaE) is encoded by the dnaE gene in which the dnaE-n encodes the first 774 amino acid residues of DnaE and the dnaE-c encodes the remaining 423 amino acid residues. These two portions are transcribed from opposite strands of the genome. On translation, two chimeric polypeptides are produced: the N-terminus of the DnaE (N-extein) with an intein portion, and the complementary intein portion with the C-terminus (C-extein) of the polymerase. The inteins undergo self excision and concomitant ligation of the exteins as a result of the protein trans-splicing reaction. Thus the Ssp DnaE intein catalyzes trans-splicing of a protein in Synechocystis (Evans ,2000; Liu, 2000). Similarly, the Ssp DnaB mini-intein of Synechocystis is also capable of protein trans-splicing (Sun *et al.,* 2006). Hence, it might be a useful approach to split a gene into halves and integrate these separately into the plastid and nuclear genomes of a cell to reduce risk of genetic outcrossing.

Two genes encoding fragments of a herbicide-resistant form of EPSPS were placed into separate genomes, the nuclear and chloroplast genome. Full-length EPSPS protein was generated in the chloroplast after translocation of the N-terminal EPSPS intein fusion protein fragment to the plastid followed by trans splicing. Gene flow from the above-mentioned transgenic plants to wild or weedy relatives would transmit only a portion of the full-length gene, correlating to reduced environmental impact. Also, the activity of protein-splicing elements in plant cells opens the possibility of protein engineering, such as intein-based protein cyclization or polymerization, in plant tissues (Chin *et al.,* 2003).Transgene expression levels in the transplastomic plants were much higher than nuclear transformed plants because of the copy number of the transgenes in the cells. After 4 days of treatment with 0.2 mM glyphosate, the wild-type plants were bleached, whereas the trans-splicing lines remained green. This result was similar to that of transgenic tobacco plant carrying an intact nuclear *EPSPS* gene and demonstrated that glyphosate resistance was conferred by the spliced EPSPS gene products (Chin *et al.,* 2003).

Deactivation by Acetylation

Recently, a new mechanism to detoxify glyphosate *in planta*, deactivation by acetylation, was developed (Castle et al., 2004). The enzyme, glyphosate acetyltransferase, converts glyphosate into the non-toxic molecule acetylglyphosate, before glyphosate can reach and inhibit the the EPSPS enzyme. The transferase gene was derived from a naturally occurring soil bacterium (*Bacillus licheniformis*). Researchers used a technology, known as gene shuffling, to optimize both the enzyme's acetylation efficiency and its specificity before incorporating the gene into plants (Castle et al., 2004; Siehl et al., 2005).

BENEFITS AND RISKS OF GLYPHOSATE –RESISTANT CROPS

Benefits

Herbicide resistance has primarily been developed to benefit farm management, whereas benefits to consumers are less obvious. Since transgenic, glyphosate-resistant crops (GRCs) were introduced over ten years ago, adoption of this technology has been dramatic. Of the 100 million hectares of transgenic crops grown annually worldwide in 2005, almost 90% were glyphosate-resistant alone or had glyphosate-resistance genes stacked with *Bt* toxin-based insect resistance genes. Although no significant negative environmental effects have been documented in the vast areas growing these crops, there is concern about possible environmental impacts of this technology. There are potential positive and negative effects of the herbicide, the transgene, and the herbicide/transgenic crop weed management system.

Glyphosate is not a low use rate herbicide; however, it is considered a low risk herbicide in terms of toxicity and environmental effects. Some studies have claimed that the volume of herbicide use is greater with GRCs. The impact of GRCs on herbicide use varies with the crop, but there are no cases so far in which use rate has substantially increased.

Glyphosate replaces herbicides that are significantly more toxic and that persist much longer in soil and water. Several studies of the total environmental impact of glyphosate versus the herbicides that it replaces have found the impact is less with GRCs. It was concluded that the major environmental advantage of growing GR sugarbeet would be lower emissions from herbicide manufacturing, transport, and field operations, thus reducing contributions to global warming, smog, ozone depletion, ecotoxicity of water, and acidification and nitrification of soil and water.

Glyphosate is not a significant soil contaminant when used at recommended doses. Glyphosate is rapidly degraded by soil microbes and is inactivated through its strong adsorption to soil, reducing its potential biological activity. The half-life of glyphosate in soil varies from just over a week to several months, which is shorter than many of the herbicides that it replaces. It is commonly metabolized to aminomethylphosphonic acid (AMPA), which has higher mobility in soil.

Potential direct effects of GRCs and their management on soil biota include changes in soil microbial activity due to direct effects of glyphosate, differences in the amount and composition of root exudates of GRCs versus non-GRCs, and alteration in microbial populations because of the effects of management practices for transgenic crops, such as changes in other herbicide applications and tillage. Most studies find no effects or very limited, transient effects of glyphosate on soil microflora. Effects that have been detected are field site and season dependent, and the changes in microbial communities associated with

GRCs are more variable and transient than those caused by other agricultural practices such as crop rotation, tillage, use of certain other herbicides, and irrigation. So far, no agriculturally significant effect of glyphosate on soil microorganisms has been documented.

Top soil loss due to tillage is perhaps the most environmentally destructive effect of row crop agriculture. GRCs facilitate reduced tillage systems, which contribute to reductions in soil erosion from water and wind, fossil fuel use, air pollution from dust, soil moisture loss, and soil compaction. Reduced tillage also improves soil structure, leading to reduced runoff and pollution of surface waters with sediment, nutrients, and pesticides.

Even though glyphosate is highly water soluble, it does not leach into ground water in most soils due to its strong adsorption to soil components. Soil and sediments in bodies of water are the main sinks for glyphosate residues from surface water, greatly reducing further transport. Once in surface water, it dissipates more rapidly than most other herbicides.

Glyphosate has a very low level of toxicity in mammals (Williams, 2000), and it is not retained to a significant extent in animal tissues. It is not a restricted use pesticide and is sold for home use. Residues of glyphosate and AMPA have been found below tolerance levels in GR soybean seed. Non-toxic levels of glyphosate and its degradation products can be expected to be found in edible parts of GRCs.

Another important and laboratory-friendly use of engineering for herbicide tolerance is its utility as a selectable marker in screening of transformants. Different high stringent concentrations can be used to differentially select high, medium and low performing transgenics in the initial screening.

Risks

The soybean nitrogen-fixing symbiont *Bradyrhizobium japonicum* possesses a glyphosate-sensitive EPSPS and glyphosate can inhibit or kill the microbe. Furthermore, glyphosate is preferentially translocated to soybean nodules. Applying glyphosate to GR soybeans can reduce nodulation, nodule size, and leghemoglobin content of nodules. In the field, effects are transient, and no evidence indicates that crop yield is affected.

Risk to aquatic organisms is negligible or small at application rates <4 kg/ha and only slightly greater at application rates of 8 kg/ha (application rates much higher than recommended rates).

Gene-flow from crops to other crops or related species is another route to the development of resistant weed populations in the field. Once the resistance gene is present in crop volunteers or related weed species then it is expected that the same weed control practices (consistent sprayings with herbicides having the same mode of action), which cause herbicide resistance to occur in naturally tolerant/resistant weed biotypes, will lead to a rapid build-up of GR-weeds and volunteers.

Transfer of genes from one population to another may lead to unwanted effects for weed management and the environment. Gene flow may enable the resistance genes to move between GR and non-GR varieties and thus pollute a crop which is considered GM-free. Or GR-genes may be stacked from years of cross-pollination of GRCs, which may result in problems for the farmer in controlling volunteer crops in the field.

Increased herbicide use is considered a risk in some parts of the world although the effects on human health or the environment are seldom specified in details, but derived effects from pesticide-use such as ground-water pollution and pesticide residues in vegetables, for example, have caused public concern. There seem to be two major explanations why herbicide use in GRCs may increase. One reason is that a high level of crop tolerance may enable the farmer to increase doses to achieve an improved weed control without harming the crop. The other reason is problems with tolerant/resistant weeds and volunteers, which require farmers to increase dose or mix herbicides with different modes of action to maintain an acceptable level of weed control.

When adopting GM crops, farmers also meet a number of constraints, namely: seed-saving is illegal, the cost of seed or of fees for technology agreement is increased, and fewer suppliers provide the input for crop production (EU, 2000). Furthermore, if the GR crop readily cross pollinates with adjacent non-GM crops, then it raises questions both about property rights of these unintentionally resistant plants and threshold levels of GM seed in harvests from non- GM crops.

GRCs have a great potential in the simplification of weed management. Handled judiciously, these crops may be beneficial to the environment by enabling no-till systems, thus reducing erosion or allowing for later weed control, which may increase biodiversity in the field. However, it must be emphasized that the risk from GRCs should be carefully evaluated prior to releasing the GRC into a cropping system, especially when the GRCs possess weedy characters or may outcross to related weeds. If this is the case, and the GRC is grown commercially, then precautions need to be taken, similar to the management strategies adopted to prevent the development of naturally-resistant weeds. Furthermore, precautions must, in particular, be taken before release into the genetic origin of the species.

Conclusion

Generalizations may not apply to all situations, as risks and benefits are geography and time dependent. In the context of the replaced herbicides and agronomic practices, the apparent health and environmental benefits of the glyphosate/GRC combination are significant in most studied cases. Glyphosate is more environmentally and toxicologically benign than many herbicides that it replaces. Perhaps the most significant contribution of GRCs to the environment has been its influence on adoption of reduced- or no-tillage agriculture. The prevalence of evidence indicates that GRC-derived feed and foods are safe and are nutritionally equivalent to their non-transgenic counterparts. Little or no direct impact or risk is expected from glyphosate resistance transgenes when they introgress into wild populations. However, when stacked with genes that impart a fitness advantage in the wild, the glyphosate-resistance trait, in the presence of glyphosate use, can increase the chances of potentially harmful gene flow. The glyphosate resistance can be developed as a selectable marker of high utility. However, environmental friendly strategies to develop herbicide resistance in a variety of crops has to go a long way.

REFERENCES

Barry, G., Kishore, G., Padgette, S., Taylor, M., Kolacz, K., Weldon, M. Re. D., Eichholtz, D., Fincher, K. and Hallas, L. (1992). Inhibitors of amino acid biosynthesis: strategies for imparting glyphosate tolerance to crop plants. In: Singh, B.K., Flores, HE. and Shannon, J.C., (eds) *Biosynthesis and molecular regulation of amino acids in plants*. American Society of Plant Physiologists, Rockville, Md, pp 139–145

Bhat S.R and Chopra V.L. (2006), Choice of technology for herbicide-resistant transgenic crops in India: examination of issues, *Curr sci*, 91: 435-438.

Birch, R. G. (1997). Plant transformation: problems and strategies for practical application. *Annu. Rev. Plant Physiol, Plant Mol. Biol*., 48: 297-326.

Carlisle, S.M. and J.T. Trevors, (1988).*Glyphosate in the Environment, Water, Soil and Air Pollution*, 39:409-420.

Castle, L.A., Siehl, D.L., Gorton, R., Patten, P.A., Chen, Y.H., Bertain, S., (2004). Discovery and directed evolution of a glyphosate tolerance gene. *Science, 304*(5674), 1151-1154.

Chaoyang Lin, (2008).A built-in strategy for containment of transgenic plants: Creation of selectively terminable transgenic rice, *PLoS ONE*,3,3,e1818(1-6)

Chin, H. G., G. D. Kim, I. Marin, F. Mersha, T. C. Evans, L. X. Chen, M.-Q.Xu, and S. Pradhan. 2003. *Protein trans-splicing in transgenic plant chloroplast:reconstruction of herbicide resistance from split genes*. Proc. Natl.Acad. Sci. USA 100:4510–4515.

Cole, D., Pallet, K., Rodgers, M. (2000). Discovering new modes of action for herbicides and the impact of genomics. *Pesticide Outlook* 11, 223–229.

Cox, C., Herbicide factsheet: Glyphosate, Part 1: Toxicology, *Journal of Pesticide Reform*, September 1995, 15:3.

Daniell ,H., Datta, R., Varma ,S., Gray, S., Lee, S-B.(1998). Containment of herbicide resistance through genetic engineering of the chloroplast genome. *Nat Biotechnol* , 16:345-348.

DeFelice, M.S. (1998). Herbicide registration dates, use rates and acute toxicity by decade. In: Hatzios KK (ed) *Herbicide handbook supplement to seventh edition*. Weed Science Society of America, Lawrence, Kans, pp 81–86.

Dill, G.M. (2005). Glyphosate-resistant crops: history, status and future. *Pest Manage Sci* 61: 219–224.

Duke, S.O. (2005). Taking stock of herbicide-resistant crops ten years after introduction. *Pest Management Science*, 61: 211–218.

Duke, S.O., A.M. Rimando, P.F. Pace, K.N. Reddy, and R.J. Smeda. (2003). Isoflavone, glyphosate, and aminomethylphosphonic acid levels in seeds of glyphosate-treated, glyphosate-resistant soybean. *J. Agric. Food Chem*. 51:340–344.

EU.2000. Economic impacts of genetically modified crops on the agri-food sector-A synthesis. Working document directorate –general for agriculture (available at *http://europa.eu.int/comm/agriculture/res/index_en.htm)43pp*

Evans, T.C. (2000) Protein trans-splicing and cyclization by a naturally split intein from the dnaE gene of Synechosystis species PCC68803. *J. Biol. Chem*. 275, 9091–9094

Franz, J.E., M.K. Mao, and J.A. Sikorski. (1997). Glyphosate, a unique global herbicide. *Amer. Chem. Soc*., Washington, DC., USA.

Gianessi, L.P. and N. Reigner. (2006). *Pesticide use in U.S. crop protection*: (2002). Crop Life Foundation, Washington, D.C. free amino acid soluble proteins and protein synthesis in inflated duckweed.

Gogarten, J.P. (2002) Inteins: structure, function and evolution. *Annu. Rev. Microbiol.* 56, 263–287

Hansen, G., and Wright, M.S. (1999). Recent advances in the transformation of plants. *Trends Plant Sci.*, 4: 226-231.

Herbicide Resistance Action Committee (2005). Classification of herbicides according to mode of action. http:==*www.plantprotection.org*= HRAC= (accessed in March 2005)

Hess ,F.D. (2000) .Light-dependent herbicides: an overview. *Weed Science* 48: 160–170

Kishore, G., and Shah, D. (1988). Amino acid biosynthesis inhibitors as herbicides. *Annual Review of Biochemistry*. 57: 627–663

LaRossa, R.A. and Falco, S.C. (1984). Amino acid biosynthetic enzymes as targets of herbicide action. *Trends in Biotechnology*, 2: 158–161

Liu, X.Q. (2000) Protein-splicing intein: genetic mobility, origin, and evolution. *Annu. Rev. Genet.* 34, 61–76

Mallory-Smith, C.A and Retzinger, E.J .Jr (2003) Revised classification of herbicides by site of action for weed resistance management strategies. *Weed Technology* 17: 605–619.

Mc.Bride, ,K.E, Svab Z, Schaaf, D.J, Hogan ,P.S, Stalker, D.M, and Maliga P. (1995).Amplification of a chimeric Bacillus gene in chloroplasts leads to an extraordinary level of an insecticidal protein in tobacco.*Bio/Technology*, 13:362-365.

Mousdale ,D.M. and Coggins, J.R. (1991). *Amino acid synthesis*. In: Kirkwood RC (ed) Target sites for herbicide action. Plenum Press, New York, pp 29–56.

Mullineaux, P.M. (1992). Genetically engineered plants for herbicide resistance. In Plant genetic manipulation for crop protection (ed. A.M.R. Gatehouse, V.A.Hilder and D.Boulter), pp.75-107. C.A.B. International, Wallingford UK.

Murtaza, F., Alibhai and William C. Stallings. (2001). *Closing down on glyphosate inhibition—with a new structure for drug discovery* [[Proceedings of the National Academy of Sciences, 98(6): 2944–2946.

OECD, (2002). Module II: herbicide biochemistry, herbicide metabolism and the residues in glufosinate-ammonium (phosphinothricin)-tolerant Herbicidal inhibitors of amino acid biosynthesis and herbicide-tolerant crops 203 transgenic plants. Series on harmonization of regulatory oversight in biotechnology No. 25 *http:==www.olis.oecd.org=olis=2002doc.nsf= LinkTo=env-jm-mono*(2002)14 (accessed in March 2005)

OECD. (1999). Consensus document on general information concerning the genes and their enzymes that confer tolerance to glyphosate herbicide. Series on harmonization of regulatory oversight in biotechnology No. 10, *http:==www.olis.oecd.org=olis=1999doc.nsf=LinkTo=env-jm-mono* (accessed in March 2005)

Ott, K., Kwagh, J., Stockton ,G.W., Sidorov, V. and Kakefuda, G (1996). Rational molecular design and genetic engineering of herbicide resistant crops by structure modeling and site-directed mutagenesis of acetohydroxyacid synthase. *J Mol Biol* 263: 359–368.

Pang, S.S., Guddat ,L,W. and Duggleby ,R.G. (2003) .Molecular basis of sulfonylurea herbicide inhibition of acetohydroxyacid synthase. *J Biol Chem* 278: 7639–7644.

Perler, F.B. (1997) Protein splicing and autoproteolysis mechanisms. *Curr. Opin. Chem. Biol.* 1, 292–299

Perler, F.B. (2002) Inbase, the intein database. *Nucleic Acids Res.* 30, 383–384

Phillips, R.L., Kaeppler, S.M., and Olhoft, P. 1994. Genetic instability of plant tissue cultures: breakdown of normal controls. *Proc. Natl. Acad. Sci.* USA, 91: 5222-5226.

Ralph Bock. (2007). Plastid biotechnology: prospects for herbicide and insect resistance, metabolic engineering and molecular farming, *Current Opinion in Biotechnology*, 18:100–106

Reade, J.P.H. and Cobb, A.H. (2002). Herbicides: modes of action and metabolism. In: Naylor REL (ed) *Weed management handbook*. Blackwell Science, Oxford, pp 134–170.

Ruhland, M., Engelhardt, G. and Pawlizki, K. (2004). Distribution and metabolism of D=L-, L- and D-glufosinate in transgenic, glufosinate tolerant crops of maize (Zea mays L ssp mays) and oilseed rape (Brassica napus L. var napus). *Pest Manage Sci* 60: 691–696.

Saari ,L.L. and Mauvais ,C.J. (1996). Chapter 8, Sulfonylurea herbicide-resistant crops. In: Duke SO (ed) *Herbicide resistant crops*. CRC Press, Boca Raton, Fla, pp 127–142

Sebastian, S.A, Fader ,G.M, Ulrich ,J.F, Forney, D.R and Chaleff RS. (1989). Semidominant soybean mutation for resistance to sulfonylurea herbicides. *Crop Science*, 29: 1403–1408

Shaner ,D.L., Bascomb, N.F. and Smith, W. (1996). Chapter 9, Imidazolinoneresistant crops: selection, characterization, and management. In: Duke SO (ed) *Herbicide resistant crops*. CRC Press, Boca Raton, Fla, pp 143–157

Shaner, D.L. and Singh, B.K. (1997) Acetohydroxyacid synthase inhibitors. In: Roe RM (ed) *Herbicide activity: toxicology, biochemistry and molecular biology*. IOS Press, Washington, DC, pp 69–110

Shaner, D.L. and Singh, B.K. (1992). How does inhibition of amino acidbiosynthesis kill plants? In: Singh BK, Flores HE, Shannon JC (eds) *Biosynthesis and molecular regulation of amino acids in plants*.

Sherman, T.D., Vaughn, K.C.and Duke, S.O. (1996). Mechanisms of action and resistance to herbicides. In: Duke SO (ed) *Herbicide resistant crops*. CRC Press, Boca Raton, Fla, pp 13–35

Siehl, D.L., Castle, L.A., Gorton, R., Chen, Y.H., Bertain, S., Cho, H.J., (2005). Evolution of a microbial acetyltransferase for modification of glyphosate: a novel tolerance strategy. *Pest Management Science*, *61*(3), 235-240.

Singh, B.K. and Shaner, D.L. (1995). Biosynthesis of branched chain amino acids: from test tube to field. *Plant Cell* 7: 935–944.

Stidham, M.A. and Singh, B.K. (1991). Imidazolinone-acetohydroxyacid synthase interactions. In: Shaner DL, O'Conner SL (eds) *The imidazolinone herbicides*. CRC Press, Boca Raton, Fla, pp 71–90.

Sun, Y. C., Y. Li, H. Zhang, H. Q. Yan, D. N. Dowling, and Y. P. Wang. (2006). Reconstitution of the enzyme AroA and its glyphosate tolerance by fragment complementation. *FEBS Lett.* 580:1521–1527.

Thompson, C.J., Movva, N.R., Tizard, R., Crameri, R., Davies, J.V., Lauwereys, M and Botterman, J., (1987). Characterization of the herbicide-resistant gene bar from *streptomyces hygroscopicus*. *EMBO J.*, 1987, 6:2519-2523.

Vaughn, K.C. and Duke SO (1991). Biochemical basis of herbicide resistance. *Chem Plant Protect* 7: 141–169.

Vencill,W.K. (ed.). (2002). *Herbicide Handbook*, 8th ed.Weed Sci. Soc. Amer., Lawrence, KS, USA.

Williams, G.M (2000) Safety evaluation and risk assessment of the herbicide roundup and its active ingredient, Glyphosate, for humans. *Regul.Toxicol.Pharmacol.*31,117-165

Wu, H. (1998) *Protein trans-splicing by a split intein encoded in a split DnaE gene of Synechocystis* sp. PCC6803. Proc. Natl. Acad. Sci.U. S. A. 95, 9226–9231.

Ye G-N, Hajdukiewicz, P.T.J., Broyles, D., Rodriguez, D., Xu ,C.W., Nehra, N., Staub ,J.M. (2001). Plastid-expressed 5-enolpyruvylshikimate- 3-phosphate synthase genes provide high level Glyphosate tolerance in tobacco. *Plant Journal*, 25:261-270.

In: Genetically Engineered Crops
Editors: Lindsay M. Grover

ISBN: 978-1-61122-526-6

Chapter 7

EFFECTS OF GENETICALLY MODIFIED CROPS ON ARBUSCULAR MYCORRHIZAL FUNGI: MECHANISMS AND IMPLICATIONS

Wen Ke Liu*

Institute of Environment and Sustainable Development in Agriculture,
Chinese Academy of Agricultural Sciences, Key Laboratory
for Agro-Environment and Climate Change, Ministry of Agriculture, Beijing , China

ABSTRACT

Impacts of genetically modified crops (GMCs) on non-target organisms and ecosystems are a key issue of risk assessment of GMCs introduction since they released. Introduction and use of genetically modified crops brought a list of changes in GMCs-soil ecosystems. The main changes include three aspects, i.e. physiological alterations of transgenic crops, rhizosphere characteristics variations of transgenic crops, and changes in management and agronomic measures, which result from the insertion of exotic genes and their biological functions. Compared to their parental line, above changes of transgenic crops would lead to a great deal of subsequently changes in soil ecosystem and the entire field ecosystem, e.g. nutrient cycling, soil microorganisms (e.g. arbuscular mycorrhizal fungi, AMF), input of agro-chemicals and so on. In this review, the changes brought by introducing GMCs in plant-soil system under field and the resultant impacts on soil ecosystem were summarized, highlighting the impact mechanisms on different scales, and ecological implications of interactions between Bt transgenic crops (BTCs) and AMF.

INTRODUCTION

In 2009, the global acreage of genetically transgenic crops (GMPs) reached 134 million hm^2, they planted great roles in input reduction of agro-chemicals, in enhancement of crop

* E-mail: liuwke@163.com

yield and economic income of farmers [1]. The 80-fold increase in biotech crop hectares between 1996 and 2009, is unprecedented, and makes biotech crops the fastest adopted crop technology in the recent history of agriculture; this reflects the confidence and trust of millions of farmers worldwide who have consistently continued to plant more biotech crops every single year since 1996, because of the multiple and significant benefits they offer. The global adoption rate for principal biotech crops, soybean, cotton, maize and canola, had reached 77%, 49%, 26% and 21%, respectively [1]. Take Bt transgenic cottons for example, Bt transgenic cotton that had been engineered to produce insecticidal toxins from *Bacillus thuringiensis* (Bt) and so to resist the pest cotton bollworm (*Helicoverpa armigera*) had been widely planted in Asia. Bt cotton not only controlled *H. armigera* on transgenic cotton designed to resist this pest but also reduced its presence on other host crops and decreased the need for insecticide sprays in general[2]. In terms of traits, GMPs are classified into four types: herbicide tolerant (HT), insect resistant (IR, mostly Bt-*Bacillus thuringiensis*), combined HT & IR, and resistant to virus diseases (VR), accounting for 63%, 15%, 22%, and <0.1% of total acreage, respectively [3]. GMPs had been expected to be widely adopted throughout the world for their consistent and substantial agronomic, economic and environmental benefits in agriculture [4-7]. However, only 25 countries in 2009 now cultivated GMPs, which reflects the strong worldwide public concerns over the potential environmental risks of GMPs. Since GMPs were released, there has been a great controversy over the potential unwanted effects on the environment and human health [8]. In recent years, the rapid development of agricultural biotechnology and release of new GMPs (species and cultivars) made ecological risk assessment of GMPs even more important and urgent. In view of the great potential uses in modern agriculture, the release of GMPs requires well-defined risk assessments on the environment and human health, especially for Bt transgenic plants (BTPs) [9、10], which constituted the absolute majority of the GMPs, and emerged in quantity in past decade.

Ecological risk evaluation has mainly focused on three issues: (a) transfer of genes from the GMPs to wild relatives and related species [11、12]; (b) resistance evolution to herbicide-tolerant crops, virus-resistant crops and insect-resistant crops[13]; and (c) impacts on non-target organisms and ecosystems [14、15、16]. Although the possible impacts of GMPs on soil ecosystems was described in the early 1990s [17、18、19], risk assessments of GMPs have mainly addressed aboveground effects in previous studies[10]. Most studies dealing with the effects of GMPs on soil ecosystems, particularly the soil microbial community, have only begun to be conducted since 2000. Taking risk probability on soil microbial communities into consideration, past studies have mainly focused on investigating the ecological risks of GMPs transferred with herbicide tolerant (HT) and insect resistant traits.

Many excellent reviews have been published addressing advances in environmental risk assessments of GMPs on soil microbial communities and biochemical processes [10、20-26]. However, few literatures cover the effects of GMPs on arbuscular mycorrhizal fungi (AMF) and relevant mechanisms. In this paper, some changes brought by introducing GMPs in field and the resultant changes in soil ecosystem were summarized, the impact mechanisms and ecological implications of interactions between Bt transgenic crops (BTCs) and AMF. Here, we take Bt transgenic crops as a example to discuss the impact mechanisms and ecological implications of interactions between Bt transgenic crops and AMF.

1. Changes in Soil-Plant System Incurred by Introducing BTCs

1.1. Physiological Changes

Previous studies showed that there were physiological changes of BTCs compared with their parental lines. Wang et al. (1998) [27] reported that Bt transgenic cotton line had a increased leaf amino acid content, but more nutrients were utilized for stem and branch growth. Breen et al. (1999) [28] reported that high nitrogen and high vegetative growth produced significantly less lint for DP90B of Bt cotton cultivar. Dong et al. (2000) [29] reported that increase of leaf NR activity and NO_3^--N enhanced boll shedding for Bt cotton cultivars. Chen et al. (2004) [30] found further that the Bt cotton cultivars were higher than their parents in leaf total nitrogen, free amino acid and soluble protein content, greater in NR and GPT activity and lower in protease activity, during peak square and boll developing period. Therefore, the nitrogen metabolism was stronger in transgenic Bt cotton cultivars. The authors suggested that keeping the balance between nitrogen and carbohydrate metabolism was important for high yield. The changed balance in Bt cotton cultivars might result dominating vegetative growth and reducing lint yield under the high nitrogen soil [31].

It is an established fact that for higher protein synthesis, greater N has to be taken up by the Bt plant. Bruns and Abel (2003) [32] found that σ-endotoxin concentration was positively associated with whole-plant N concentration. Bt hybrids required more N to synthesize Bt protein, therefore they take more N from the soil than that taken up by their non-Bt counterparts [33]. Tian et al. (2000) [34] reported that CCRI30 of Bt cotton cultivar had smaller boll than its parent because of the poor supply of assimilates. Chen et al. (2004) [30] showed that the total nitrogen reduced sharply in the bolls of the Bt cotton cultivars, this result implies that the reduction of total nitrogen caused decrease of nitrogen metabolism and limited the boll development.

1.2. Environmental Release of BT Insecticidal Proteins (BIPs)

BIPs expression level and interspecific and intraspecific variations between species and cultivars of Bt transgenic crops were widely investigated. In addition, a large amount of Bt toxins were released through biomass deposition, root secretion and so on, which may result in Bt toxin accumulation in soil systems. Moreover, large variability of soil Bt toxin level in soil may occur since the differences in tissue expression of Bt insecticidal proteins of BTCs.

Bt toxin is introduced into soil primarily through the degradation of plant biomass remnants, root exudates [35] and probably pollen during tasselling [15]. There are three major fates of BIPs into environment, including soil ecosystems, surface water ecosystems and animal digestive systems. Sims and Ream (1997) [36] estimated that a potential maximum of 1.6 mg Cry2A protein kg^{-1} soil would result from the incorporation of Bt cotton residues. Cry3Bb1 expression was highest in young leaves (228.4 μg/g dw and 35.5 μg/g fw) and lowest in pollen (3.8 μg/g fw). On the basis of the total corn biomass produced on 1 hectare, it was estimated that up to 905 g of Cry3Bb1 is produced per hectare Bt corn MON88017 [37]. Root exudation, root and shoot residues, pollen to surface water during tasseling, crop residue via runoff into aquatic systems, migration with eroded sediment by wind, rain or snow melt[38].

1.4. Rhizosphere Characteristics Changes

Root expression levels of modified genes, root exudation, soil persistence of BIPs, and the unintentional rhizosphere changes of transgenic plants are the main aspects that have been acknowledged. Expression and release of Bt toxins from GMPs is the foremost and original changes compared to their parental lines. Biological activity and soil persistence of Bt toxins are crucial parameters for assessing potential risks and environmental fates in soil ecosystems [35]. A large variability in Bt toxin expression, rhizosphere release and soil persistence between BIPs species and cultivars has been found. In general, the expression level of transferred genes in leaves is higher than that in roots for the same BTPs cultivar. For example, the Cry1Ab protein contents in the shoot and root of KMD were 3.23-8.22 and 0.68-0.89 mg/g (fresh weight), respectively [39]. For MON863 maize, Bt crystal protein expression in root ranged from 25 mg/g to 56 mg/g, but it was 65 mg/g to 93 mg/g in leaf. GK12 and ZK30, two Bt transgenic cottons, showed 21.2-44.7 ng/g in root tissue, which was lower than that of the shoot at 103.5-134.1 ng/g[40]. A higher level of CryIAb toxin expression was reported for Bt 176 than Bt 11 (80.63 and <0.55 μg CryIAb/g protein, respectively) [41] (Turrini et al. 2004a). High levels of proteins expressed in roots of GMPs may pose potential damage to the growth and development of the intraradical part of AMF. Dong et al. (2006) [42] showed new evidence that the Bt crystal proteins were mostly located in the cytoplasm and intercellular space of Bt transgenic cotton GK-12 cotyledon mesophyll cells, which provides indirectly evidence for the above intraradical damage mechanism. Knox et al. (2007) [43] used ELISA to quantify *in vitro* expression of the Cry1Ac and Cry2Ab proteins in mucilage, root border cells and root tips in five transgenic cultivars of cotton compared to conventional cultivar Sicot 189. Expression of Cry proteins in roots and border cells of the transgenic cotton cultivars was constitutive and at detectable levels, with Cry1Ac and Cry2Ab protein expression ranging from <20 ppb to >100 pbb.

In rhizosphere soil of Bt transgenic cotton (Xincai1), a maximal 56.14mg/kg Bt toxins (Cry1Ac) were detected [44]. During the entire growth period, Bt cotton varieties (NuCOTN99B and SGK321) excreted Bt toxins into the rhizosphere at a rate of up to 0.3 mg/g, but they varied temporally and disappeared to an undetectable level [45]. Rui et al. (2006) [46] found that, as the transgenic CpTI/Bt cotton cultivar SGK321 developed, the residue of CpTI in the rhizosphere increased, reaching a peak at the topping stage. After this stage, the residue began to decrease and was nil the following year. The rhizoshere contents of the Cry1Ab toxin proteins of Bt transgenic rice via root exudation were 1.66-48.02ng/(individual/d) [47、48]. Sun et al. (2007) [40] incubated soil amended with the leaves and stems of two transgenic cotton varieties. The results showed that, after Bt cotton tissue amendment, the Bt toxin was introduced into soil upon decomposition; about 50% of the introduced Bt toxin persisted in soil for at least 56 days. Straw degradation, root residue decomposition and root exudation can increase Bt toxins to a high level at different times. The persistence and activity of Bt toxins is closely related to soil pH, clay content, soil water content and so on, and the Bt toxins could maintain their activity for 180-234 days

[49、50、51、52]. Furthermore, in comparison to conventional lines, GMPs produced great changes in certain properties, such as lignin content, which confirmed their experimental impacts on mycorrhizal colonization, and maybe AMF community and biodiversity. To date, lignin content, root exudates changes and photosynthetic changes have been confirmed [45、53]. Moreover, some development stages (germination, intraradical development) of AMF depend highly on these physiological processes.

1.5. Changes in Chemicals Input and Agronomic Management

Updated global impact assessments for biotech crops indicate that for the period 1996 to 2008 economic gains of 51.9 billion US$ were generated from two sources, firstly, reduced production costs (50%), and secondly, substantial yield gains (50%) of 167 million tons. The latter would have required 62.6 million additional hectares had biotech crops not been deployed. Hence, biotech crops are an important land saving technology. During the same period, 1996 to 2008, pesticide reduction was estimated at 356 million kg of active ingredient, a saving of 8.4% in pesticides. In 2008 alone, the CO_2 savings from biotech crops through sequestration was 14.4 billion kg of CO_2 equivalent to removing 7 million cars from the road [54]. Requirement alterations in pesticides, fertilizers and water of BTCs brought a series of changes in agronomic management, such as fertilization rate and fertilizer species, pesticide usage and species, degree of human intervention and so on. With respect to AMF, the subsequent key changes focused on the rhizosphere environment and root susceptibility of Bt transgenic crops. Above changes happened in BTC-soil ecosystems will affect the formation and symbiosis of AMF significantly. Some changes are beneficial, but some are detrimental. Nowadays, there lacks evidences from the field, especially from long-term experimental investigations. For us, it is necessary to analyze the components incurred by introduction of BTCs at the level of agro-ecosystems, and to make a long-term field experiments to obtain the terminal results about interactions between BTCs and AMF.

After insertion of exotic antimicrobial or insecticidal genes into these plants, some changes occurred including expression and exudation of anti-fungal or insecticidal proteins and derivative changes in chemical composition. In addition, introduction of GMPs does induce some changes in agro-ecosystems, including chemical input (fungicides, pesticides and even fertilizer), biochemical processes in rhizosphere ecosystem and so on, which may affect AMF growth, development and even community structure. Life cycle of AMF consists of several processes, involving two part development of AMF (intraradical and extraradical biomass formation). All processes of AMF symbiotic development are subjected to changes in environmental factors, including germination of propagules, pre-symbiotic hyphal growth, colonization, intraradical and extraradical fungal development (arbuscular, hyphae and vesicular), and sporulation. A budget analysis by Pray et al. (2001) [55]shows that while there is no significant difference in fertilizer and machinery uses between Bt and non-Bt cotton production, significant reductions were recorded in pesticide and labour use (labour used to spray pesticides). Huang et al. (2002b) [56] revealed that Bt cotton adopters spray 67% fewer times and reduce pesticide expenditures by 82%. Because the reduction of the farmers pesticide-spraying frequency (from an average of 20 times during one crop season to eight times), Bt cotton technology is also considered a labour-saving technology.

2. Arbuscular Mycorrhizal Fungi: Vital Roles and Sensitivity in Ecosystems

2.1. Key Roles of AMF in Ecosystems

AMF are ubiquitous soil microorganisms, existing in almost all kinds of soil ecosystems [57], and forming symbioses with more than 80% of plant species [58]. Based on their obligate biotrophy, AMF have to form symbioses to exchange carbohydrates necessary for completion their life cycle. Arbuscular mycorrhizal association appears to have evolved as a survival mechanism for both the fungi and higher plants, allowing each to survive natural stresses such as soil impoverishment, diseases, pollution, drought, and low temperatures. It is well known that AMF play important roles in natural and agro-ecosystems based on their multi-functional mechanisms[59]. In terms of benefits to mycorrhizal plants, AMF are crucial for plant nutrition[60、61], disease tolerance[62], and productivity and diversity of the plant community[63]. Additionally, the microbial communities in soil ecosystems responsible for soil structure and nutrient cycling [64] are strongly dominated by mycorrhizal fungi. Currently, the most well-known and important benefit to host plants of arbuscular mycorrhiza concentrated on improvement in mineral nutrition status, which mainly depends on the nutrient uptake by the extended hyphae of AMF beyond the depleted zone in the soil. In essence, both individual and community AMF are fundamental for many of the biochemical processes that occur in soil, such as transformation and transfer of P and N compounds. Furthermore, hyphal bridge formation between two mycorrhizal plants allows substances (P and N etc.) exchange, which establishes an underground hyphal web ensuring the stability of the ecosystem. AMF have great potential use in low-input agriculture and organic agriculture based on their symbiotic functions, ubiquitous distribution and low specificity for host plant species. Numerous studies have been conducted to isolate and select efficient AMF for use in agriculture [65]. In low-input agricultural systems, AMF can improve the fitness of host plants by absorbing more soil nutrients and preventing from diseases, which substantially decreases the need for chemical fertilizer and insecticides & fungicide. In summary, maintaining a suitable community of AMF allows for more sustainable development of agro-ecosystems.

2.2. Sensitivity to Environmental Alterations of AMF

It is well documented that AMF are sensitive to four aspects of environmental alterations in agro-ecosystems, including plant (species, cultivars and diversity) [66], soil (soil type and fertility level, especially phosphorus) [67], climate (temperature, CO_2 concentration etc.) [68] and agronomic practices (tillage, fertilization, irrigation etc.) [69、70、71、72]. The sensitive response of AMF to changes of above four aspects presented variation at root colonization, fungal biomass and sporulation, which, in turn, determined the magnitude of AMF propagule reserve. Host plant species is the primary factor that determines colonization level, spore production number, community structure [66], and diversity of AMF and plant diversity in a special site correlated closely [63]. Plants themselves can impact AMF in multiple ways, including changes in root exudates (signal substances determine the recognition colonization), constitution of straws/residues, and expression of anti-fungal/Bt proteins. The first three species of GMPs, soybeans, maize, and cotton, are heavily mycorrhizal-dependant species, obtaining numerous benefits from symbiont AMF during symbiosis. However, after insertion of exotic antimicrobial or insecticidal genes into these plants, some changes occurred including expression and exudation of anti-fungal or insecticidal proteins and derivative changes in chemical composition. In addition, introduction of GMPs induces input changes in agro-ecosystems, including chemical input (fungicides, pesticides and even fertilizer), biochemical processes in rhizosphere ecosystem and so on, which may affect AMF growth, development and even community structure. Life cycle of AMF consists of several processes, involving two part development of AMF (intraradical and extraradical biomass formation). All processes of AMF symbiotic development are subjected to changes in environmental factors, including germination of propagules, pre-symbiotic hyphal growth, colonization, intraradical and extraradical fungal development (arbuscular, hyphae and vesicular), and sporulation. With respect to the above considerations, many documents have assumed that AMF are a sensitive non-target microbial community that should be monitored in ecological risks assessments of the released GMPs due to the fact that AMF symbiotic development is strongly influenced by agronomic practices, soil fertility and plant species and cultivars [73、74、75].

3. Impact Mechanisms of BTCs on AMF and the Ecological Implications

3.1. Impact Mechanisms of BTCs on AMF: A Hypothesis

Generally, symbiotic AMF consist of an internal phase inside the root and an external phase, which forms an extensive network within the plant-soil system. Spore, hyphae and infected root segments are effective propagules for the survival of AMF. The entire environmental conditions for AMF life changed when the conventional lines were replaced by GMPs transgenic ones. Taking BTPs for example, in the rhizosphere, Bt toxins might accumulate to a high level that directly impacts extraradical mycelium growth, development, and sporulation. Moreover, Bt toxins are constitutively expressed in the roots at a high level (at least for some cultivars), to which the important organs (e.g., the arbuscular and hyphae of AMF) are exposed, which may directly influence their intraradical development. In summary, AMF are thought to be exposed to Bt toxins both for their intraradical and extraradical tissues. Furthermore, Bt toxin exposure will be experienced throughout the life cycle of AMF and even longer for continuity of Bt exudation in entire growth period [45、46]. Therefore, BTPs may impact AMF development over their whole life cycle by Bt exudation [26]. We suggest that there is a temporal-spatial relevance between distribution of Bt proteins and fungal development of AMF. Temporally, BTPs express and release Bt proteins accompany entire stages of the AMF symbiotic growth & development processes of AMF. Spatially, extraradical and intraradical parts of AMF are both exposed to Bt toxins expressed in the roots or exuded into the rhizosphere. Bt toxins may inhibit the specialized structure formation of AMF, which has a negative effect on normal functioning. In the past decade, few studies have examined the impact of BTPs on AMF. Here, we presumed that AMF are more susceptible to GMPs release than other soil microbial communities, in particular to BTPs. Above hypothesis based on morphological and anatomical characteristics of symbiotic AMF is adequate for all trait GMPs, which differ in transferred genes, and their expression and release.

3.2. Impacts of BTCs on AMF

BTPs are the most important GMPs today, and they are modified to express the Bt insecticidal proteins (Cry1Ab, Cry1Ac etc.) from a subspecies of the bacterium Bacillus thuringiensis (Bt) in order to kill lepidopteran pests that feed on cotton, rice, and maize and coleopteran pests that feed on potato [53]. To date, some published data have revealed the significant negative effects of BTPs on root colonization and the development of AMF. Most results focused on Bt transgenic maize [76、77] and rice [78、79]. Turrini et al. (2004a)[41]provided evidence of the effects of Bt maizes on pre-symbiotic hyphal growth and colonization. Bt176 (with Cry1Ab gene) maize significantly reduced pre-symbiotic hyphal growth and impeded the development of appressoria. Many appressoria that formed on Bt 176 maize roots were not able to develop infection units since the infection pegs became sepatate and devoid of proplasm. The data showed that there were significant differences between Bt maize line Bt 176 and Bt 11 in exerting negative effects, which might relate to the level of Cry1Ab toxin expression (Bt 176 and Bt 11 express 80.63 and <0.55μg/g protein, respectively). Moreover, Bt maize residues, plowed under at harvest and mixed with soil for up to four months still affected mycorrhizal establishment by indigenous AMF [76]. Ren (2006) [77] and Yang et al. (2002) [78] found that root colonization of Bt transgenic rice and spore density in rhizosphere soil were lower than those of the parental lines. On the contrary, there were also some data presented no impacts of BTPs on AMF [79、80]. de Vaufleury et al.(2007) [80] found that the tested the Bt-maize (MEB307 expressing the insecticidal Cry1Ab protein) and the near-isogenic non-Bt variety (Monumental) gave similar mycorrhizal colonization of roots (frequency of mycorrhizal roots was 88.7% and 83.3% respectively). In another experiment, the observed pattern of colonization was virtually identical among both conventional and GM cultivars of cotton (Cry1Ac and Cry2Ab), clearly indicating that colonization by AM fungi were not affected by the expressed transgenic traits[79]. The above documents gave us fragmentary information regarding how BTPs affected colonization and symbiotic development of AMF, and there were significant differences in mycorrhization between Bt crop lines. However, these studies only took snapshots of the effects of BTPs on AMF, and many questions still remain. No study has yet provided evidence for effects of BTPs on other stages of the life cycle of AMF except colonization, especially at the top stage of Bt toxin expression of BTPs, which coincides with the rapid period for AMF morphological establishment. This is an urgent issue, especially because more and more Bt transgenic varieties are being bred and released into agro-ecosystems. We presume that, if AMF were inhibited or prevented from colonizing BTPs, even in only a few cases, this would deleteriously affect the AMF community as a whole. However, investigation of the responses of AMF to BTPs has been lacking up to this point.

Unfortunately, the ultimate outcome of these practices may result in the disappearance of some AMF unable to adequately adapt to the increased stress caused by BTPs.

3.3. Ecological Implications of Impacts of Gmps on AMF

As early as 1993, Miller predicted that the release of GMPs would cause ecological risks on soil microbial communities, particularly AMF. One ecological consequence of reducing mycorrhizal colonization was a decrease in the soil's mycorrhizal propagule reserve, which diminished the next crop's production, especially under low-input cropping practices [81]. A further consequence with both ecological and evolutionary consequences was the escape of the transgene for improved pathogen resistance into wild populations. Gene transfer would decrease the affinity of GMPs and their relatives with AMF, which might further aggravate the ecological risks of GMPs on AMF. Today, we do not know whether GMPs have significant effects on AMF formation and functions in ecological systems [82]. However, we have surprisingly little evidences of environmental risks to AMF derived from the current use of GMPs. Much less, some studies gave no effects of GMPs on AMF [79、80]. Our understanding of the interactions between GMPs and AMF are limited and rudimentary. GMPs determine the release and exudation of inserted genetic products (e.g. Bt toxins), then the structure and function of soil microbial communities, and finally the activity of soil enzymes and related nutrient cycling and transformation processes. There is a strong case to be made for reassessing the environmental risks using a small number of large-scale, appropriately powered field studies on a trait-by-trait basis [83]. The cumulative effects of long-term GMPs planting on AMF are important topics that deserve more attention.

Many studies about the impacts of GMPs on soil ecosystems (population and diversity of soil microbial community, activity of soil enzymes, AMF colonization, etc.) have been published and represent two conclusions: significant effect and no effect. Related literatures were cited in Bruinsma et al. (2003) [10]. Simultaneously, numerous recent studies have presented ambivalent data about the ecological risks of GMPs on soil ecosystems. What is the cause of that? In part, the complexity of ecosystems presents a considerable challenge for experiments to assess the risks of GMPs, that varies spatially, temporally, and according to trait and cultivars [9]. In fact, AMF may play important roles in interactions between GMPs and soil microorganisms. In natural ecosystems, AMF have important link functions between GMPs and soil ecosystems (the rhizosphere in particular). Some data have given us some strong indication that they may be harmed severely after GMPs are released without limits based on some of the mechanisms mentioned in this review. In light of the roles of AMF in agro-ecosystems and their sensitivity to environmental changes, it is suggested that AMF should bc fully considered in ecological risk assessments of GMPs [26]. AMF may alleviate injuries caused by Bt toxins in soil through an accelerating degradation process or physical shield. The detrimental effects of GMPs on AMF are certain to impact the structure and function of soil microbial communities, which may be represented by changes in the activities of soil enzymes since they are the most sensitive biochemical index for measuring changes in soil ecosystems. Consequently, nutrient transformation and cycling will be influenced substantially. Surely, Bt toxins may directly affect soil microbial communities. In addition, the presence of AMF may affect the degradation rate of Bt toxins in rhizosphere and mycorrhizosphere soil to some extent, despite the opacity of the mechanisms discussed above. Taking the above roles of symbiotic mechanisms in the rhizosphere into consideration, symbiotic mechanisms may respond significantly to BTPs both directly and indirectly, which then generates feedback from other rhizosphere microorganisms. That is to say, BTPs may significantly affect general microorganisms, but no similar data have been observed for the effects of symbiotic mechanisms.

4. Conclusions and Prospects

After insertion of exotic antimicrobial or insecticidal genes into these plants, some changes occurred including expression and exudation of anti-fungal or insecticidal proteins and derivative changes in chemical composition. In addition, introduction of GMPs induces input changes in agro-ecosystems, including chemical input (fungicides, pesticides and even fertilizer), biochemical processes in rhizosphere ecosystem and so on, which affect AMF growth, development and even community structure. However, the impacts and relevant mechanisms of GMPs on AMF are limited and unclear because the complexity of interactions between GMPs and AMF in agro-ecosystems.

There is combination complexity with regard to ecological regions, crop species, varieties and Bt genes transferred. The heterogeneity of soil ecosystems, wide variety of soil microbial communities and complex components make it difficult to understand overall the changes in biochemical processes caused by GMPs. In soil ecosystems, the components are complex and dynamic, including physical, chemical and biochemical factors and processes. Many factors in soil ecosystems are interdependent and interactive. Therefore, more multi-site experiments should be conducted in order to further explore the effects of BTPs on soil ecosystems and their interaction mechanisms. Furthermore, the current results were conducted almost entirely at a special site during a short period (one growth season), so long-term monitor experiments at different sites covering all GM plants are seriously lacking. Today, knowledge about the three above-mentioned issues is surprisingly limited. More experiments are needed to systematically examine the relationship among Bt toxins, soil microorganisms and soil biochemical properties. In order to accelerate research progress, three measures should be taken into consideration. First, a series of proper model BTPs should be selected in terms of Bt toxin expression level. Second, BTPs may have a greater hazard potential than other GMPs such as antimicrobial species, which should be highlighted in future risk assessments of GMPs. Third, more multi-site field trials should be conducted. Interactions between GMPs and AMF should be paid mire attention on. Anyway, diverse soil microbial communities, especially beneficial microorganisms, are fundamental for the health and stability of agro-ecosystems.

Acknowledgments

I want to thank the financial support of the Open Fund of Key Laboratory of Crop Nutrition and Fertilization (2009), Ministry of Agriculture.

REFERENCES

[1] James, C. *Global Status of Commercialized Biotech/GM Crops*: 2009. ISAAA Brief No. 41. 2009. ISAAA: Ithaca, NY.

[2] Wu, KM; Lu, YH; Feng, HQ; Jiang, YY; Zhao, JZ. Suppression of cotton bollworm in multiple crops in China in areas with Bt toxin-containing cotton. *Science*, 2008, 321,1676-1678.

[3] James, C. *Global status of commercialized Biotech/GM crops*: 2008, ISAAA Briefs No. 39. 2008. ISAAA, Ithaca, NY.

[4] Huang, JK; Hu, FR;Pray, C; Qiao, FB; Rozelle, S. Biotechnology as an alternative to chemical pesticides: a case study of Bt cotton in China. *Agricultural Economics*, 2003, 29, 55-67.

[5] Huang, JK; Rozelle, S; Pray, C; Wang QF. Plant biotechnology in China. *Nature*, 2002a, 295,674-676.

[6] Brookes, G.; Barfoot, P. Global impact of biotech crops: socio-economic and environmental effects in the first ten years of commercial use. *AgBioForum*, 2006, 3,139-151.

[7] Ferry, N; Edwards, MG; Gatehouse, J; Capell, T; Christou, P; Gatehouse AMR. Transgenic plants for insect pest control: a forward looking scientific perspective. *Transgenic Research*, 2006, 15, 13-19.

[8] Liu, B; Zeng, Q Yan, F.M Xu, HG; Xu CR. Effects of transgenic plants on soil microorganisms. *Plant and Soil,* 2005, 271, 1-13.

[9] Wolfenbarger, L.L.and P.R.Phifer. (2000) The ecological risks and beneficial engineered plants. *Science* 290: 2088-2093.

[10] Bruinsma, M; Kowalchuk, GA; van Veen, JA. Effects of genetically modified plants on microbial communities and processes in soil. *Biol. Fertil. Soils*, 2003,37, 329–337.

[11] Hails, RS. Genetically modified plants-the debate continues. *Trends in Ecology and Evolution*, 2000, 15, 14-18.

[12] Snow, A; Palma PM. Commercialization of transgenic plants: potential ecological risks. *BioScience*, 1997, 47, 86–96.

[13] Tabashnik, BE. Evolution of resistance to Bacillus thuringiensis. *Annual Review of Entomology*, 1994,39, 47–49.

[14] Hilbeck, A; Moar,,W J; Pusztai-Carey, M, Filippini, A; Bigler F. Toxicity of *Bacillus thuringensis* Cry1Ab toxin to the predator Chrysoperla carnea. *Environmental Entomology*,1998, 27, 1255–1263.

[15] Losey, JE; Rayor, LS;.Carter ME Transgenic pollen harms monarch larvae. *Nature*, 1999, 399,214.

[16] Watrud, LS; Seidler, RJ; Huang, PM; Adriano, DC; Logan, TJ; Checkai. RT. *Non-target ecological effects of plant, microbial and chemical introductions to terrestrial* system. In Soil Chemistry and Ecosystem Health Madison, 1998. Wisconsin: Soil Science Society of America. pp. 313-340.

[17] Angle, JS. Release of transgenic plants: biodiversity and population-level considerations. *Molecular Ecology*, 1994, 3,45–50.

[18] Jepson, PC; Croft, BA; Pratt GE. Test systems to determine the ecological risks posed by toxin release from *Bacillus thuringiensis* genes in crop plants. *Molecular Ecology*,1994,3, 81-89.
[19] Morra, MJ. Assessing the impact of transgenic plant products on soil organisms. *Mol. Ecol.*, 1994, 3,53-55.
[20] Saxena, D; Stotzky G.. Fate and effects in soil of insecticidal toxins from *Bacillus thuringiensis* in transgenic plants. In Collection of biosafety reviews, vol. 1. *International Centre for Genetic Engineering and Biotechnology*, 2003. Trieste, pp 7–83.
[21] Dunfield, KE; Germida JJ. Seasonal changes in the rhizosphere microbial communities associated with field-grown genetically modified canola (*Brassica napus*). *Appl. Environ. Microbiol.* ,2003, 69, 7310-7318.
[22] Motavalli, PP; Kremer, RJ; Fang, M; Means NE.. Impact of genetically modified crops and their management on soil microbially mediated plant nutrient transformations. *J. Environ. Quality*, 2004, 33, 816-814.
[23] Velkov, V; Medvinsky, A; Sokolov, M; Marchenko A. Will transgenic plants adversely affect the environment? *J. Biosci.*, 2005,30, 515-548.
[24] Giovannetti, M; Sbrana, C; Turrini, A. The impact of genetically modified crops on soil microbial communities. *Biology Forum*, 2005, 98, 393-418.
[25] Liu, WK. Effects of Bt transgenic crops on soil ecosystems: a review of a ten-year research in China. *Frontiers of Agriculture in China*, 2009, 3, 190-198.
[26] Liu, WK; Du LF. Interactions between Bt transgenic crops and arbuscular mycorrhizal fungi: a new urgent soil ecology issue in agro-ecosystems. *Acta Agriculturae Scandinavica Section B-Soil and Plant Science*,2008, 58, 187-192.
[27] Wang, YQ; Pan, XB; Cui, XW; Pan, SM. Research on the physiological traits of Bt transgenic cotton R93-4. *China Cottons*, 1998, 25, 14-16.
[28] Breen, P.J; Flynn, R; Ellers, KC; French, C; Dugger, P. *Effect of nitrogen and vegetative growth on plant resistance to bollworm, Helicoverpa zea, in selected Bt cotton varieties*. In: Richter, D. (Ed.), 1999. American Proceedings Beltwide Cotton Conferences, vol. 2. pp. 1234–1236.
[29] Dong, ZQ; He, ZP; Zhai, XJ. The metabolic characters of nitrogen in leaves of Bt transgenic cotton Nucoton 333B and its regulation. *Acta Gossypii Sin.*, 2000, 12, 113–117.
[30] Chen, DH; Ye, G.Y; Yang, CQ; Chen, Y; Wu, YK. Effect after introducing *Bacillus thuringiensis* gene on nitrogen metabolism in cotton. *Field Crops Res*, 2004, 87, 235–244.
[31] Chen, DH; He, ZP; Wu, YK. Study on the relationship between endogenous hormone of leaf and accumulation of nitrogen, phosphorus and potassium of plant as well as the cause of boll weight raised in high yield cotton. *Acta Agron. Sinca*, 2000, 6, 659–665.
[32] Bruns, AH; Abel, CA. Nitrogen fertility effects on Bt δ-endotoxin and nitrogen concentration of maize during early growth. *Agronomy Journal*, 2003, 95,207-211.
[33] Subedi, KD; Ma, BL. Dry matter and nitrogen partitioning patterns in Bt and non-Bt near-isoline maize hybrids. *Crop Science*, 207, 47:1186-1192.
[34] Tian, XL; He, ZP; Wang, BM. Boll development and yield components of Bt cotton CCRI30 influenced by flowering date. *Cotton Sci.,* 2000,6, 306-309.

[35] Icoz, I; Stotzky, G. Fate and effects of insect-resistant Bt crops in soil ecosystems. *Soil Biol Biochem*, 2008, 40:559–586
[36] Sims, SR; Ream, JE. Soil inactivation of the insecticidal protein within transgenic cotton tissue: laboratory microcosms and field studies. *Journal of Agricultural and Food Chemistry*, 1997, 45, 1502-1505.
[37] Nguyen, HT; Jehle, JA. Expression of Cry3Bb1 in transgenic corn MON88017. *J. Agric. Food Chem.*, 2009, 57, 9990-9996.
[38] Prihoda, KR; Coats, JR. Aquatic fate and effects of *Bacillus thuringiensis* cry3bb1 protein: toward risk assessment. *Environmental Toxicology and Chemistry*, 2008, 27, 4:793–798.
[39] Wang, H; Ye, Q; Wang, W; Wu, L; Wu, W. Cry1Ab protein from Bt transgenic rice does not residue in rhizosphere soil. *Environmental Pollution*, 2006, 143, 449–455.
[40] Sun, CX; Chen, LJ; Wu, ZJ; Zhou, LK. Soil persistence of *Bacillus thuringiensis* (Bt) toxin from transgenic Bt cotton tissues and its effect on soil enzyme activities. *Biology and Fertility of Soils*, 2007, 43,617–620.
[41] Turrini, A; Sbrana, C; Nuti, MP. Development of a model system to assess the impact of genetically modified corn and aubergine plants on arbuscular mycorrhizal fungi. *Plant and Soil* ,2004, 266, 69-75.
[42] Dong, ZQ; Zhao, M; Shu, WH; Zhang,BM; Hao, HJ.. (2006) The subcellular localization of the Bt crystal protein in transgenic Bt cotton cell. *Acta Agronomica Sinica* ,32,1924-1926.
[43] Knox, OGG; Gupta, VVSR; Nehl, DB; Stiller WN. Constitutive expression of cry proteins in roots and border cells of transgenic cotton. *Euphytica*, 2007, 154, 83-90.
[44] Zhang, MD; Yang WD. Effect of transgenic Bt cotton planting on biological characteristics and nutrient content in rhizosphere soil. *Plant Nutrition and Fertilizer Science*, 2008, 14,162-168.
[45] Rui, YK; Yi, GX; Zhao, J; Wang, BM; Li, ZH; Zhai, ZX; He, ZP;. Li. QX Changes of Bt Toxin in Rhizosphere of Transgenic Bt-Cotton and Its Influence on Soil Functional Bacteria. *World Journal of Microbiology and Biotechnology*, 2005, 21,1279–1284.
[46] Rui, YK; Yi, G.X; Guo, J; Guo, X; Luo, YB; Wang, BM;.Li ZH. Transgenic cotton could safely be grown since CpTI toxin rapidly degrades in the rhizosphere soil. *Acta Agriculturae Scandinavica Section B-Soil and Plant Science*, 2006, 57,122-125.
[47] Wu, WX; Ye, QF; Min H. Effect of straws from Bt-transgenic rice on selected biological activities in water-flooded soil. *Eur. J. Soil Biol.*, 2004a, 40, 15–22.
[48] Wu, WX; Ye, QF; Min, H; Duan,XJ; Jin WM. Bt transgenic rice straw affects the culturable microbiota and dehydrogenase and phosphatase activities in a flooded paddy soil. *Soil Biol. Biochem.*,2004b,36, 289-295.
[49] Crecchio, C; Stotzky G.. Insecticidal activity and biodegradation of the toxin from *Bacillus thuringiensis subsp. Kurstaki* and *subsp. Tenebrionis* adsorbed and bound on pure and soil calys. *Appl. Environ. Microbiol.* 1998, 61,1786-1790.
[50] Tapp, H; Stotzky G.. 1Insecticidal activity of the toxin from *Bacillus thuringiensis subsp. Kurstaki* and *subsp. Tenebrionis* bound to humic acids from soil. *Soil Biol. Biochem.*, 1995, 30,463-470.
[51] Stotzky, G.. Persistence and biological activity in soil of insecticidal proteins from *Bacillus thuringiensis* and of bacterial DNA bound on clays and humic acids. *J. Environ. Qual.*, 2000, 29,691-705.

[52] Saxena, D; Stotzky G.. Bt toxin uptake from soil by plants. *Nature Biotechnology*,2001, 19,199.

[53] Flores, S., Saxena, D. and G. Stotzky. (2005). Transgenic Bt plants decompose less in soil than non-Bt plants. *Soil Biology & Biochemistry* 37,1073-1082.

[54] Brookes, G.; Barfoot, P. *GM crops: global socio-economic and environmental impacts* 1996-2008, 2010. PG Economics Ltd, Dorchester, UK.

[55] Pray, C; Ma, D; Huang, J; Qiao, F. Impact of Bt cotton in China. *World Development*, 2001, 29,813-825.

[56] Huang, JK; Hu, R; Pray, C; Rozelle, S; Qiao, F. Smallholders, transgenic varieties, and production efficiency: the case of cotton farmers in China. *Australian Journal of Agricultural and Resource Economics*, 2002b, 46(3):1-21.

[57] Abbott, LK; Robson AD. Factors influencing the occurrence of vesicular-arbuscular mycorrhizas. *Agriculture, Ecosystems and Environment*, 1991, 35,121-150.

[58] Smith, SE; Read DJ. *Mycorrhizal symbiosis*, 2nd ed. 1997. Academic Press, San Diego, CA.

[59] Newsham, KK; Fitter AH; Watkinson. AR Multifunctionality and biodiversity in arbuscular mycorrhizas. *Trends in Ecology Evolution*, 1995, 10,407-411.

[60] Li, XL; Geoege, E; Marschner H. Extension of the phosphorus depletion zone in VAM white clover in a calcareous soil. *Plant and soil* , 1991a, 136,41-48.

[61] Li, XL; George, E; Marschner H. Acquisition of phosphorus and copper by VAM hyphae and root to shoot transport in white clover. *Plant and Soil*, 1991b, 136, 49-57.

[62] Selosse, MA; Baudoin, E; Vanderkoornhuyse P. Symbiotic microorganisms, a key for ecological success and protection of plants. *Cr Acad Sci III Vie*, 2004,327, 639–648.

[63] Van der Heijden, MGA; Klironomos, JN; Ursic M. Mycorrhizal fungal diversity determines plant biodiversity, ecosystem variability and productivity. *Nature*, 1998, 396, 69-72.

[64] Rillig MC; Mummey DL. Mycorrhizas and soil structure. *New Phytol.* , 206, 171:41-53.

[65] Gai, JP; Christie, P; Feng, G; Li XL. Twenty years of research on community composition and species distribution of arbuscular mycorrhizal fungi in China: a review. *Mycorrhiza*, 2006,16,229-239.

[66] Liu, RJ; Wang FY. Selection of appropriate host plants used in trap culture of arbuscular mycorrhizal fungi. *Mycorrhiza*,. 2003, 13, 123-127.

[67] Liu, WK. N, P contribution and soil adaptability of four arbuscular mycorrhizal fungi. *Acta Agriculturae Scandinavica Section B-Soil and Plant Science*, 2008, 58, 285-288.

[68] Gavito, ME; Schweiger, P; Jakobsen I. P uptake arbuscular mycorrhizal hyphae: effect of soil temperature and atmospheric CO_2 enriched. *Global Change Biology*, 2003, 9,106-116.

[69] Johnson, NC;.Pfleger FL *Vesicular-arbuscular Mycorrhizae and cultural stresses*. In Mycorrhizae in Sustainable Agriculture. 1992. ASA Special Publication no. 54 Eds. GJ Bethenfalvay and RG Linderman. Madison, EI. pp 71–99.

[70] Helgason, T; Daniell, TJ; Husband, R; Fitter, AH; Young JPW. Ploughing up the wood-wide web. *Nature*, 1998, 394-431.

[71] Fontanet, X., Estaun, V., Camprubi, A. and C. Calvet. (1998) Fungicides added to potting substrate affect mycorrhizal symbiosis between a peach-almond rootstock and *Glomus sp. Hortscience* 33, 1217–1219.

[72] Giovannetti, M; Avio L. *Biotechnology of Arbuscular Mycorrhizas*. In Applied Mycology and Biotechnology. Vol.2. Agriculture and Food Production. Eds. GG Khachatourians and DK Arora. 2002. Elsevier Science, Amsterdam. pp 275–310.

[73] Ferreira, LHPL; Molina, JC; Brasil, C; Andrade G.. Evaluation of *Bacillus thuringiensis* bioinsecticidal protein effects on soil microorganisms. *Plant and Soil*, 2003, 256,161-168.

[74] O'Callaghan, M; Glare, TR; Burgess, EPJ; Malone LA. Effects of plants genetically modified for insect resistance on non-target organisms. *Annual Review of Entomology*, 2005, 50, 271-292.

[75] Liu B; Zeng Q; Yan FM; Xu HG; Xu CR.Effects of transgenic plants on soil microorganisms. *Plant and Soil*, 2005, 271,1-13

[76] Castaldini, M; Turrini, A; Sbrana, C; Benedetti, A; Marchionni, M; Fabiani, A; Landi, S; Santomassimo, F; Pietrangeli, R; Nuti, MP; Miclaus, N; Giovannetti M. Impact of Bt corn on rhizospheric and soil eubacterial communities and on beneficial mycorrhizal symbiosis in experimental microcosms. *Applied and Environmental Microbiology*, 2005, 71,6719-6729.

[77] Ren, X. Effect of Bt Transgenic Rice (KMD) on Soil Bacterial Community and Rhizosphere AM Fungi. 2006. Master degree thesis of Zhejiang University. HangZhou.

[78] Yang, YF; Yuan, HX; Liu, YL; Xu, XP; Li BJ. Research on root microorganism community of "RCH" transgenic rice. *Chin. J. Eco-Agr.*, 2002, 10, 29–31.

[79] Knox, OGG; Nehl DB; Mor T; Roberts G.N; Gupta VVSR. Genetically modified cotton has no effect on arbuscular mycorrhizal colonisation of roots. *Field Crops Research*, 2008, 109, 57-60.

[80] de Vaufleury, A; Kramarz, P.E; Binet, P; Cortet, J; Caul, S; Andersen, MN; Plumey, E; Coeurdassier, M; Krogh PH. Exposure and effects assessments of Bt-maize on non-target organisms (gastropods, microarthropods, mycorrhizal fungi) in microcosms. *Pedobiologia*, 2007,51,185-194.

[81] Miller, RM. Nontarget and ecological effects of transgenically altered disease resistance in crops-possible effects on the mycorrhizal symbiosis. *Mol. Ecol.*, 1993, 2,327-335.

[82] Liu WK. Do genetically modified plants impact arbuscular mycorrhizal fungi? *Ecotoxicology*, 2010, 19,229–238.

[83] Firbank, L; Lonsdal M;. Poppy. G. Reassessing the environmental risks of GM crops. *Nature Biotechnology*, 2005,23, 1475- 1476.

In: Genetically Engineered Crops
Editors: Lindsay M. Grover
ISBN: 978-1-61122-526-6

Chapter 8

PLANTS UNDER ATTACK: PHYTOALEXINS IN DEFENSE MECHANISM

V. Nagaveni, A. Mahadeva and Rohini Sreevathsa[*]
Department of Crop Physiology, University of Agricultural Sciences, GKVK, Bangalore, India

ABSTRACT

Plants are endowed with diverse mechanisms including both passive (preformed) and active (inducible) defense responses that protect them from microbial attack. Inducible plant defense responses include phytoalexins, lytic enzymes, oxidizing agents, cell wall lignification, PR proteins etc. Phytoalexins are a chemically heterogeneous group of low-molecular-weight secondary metabolites with antimicrobial properties that play a key role in defense system. The concept of phytoalexins in plants was first developed in 1940. It has taken more than 70 years to gain a basic perceptive of how a small part of nature's plant antimicrobial compound is manufactured. However, role of these compounds in defense has been revealed through several studies *viz.,* rate of phytoalexins synthesis in relation to cessation of pathogen development, quantification of phytoalexins at the infection site and relationship of pathogen virulence to the phytoalexin tolerance. The application of molecular and recombinant DNA technologies has led to significant advances in host-pathogen interactions and disease resistance mechanisms. In addition, latest successes in genetic engineering of plants afford new strategies for directly manipulating these interactions and enhancing disease resistance. In this review we discuss engineering novel phytoalexins in heterologous plants that provide a unique strategy to develop plant innate defense systems.

INTRODUCTION

Disease is caused whenever the vital functions of a plant are disrupted by a biotic stimulus. The biochemical dynamics of parasitism and pathogenesis are triggered and

* Corresponding Author E-mail: rohinisreevathsa@rediffmail.com, Telephone: 91-080-23636713 extn: 47

controlled by a series of interactions between the host and the pathogen. Plants have their own built-in defense mechanism(s) against pathogen(s). Once the pathogens are recognized by the host, a series of alarm signals are sent out to host cell proteins to produce substances inhibitory to the pathogen and to mobilize themselves or their products towards the point of cell attack by the pathogen. The defensive strategy of plants exists in two stages. The first is assumed to involve the rapid accumulation of phenols at the infection site, which functions to slow or even halt the growth of the pathogen and to allow for the activation of secondary strategies that would more thoroughly restrict the pathogen. The second is the activation of specific defense such as *de novo* synthesis of phytoalexins or other stress-related substances (Fig. 1) (Nicholson and Hammerschmidt, 1992; Agrios, 2004).

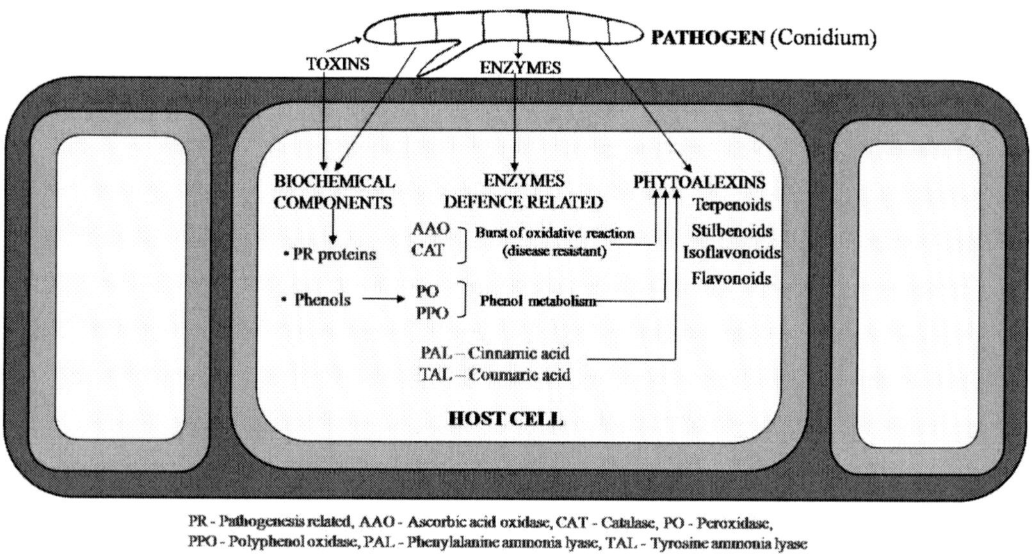

Figure 1. Schematic representation of host-pathogen interaction in synthesis of phytoalexins.

Phytoalexin synthesis is an important chemical barrier of post-infectional defense mechanism, which parallels the mechanism of defense through production of antibodies in the animal system. Phytoalexins are antimicrobial, low-molecular-weight secondary metabolites produced in plants in response to infection by pathogens.

The phytoalexin theory of disease resistance in plants was proposed by Muller and Borger (1940) as a working hypothesis to explain the dynamic nature and toxicological aspects of host-pathogen interactions. For the first time, he propounded the term 'Phytoalexin' (PA) (Phyton=plant ; alexin=warding off substance) to describe the substance that inhibited fungus development and formed when living plant tissues (potato tuber) were invaded by the parasite (*Phytophthora infestans*). After that, many experiments have been carried out in different crop plants belonging to different families. It has taken 70 years to gain a basic understanding of how a small part of nature's plant antimicrobial substance is manufactured. PA has been implicated as an important component of natural resistance.

ROLE OF PHYTOALEXINS IN DISEASE RESISTANCE

Cruickshank (1980) proposed seven primary conclusions of Muller and Borger (1940) which constituted the 'Phytoalexin theory'. They are listed below :-

1. A principle designated as 'Phytoalexin' (PA) that inhibits the development of the fungus in hypersensitive tissue, is formed or activated only when the host cells come in contact with the pathogen.
2. The defense reaction occurs only in living cells.
3. The inhibitory material is a chemical substance and may be regarded as the product of necrobiosis of the host cell.
4. The PAs are non-specific in their toxicity towards fungi.
5. The basis of differentiation between resistant and susceptible host lies in the speed of formation of PA.
6. The defense reaction is confined to the tissue colonized by the fungus and its immediate neighborhood.
7. The resistant state is not inherited. It is developed after the fungus has attempted infection. The sensitivity of the host cell, which determines the speed of the host reaction, is specific and genetically determined.

Many investigators have worked on PAs in many crop plants and suggested the possible role of PAs in disease resistance. They are very briefly mentioned below :-

- PAs are synthesized by both necrotic (dead) as well as living cells (Varns et al. 1971).
- Incompatible pathogens induce more PA production (Teasdale et al. 1974).
- PAs are highly inhibitory to spore germination of fungi (Skipp and Bailey, 1976).
- PAs accumulate very quickly at the site of infection following penetration by the microbe, in sufficiently high concentration and inhibit bacteria (Lyon and Wood, 1975), nematodes (Smith, 1976) as well as fungi (Bailey et al. 1980).
- PAs are produced more rapidly in resistant reactions (Mayama et al. 1982; O'neill and Mansfield, 1982).
- The term PA is usually restricted to antibiotic compounds which require *de novo* expression of enzymes involved in their biosynthetic pathways (Anderson, 1991).
- PAs are produced in response to infection by pathogenic fungi but are absent in healthy plants (Grayer and Kokubun, 2001).

Keen (1981) described several lines of evidence that support the role of PAs in disease resistance. They are documented below :-

- Localization and timing of PA accumulation in infected tissue in relation to pathogen development.
- Strong positive correlation of rapid PA production with incompatible interactions in gene-for-gene plant-pathogen systems.

* Association of rapid PA accumulation with resistance genes that condition rapid restriction of pathogen development.
* Use of metabolic inhibitors that enhance susceptibility and block PA production.
* A positive relationship between pathogen virulence and tolerance of PAs.
* An increase of plant tissue resistance by stimulation of PA production prior to inoculation.

Misaghi (1982) listed out some salient features of PAs. They are as follows :-

- PAs may accumulate not only in response to wounding and infection, but also in response to a large number of PA inducers viz., chemicals, physiological stimuli and microbial metabolites.
- A single PA may be responsible for resistance to all diseases in any particular plant.
- In the same tissue, one PA may accumulate fairly over a limited range and reduce thereafter.
- The production and activity of PA in the disease system is controlled by complex elicitor - repressor regulation, which is apparently different in different plants.

Table 1. Production of phytoalexins in plants in response to pathogen attack.

Host	Pathogen	Phytoalexin isolated	Class of Phytoalexin	References
Potato	*Erwinia carotovora, Erwniia carotovora var. Atroseptica*	Rishitin and Phytuberin	Terpenoids	Ghanekar et al. (1984) ; Lyon (1973)
Potato	*Phytophthora infestans*	Phytuberin, Rishitin, Lubimin, Solavetivone		Andreu *et al.* (2001)
Grapevine	*Botrytis cinerea or Plasmopara viticola*	Resveratrol (trans-3,4¢, 5-trihydroxystilbene)	Stilbenoids	Blaich and Bachmann (1980); Langcake (1981); Dercks and Creasy (1989)
Alfalfa	*Phoma medicaginis*	Resveratrol		Hipskind and Paiva (2000)
Tobacco	*Botrytis cinerea*	Resveratrol		Yu (2006)
Soybean	*Phytophthera megasperma* f sp. *glycinea*	Glyceollin	Isoflavonoids	Graham and Graham (1991)
Chickpea	*Ascochyta rabiei*	Isoflavones and Pterocarpan		Helmut and Barz (1987)
Pea	*Nectria haematococca*	Pterocarpans, Maackiain and Pisatin		Dixon and Steele (1999); Wasmann and Van Etten (1996); Enkerli et al. (1998)
Soybean	*Phytophthora sojae*	Isoflavone		Senthil et al. (2005)
Lotus	*Agrobacterium rhizogenes*	Vestitol (7,2'-dihydroxy-'methoxyisoflavan) Sativan (7-hydroxy-2',4'dimethoxyisoflavan)		Robbins et al. (1991)
Bean	*Fusarium solani* f. sp.*phaseoli*	Phaseollin Kievtone, phaseollinisoflavan Phaseollin, 2 methoxy-phaseollinisoflavan		Perrin (1964) Bailey and Burden (1973) Perrin et al. (1972) Van Etten (1976)
Bean	*Botrytis cinera*	Phaseollin, phaseollidin, phaseollinisoflavan, 6a-hydroxyphaseollin		Van Den Heuvel and Grootveld (1980)
Sugarcane	*Colletotrichum falcatum*	3-deoxyanthocyanidins	Flavonoids	Malathi et al. (2008)
Sorghum	*Colletotrichum sublineolum*			Snyder and Nicholson (1990); Lo et al. (1998); Du (2010)

PAs are only a component of the complex mechanisms for disease resistance in plants. Studies on PAs alone have contributed a great deal to plant biochemistry and molecular biology. PAs are low molecular mass secondary metabolites with antimicrobial activities produced in a broad range of plant species. These are produced in plants after pathogen attack and also under abiotic stresses. Production of these phytoalexins to develop fungus and bacterial resistant plants has been exploited in many plant species and few of them are listed in Table 1.

PAs are a chemically heterogeneous group of compounds of phenylpropanoid metabolism and the biosynthetic pathway of phenylpropanoid phytoalexins has been elucidated in several plants (Dixon, 1986). The products of the pathway are common precussors of many classes of natural plant products including flavonoids, isoflavonoids, stilbenes, lignins and other phenolics (Ebel, 1986) (Fig. 2).

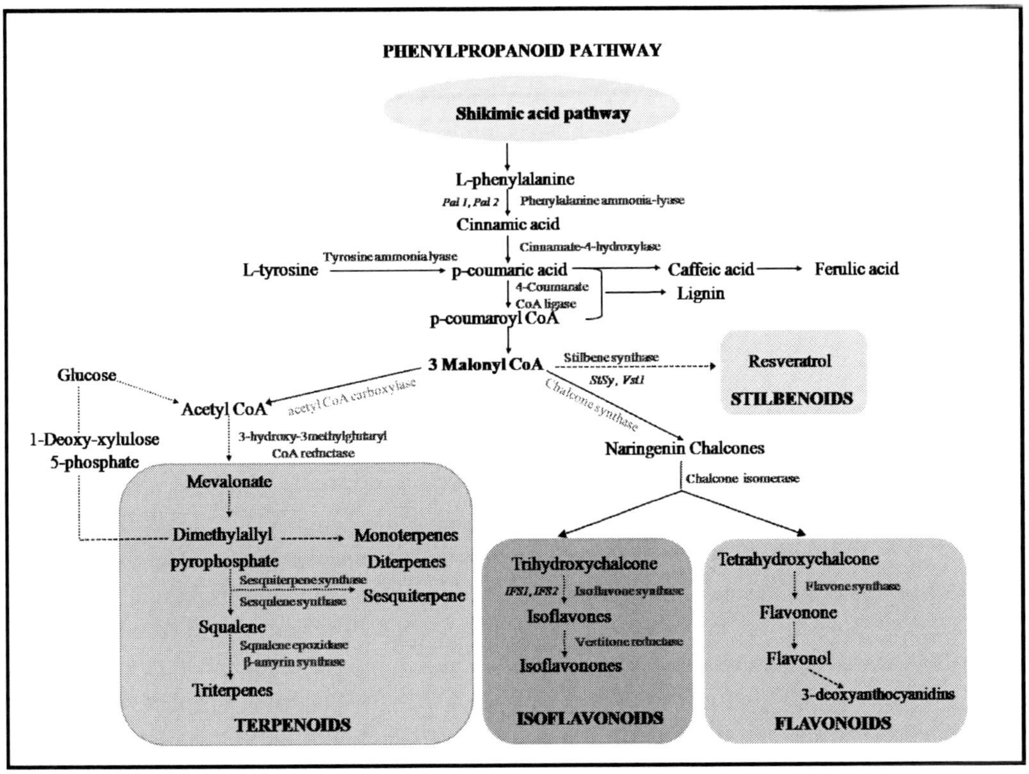

Figure 2. Synthesis of different classes of phytoalexins in phenylpropanoid pathway.

Phenylpropanoids belong to the largest group of secondary metabolites produced by plants, mainly, in response to biotic or abiotic stresses. Several phenylpropanoid compounds have been identified from various species, making them one of the largest natural product families in the world. Many of these compounds are essential for normal plant growth and development. They also play important roles in defense against plant pathogens.

Phenylpropanoids are plants phenols produced through shikimic acid pathway and are natural products which are essential for plants. They are derived from trans-cinnamic acid formed by deamination of L-Phe by L-Phe ammonia-lyase (PAL; EC 4.3.1.5). PAL is a tetrameric enzyme whose subunits are encoded by multigene families in most species studied

(Cramer, et al: 1989; Nagai, et al. 1994; Wanner, et al. 1995). PAL catalyses the first committed step of the core pathway of general phenylpropanoid metabolism. Branch pathways lead to the synthesis of compounds that have diverse defensive functions in plants such as cell wall strengthening and repair (lignin and suberin), antimicrobial activity (flavonoids, isoflavonoids, terpenoids and stilbenes phytoalexins). In tobacco (*Nicotiana tabacum), PAL* genes have a simple organization consisting of two families, each with two very closely related genes (Nagai et al. 1994; Pellegrini et al. 1994). Hydroxylation of cinnamic acid in the 4-position leads to 'p'-coumaric acid, which can be further modified into hydroxylated derivatives and 'p'-coumaric acid. 4-coumarate is converted to 4-coumaroyl-CoA by an enzyme 4-coumarate:CoA ligase. Biosynthesis of lignin (barrier to infection in the peripheral cell) occurs via the phenylpropanoid pathway, in which the enzyme 4-coumarate:CoA ligase plays a key role by catalyzing the formation of hydroxycinnamoyl-CoA esters.

Stilbenoids, such as resveratrol, are hydroxylated derivatives of stilbene and *StSy* and *Vst1* genes encodes stilbene synthase. They are formed through an alternative cyclization of 4-coumaroyl-CoA and 3 malonyl-CoA. Isoflavonoids and flavoniods are derived from a chalcone precursor, the product of the condensation of 4-coumaroyl-CoA and three molecules of 3 malonyl-CoA by the enzyme chalcone synthase. Naringenin chalcone is rapidly converted to trihydroxychalcone and tetrahydroxychalcone by the enzyme chalcone isomerase. Subsequently trihydroxychalcone and tetrahydroxychalcone are converted into isoflavonoids and flavonoids by the enzymes isoflavone synthase (encoded by genes *IFS1* and *IFS2*) and flavones synthase respectively. Whereas terpenoids are enzymatically synthesized from acetyl-CoA through 3 malonyl-CoA of phenylpropanoid pathway by acetyl-CoA carboxylase which is then reduced to form mevalonate. The synthesis of dimethylallyl pyrophosphate is by the phosphorylation of mevalonate that results in triterpenoids and some sesquiterpenoids. Deoxy-xylulose 5-phosphate pathway leads to the synthesis of monoterpenes, diterpenes and some sesquiterpenes (Fig. 2).

Molecular engineering provides a tool to determine the complex biochemical pathways involved in the synthesis and regulation of phenylpropanoids and to manipulate pathways to increase or initiate the production of economically desirable compounds (Dixon et al. 1999). Expression of plant genes in heterologous plant systems can lead to improved or novel synthesis of valuable compounds (Yun et al. 1992) and improved disease resistance (Hain et al. 1993). The overexpression or downregulation of enzymes involved in phenylpropanoid and lignin biosynthesis has demonstrated that it is possible to alter the content and properties of lignin and its associated phenolics (Boudet, 1998). This finding has important implications for the manipulation of plant quality with respect to pulping, forage digestibility, texture and defense responses (Campbell and Sederoff, 1996).

The phytoalexin biosynthesis in plant requires activation of the genes encoding enzymes involved in the synthetic pathway (Lamb et al. 1989). All affords in molecular biology and biotechnology is to introduce new approaches into disease control for friendlier environment. In recent years, genetic engineering is offering a means to enhance the production of important metabolites and/or to establish the production of a new metabolite in the plant. As the knowledge of the enzymes that function in the secondary metabolism is accumulating, the transfer of whole biosynthetic pathways is becoming feasible. The major biosynthetic steps in phenylpropanoids pathway and to some extent the PAs have been unraveled and many of the enzymes and genes related to these compounds have not been characterized. Thus phenylpropanoid pathway is a good target for genetic manipulation because it is among the pathways most frequently induced by pathogens or pathogen elicitors. Plants respond to infection by reinforcing the cell wall (particularly the accumulation of various wall-bound phenolic compounds) and by forming a variety of antimicrobial compounds (phytoalexins) at the infection sites; many of them being phenolic in nature (Fry, 1987; Dixon, 2001; Lozovaya, et al. 2004).

Though, genetic engineering has opened new avenues to modify crops by providing new solutions to solve specific needs, genetic manipulation of secondary metabolites is an area of molecular biology that remains in its infancy. Basic metabolic processes and pathway interactions will help in developing more powerful strategies for improving disease resistance and other processes via genetic engineering. Manipulating metabolic events, leads to production of anti-microbial compounds or strengthening the cell wall barrier to pathogen entry. Most genes for the phenylpropanoid pathway enzymes have been cloned from various species over the last 20 years and can be used now for genetic transformation of plants. The transgenic plants can be made capable of expressing or overexpressing genes of interest at an early stage, preferably with constitutive expression throughout the plant body system to combat any pathogen invasion.

Transgenic Plants Producing Phytoalexins

Genetic engineering of disease-resistance through transfer of plant defense-related genes into crops is valuable in terms of cost, efficacy and reduction of pesticide usage (Lin et al. 2004). Genetic transformation of plants to introduce new disease resistance factors, such as heterologous genes coding for enzymes to produce different class of phytoalexins, and modify the plant biosynthetic pathways may represent a choice to classical method for pathogen resistance.

Improvement of disease resistance is one of the aims of transgenic technology. The expression of bacterial genes in transgenic plants also has proven to be effective, both in introducing new pathways to increase the accumulation of desired compounds (Fecker et al. 1993) and in facilitating understanding of the plant defense response (Delaney et al. 1994). Therefore, overexpression of such genes in plants may be fruitful to achieve the goal. Several groups of researchers have developed fungal resistant plants by overexpressing gene of interest. Targeted transgenic approaches to directly altered phytoalexin profiles have been separately undertaken in the case of stilbenoids, terpenoids, isoflavonoids and flavoniods.

Stilbenoids

Stilbenes represent a group of natural phenolic compounds occurring in a number of plants families, including Fagaceae, Liliaceae, Moraceae, Myrtaceae, Papilionaceae, Pinaceae and Vitaceae (Sotheeswaran and Pasupathy, 1993). The stilbenes biosynthesis requires the presence of stilbene synthase (*StSy*), a phenylpropanoid branch pathway-specific enzyme (Hain et al. 1990; Bavaresco and Fregoni, 2001). Stilbene synthase activity is known to be strongly induced by biotic, as well as by abiotic stress factors (Langcake, 1981; Rosemann et al. 1991), and the afterwards synthesized phytoalexins have antifungal activities and play an important role in plant defense (Hart, 1981).

Stilbene synthase synthesizes the phytoalexin resveratrol a substance with fungicidal potential and are highly bioactive substances of pharmaceutical interest (Langcake and Pryce, 1976). Resveratrol is synthesized by the enzyme trihydroxystilbene synthase (stilbene synthase or resveratrol synthase) (EC 2.3.1.95), using one molecule of 4-coumaroyl-CoA three molecules of malonyl-CoA (Rupprich and Kindl, 1978) as substrates. Since these two substrates are commonly present in plants, introduction of a gene encoding stilbene synthase may be sufficient to synthesize resveratrol in heterologous plant species for increased resistance against fungal pathogens.

Several defense related *StSy* genes have been isolated and characterized (Schro¨der et al. 1989; Schwekendiek, et al. 1992; Sparvoli et al. 1994; Wiese et al. 1994; Preisig-Muller et al. 1999; Kobayashi et al. 2000; Kodan et al. 2001) but only a few studies have been reported, describing their use in plant genetic engineering. The stilbene synthetase gene (*StSy*) from grapevine was introduced into tobacco plants. Expression of the transgene resulted in production of resveratrol and the transgenic plants exhibited enhanced resistance against *Botrytis cinerea*. However, the level of resistance obtained was not sufficient for commercial exploitation. Production of phytotoxin metabolites, such as mycotoxins and oxalic acid by fungi has been shown to facilitate infection of host tissues followed by cell death (Hain et al. 1993). Degradation of these compounds by enzymes expressed in the transgenic plants could provide an opportunity to enhance resistance against diseases.

Thomzik (1997) introduced grapevine stilbene synthase gene in tomato plants and found accumulation of the phytoalexin resveratrol (the product of stilbene synthase) shortly after fungal inoculation. The transformants showed enhanced resistance against *Phytophthora infestans*. No significant increase in resistance was observed in transgenic tomato plants although accumulation of resveratrol occurred after they were inoculated with *Botrytis cinerea* and *Alternaria solani*. Lorenzen et al. (1997) reported the transfer of a grapevine stilbene synthase gene to japonica rice variety, Nipponbare, using PEG-mediated direct gene transfer and demonstrated the effectiveness of the grapevine *StSy* gene resistance to blast in monocotyledons. Zhong et al. (1998) transferred *StSy* gene from peanut into rice. As a result the ability of resistance to pathogens causing blast and bacterial blight of rice was greatly enhanced. Giorcelli *et al.* (2004) transformed a construct containing a cDNA insert encoding stilbene synthase (*StSy*) from grapevine (*Vitis vinifera* L.) under the control of the cauliflower mosaic virus (CaMV) 35S promoter, and a chimeric kanamycin resistance gene to white poplar (*Populus alba* L.) and found high accumulation of the antioxidant resveratrol glucosides. Zhu *et al.* (2004) showed the expression of the grapevine stilbene synthase gene *Vst1* in papaya (*Carica papaya* L.) which provided an increased resistance against fruit rot diseases caused by *Phytophthora palmivora*.

Other than these plants, stilbene synthase genes isolated from grapevine have also been transformed, with variable results in, oil-seed (Thomzik, 1993), kiwi (Kobayashi *et al.* 2000), alfalfa (Hipskind and Paiva, 2000), apple (Szankowski *et al.* 2003), and a grapevine rootstock (Coutos-Thevenot *et al.* 2001).

TERPENOIDS

Terpenoids constitute one of the most diverse families of natural products, with over 40,000 different structures of terpenoids discovered so far. Other than defense mechanism many of the terpenoids produced are non-volatile and are involved in important plant processes such as membrane structure (sterols), photosynthesis (chlorophyll side chains, carotenoids), redox chemistry (quinones) and growth regulation (gibberellins, abscisic acid, brassinosteroids). Terpenoids are synthesized de novo from acetyl-CoA in plastids and the cytoplasm. Despite their diversity, all terpenoids derive from the dimethylallyl diphosphate (Croteau *et al.* 2000). In plants, dimethylallyl pyrophosphate is synthesized via mevalonate pathway, which is active in the cytosol. The volatile terpenoids derived from deoxy-xylulose 5-phosphate pathway viz., monoterpenoids, sesquiterpenoids and some diterpenoids are involved in host pathogen interactions affording defense or stress responses (Dudareva *et al.* 2004; Pichersky and Gershenzon, 2002; Pichersky *et al.* 2006). Metabolic engineering of the terpenoid pathway can be used for the fundamental studies in terpenoid biosynthesis.

FLAVONOIDS

Flavonoids are secondary metabolites produced by plants, and are mainly present in flowers as precursors of anthocyanins (Holton and Cornish 1995). But, flavonoids are also known to be the precursors of phytoalexins which are synthesized as part of the hypersensitive response to incompatible pathogens (Dixon and Paiva 1995). The biosynthesis of flavonoid phytoalexins has not been intensively studied, however, few studies are reported. In rice (*Oryza sativa*) and sorghum (*Sorghum bicolor*), flavonoid compounds have been shown to act as phytoalexins against *Magnaporthe grisea* and *Colletotrichum spp.*, respectively (Kodama et al. 1992; Snyder and Nicholson 1990). In sorghum leaves, a suite of reddish brown flavonoid compounds are induced in the epidermal cells at the site of attempted fungal ingress (Snyder and Nicholson 1990). These pigments belonged to the 3-deoxyanthocyanidin class which included luteolinidin, 5-methoxy-luteolinidin, apigeninidin, caffeic acid ester of arabinosyl 5-oapigeninidin and 7-methoxyapigeninidin (Lo et al. 1996; Wharton and Nicholson 2000). Biosynthesis of these flavonoid compounds was induced by inoculation of seedlings with *C. heterostrophus* fungus. Attempted penetration by this nonpathogenic fungus of sorghum leads to extremely rapid induction of 3-deoxyanthocyanidins (Aguero et al. 2002).

The flavonoid pathway is best characterized in plants and is therefore an excellent target for genetic engineering. Heather *et al.* (1997) investigated alfalfa's response to *R. meliloti* inoculation by monitoring the expression of chalcone synthase, chalcone isomerase and isoflavone reductase genes, the genes which encode enzymes of the phenylpropanoid biosynthetic or "defense" pathway. They found an increase in the production of flavanone mediated by chalcone isomerase. Dixon and Steele (1999) reported that manipulation of flavonoid biosynthesis can be approached via several strategies, including sense or antisense manipulation of pathway genes, modification of the expression of regulatory genes, or generation of novel enzymatic specificities by rational approaches based on emerging protein structure data. The expression of phytoalexins in transgenic plants may be difficult due to complexities involved in their biosynthesis. However, Ibraheem *et al.* (2010) demonstrated that the accumulation of 3-deoxyanthocyanidin flavonoid phytoalexins in sorghum resistance to *Colletotrichum sublineolum* (anthracnose leaf blight) required a functional *y1* gene. Thus, genetic engineering can be used to change the metabolic pathways to increase the amounts of various flavonoids, which play an important role in host-plant resistance to diseases.

ISOFLAVONOIDS

Isoflavonoids are a group of secondary metabolites that play a major role in different types of plant–microbe interactions. The isoflavonoid pathway is one of many components of the general phenylpropanoid pathway that are subjected to complex regulation in different tissue types and in response to different environmental stimuli (Dixon and Paiva, 1995; Bernards *et al.* 2000). Isoflavone synthase (IFS) catalyzes the key entry point step of isoflavone biosynthesis from the general phenylpropanoid pathway. A number of isoflavonoids are identified as precursors to the major phytoalexins in legumes, thus playing key roles in nonspecific plant defense against bacterial and fungal pathogens (Dixon and Steele, 1999; Grayer and Kokubun, 2001). Coordinated induction of isoflavonoid metabolic genes has been observed in response to pathogenic infection in alfalfa (Paiva et al. 1994). Defense induced expression of IFS and several other genes involved in isoflavone metabolism in leaves of alfalfa has been reported previously (He and Dixon, 2000). Another root-specific and pathogen-inducible promoter is that of the alfalfa *IFR* gene (Oommen et al. 1994). In licorice cell cultures, induction of IFS by yeast extract has been reported (Akashi et al. 1998). Similarly, Overkamp et al. (2000) also observed the induction of a P450 cDNA with 80% homology to IFS1 in chickpea cell cultures challenged with elicitor from the fungus *Phytophthora sojae*. However, there has been little or no evidence for the tissue-specific distribution of the key isoflavonoid biosynthetic enzyme IFS and its regulation by developmental and environmental cues.

Overexpression of isoflavone O-methyltransferase resulted in the coordinated overexpression of other genes involved in the synthesis of the phytoalexin medicarpin. This occurred in alfalfa but only in response to *Phoma medicaginis* infection indicating the feasibility of engineering an inducible phytoalexin response (He and Dixon, 2000). Durango et al. (2002) noticed the accumulation of isoflavonoid phytoalexins in several Colombian bean cultivars resistant and susceptible to *Colletotrichum lindemuthianum* fungus, the causal agent of anthrachnose disease. Senthil et al. (2004) cloned and characterized the 5¢ upstream elements of the *IFS1* and *IFS2* genes from soybean in order to generate promoter:reporter constructs and examine their expression in soybean transgenic plants. Lozovaya et al. (2005) have transformed soybean hairy roots with two important genes of phenylpropanoid metabolism: chalcone synthase and isoflavone synthase driven by either a constitutive cassava vein mosaic virus (CsVMV) or root-specific rolD promoter. Thus, metabolic engineering of isoflavonoid pathway can be a favourite target to increase the production of isoflavonoid phytoalexins in plants.

CONCLUSION

Diseases are caused by a wide range of plant pathogens including fungi and their management requires the use of techniques in transgenic technology. Recent advances in transgenic technology, are leading to a more complete understanding of the enzymatic machinery that underlies the often complex pathways of plant natural product biosynthesis. Transgenic expression of natural or genetically modified enzymes in plants has applications in engineering defense compounds. A genomics based approach involves a search for homology with known genes from other species and/or an analysis of gene expression under differing environmental conditions. Synthesis of phytoalexins requires the activity of numerous enzymes, highly coordinated signaling events in the attacked cells to establish defense response successfully. Interestingly, common sequence elements can be identified in the promoters of several genes encoding enzymes required for different steps in the biosynthesis of phytoalexins. Increasing production of an endogenous antimicrobial compound/phytoalexins through overexpression of a rate-limiting enzyme is conceptually a simple strategy. But other enzymes in the phenylpropanoid pathway can be ventured to improve plant disease resistance. The future perspective therefore lies in exploitation of the rate limiting enzymes for phytoalexins synthesis specific to the plant, either in combination with other PR proteins or singly through transgenic technology.

REFERENCES

Agrios, N. *Plant Pathology*. Academic Press, India at Replika Press Pvt. Ltd., Delhi. 2004, 93-136.

Aguero, M.E.; Gevens A.; Nicholson, R.L. Interaction of *Cochliobolus heterostrophus* with phytoalexin inclusions in *Sorghum bicolor*. *Physiol.. Mol. Plant Pathol.*, 2002, 61, 267-271.

Akashi, T.; Aoki, T.; Ayabe, S. Identification of a cytochrome P450 cDNA encoding (2S)-flavanone 2-hydroxylase of licorice (*Glycyrrhiza echinata* L.; Fabaceae) which represents licodione synthase and flavone synthase II. *FEBS Letters*, 1998, 431, 287–290.

Anderson, A. J. *Phytoalexins and plant resistance.* In: Mycotoxins and Phytoalexins. (Eds.) Sharma, R.P. and Salunke, D.K., CRC Press, Boca Raton, 1991, 569-594.

Andreu, A.; Oliva, C.; Distel, S.I.; Daleo, G. Production of phytoalexins, glycoalkaloids and phenolics in leaves and tubers of potato cultivars with different degrees of field resistance after infection with *Phytophthora infestans*. *Potato Res.*, 2001, 44, 1 – 9.

Bailey, J. A.; Burden, R.S. Biochemical changes and phytoalexin accumulation in *Phaseolus vulgaris* L. following cellular browning caused by tobacco necrosis virus. *Physiol. Plant Pathol.*, 1973, 3, 171-77.

Bailey, J. A.; Rowell, P.M.; Arnold, G.M. The temporal relationship between host cell death, phytoalexin accumulation and fungal inhibition during hypersensitive reactions of *Phaseolus vulgaris* to *Colletotrichum lindemuthianum*. *Physiol. Plant Pathol.,* 1980, 17, 329.

Bavaresco, L.; Fregoni, C. *Physiological role and molecular aspects of grapevine stilbenic compounds.* In: Roubelakis-Angelakis KA (Eds), Molecular Biology and Biotechnology of the Grapevine Kluwer Academic Publishers, Dordrecht, Netherlands, 2001, pp. 153–182.

Bernards, M.A.; Susag, L.M.; Bedgar, D.L.; Anterola, A.M. and Lewis, N.G. Induced phenylpropanoid metabolism during suberization and lignification: a comparative analysis. *J. Plant Physiol.*, 2000, 157, 601–607.

Blaich, R.; Bachmann, O. Die Resveratrolsynthese bei Vitaceen: Induktion und Zytologische Beobachtungen. *Vitis*. 1980, 19, 230-240.

Boudet, A.M. A new view of lignification. *Trends Plant Sci.*, 1998, 3, 67–71.

Campbell, M.M.; Sederoff, R.R. Variation in lignin content and composition: Mechanisms of control and implications for the genetic improvement of plants. *Plant Physiol.*, 1996, 110, 3–13.

Coutos-Thevenot, P.; Poinssot, B.; Bonomelli, A.; Yean, H.; Breda, C.; Buffard, D.; Esnault, R.; Hain, R.; Boulay, M. *In vitro* tolerance to *Botrytis cinerea* of grapevine 41B rootstock in transgenic plants expressing the stilbene synthase *Vst1* gene under the control of a pathogen-inducible PR 10 promoter. *J. Exp. Bot.*, 2001, 52, 901–910.

Cramer, C.L.; Edwards, K.; Dron, M.; Liang, X.; Dildine, S.L.; Bolwell, G.P.; Dixon, R.A.; Lamb, C.J.; Schuch, W. Phenylalanine ammonia-lyase gene organization and structure. *Plant Mol. Biol.,* 1989, 12, 367–383.

Croteau, R.; Kutchan, T.M.; Lewis, N.G. Natural products (secondary metabolites). In Biochemistry and Molecular Biology of Plants, Buchanan, B.; Gruissem, W. and Jones, R., (Eds). Rockville, MD*: American Society of Plant Physiologists*, 2000, pp. 1250–1318.

Cruickshank, I.A.M. *Defenses triggered by the invader: Chemical Defenses.* Plant Defenses, Academic Press, Inc., 1980, 5, 247-267.

Delaney, T.P.; Uknes, S.; Vernooij, B.; Friedrich, L.; Weymann, K.; Negrotto, D.; Gaffney, T.; Gut-Rella, M.; Kessmann, H.; Ward, E.; Ryals, J. A central role of salicylic acid in plant disease resistance. *Sci.*, 1994, 266, 1247–1250.

Dercks, W.; Creasy, L.L. The signifcance of stilbene phytoalexins in the *Plasmopara iticola* - grapevine interaction. *Physiol. Mol. Plant Pathol.*, 1989, 34, 189-202.

Dixon, R.A. The phytoalexin response: Eliciation, signalling and control of host gene expression. *Biol. Rev.*, 1986, 61, 239-291.

Dixon, R.A. *Isoflavonoids: biochemistry, molecular biology, and biological functions, in Comprehensive Natural Products Chemistry* (Vol. 1) Sankawa, U. (Ed.), *Elsevier,* 1999, pp. 773–823.

Dixon, R.A. Natural products and plant disease resistance. *Nature*, 2001, 411, 843-847.

Dixon, R.A.; Paiva, N.L. Stress-induced phenylpropanoid metabolism. *Plant Cell*, 1995, 7, 1085- 1097.

Dixon, R.A.; Steele C.L. Flavonoids and isoflavonoids - a gold mine for metabolic engineering. *Trends Plant Sci.*, 1999, 4(10), 394-400.

Du, Y.; Chu, H.; Wang, M.; Chu, I. K.; Lo, C. Identification of flavone phytoalexins and a pathogen inducible flavone synthase II gene (*SbFNSII*) in sorghum. *J. Expt. Bot.*, 2010, 61(4), 983–994.

Dudareva, N.; Pichersky, E.; Gershenzon, J. Biochemistry of plant volatiles. *Plant Physiol.,* 2004, 135, 1893–1902.

Durango, D.; Quiñones, W.; Torres; F.; Rosero, Y.; Gil, J.; Echeverri, F. Phytoalexin accumulation in Colombian bean varieties and aminosugars as elicitors. *Molecules*, 2002, 7, 817-832.

Ebel, J. Phytoalexin synthesis: the biochemical analysis of the induction process. *Annu. Rev. Phytopathol.*, 1986, 24, 235–264.

Enkerli, J.; Bhatt, G.; Covert, S.F. Maackiain detoxification contributes to the virulence of *Nectria haematococca* MP VI on chickpea, *Mol. Plant–Microbe Interact.* 1998, 11, 317–326.

Fecker, L.; Riigenhagen, C.; Berlin, J. Increased production of cadaverine and anabasine in hairy root cultures of *Nicotiana tabacum* expressing a bacterial lysine decarboxylase gene. *Plant Mo1 Biol.*, 1993, 23, 11-21.

Fry, S.C. Intracellular feruloylation of pectic polysaccharides. 1987, *Planta* 171, 205 – 211.

Ghanekar, A.S.; Padwal-Desai, S.R.; Nadkarni, G.B. The involvement of phenolics and phytoalexins in resistance of potato to soft rot. *Potato Res.*, 1984, 27, 189 199.

Giorcelli, A.; Sparvoli, F.Y.; Mattivi, F.; Tava, Y.A.; Balestrazzi, A.; Vrhovsek, U.; Calligari P.; Bollini, R.; Confalonieri, M. Expression of the stilbene synthase (*StSy*) gene from grapevine in transgenic white poplar results in high accumulation of the antioxidant resveratrol glucosides. *Transgenic Res.*, 2004, 13: 203–214.

Graham, T.L.; Graham, M.Y. Glyceollin elicitors induce major but distinctly different shifts in isoflavonoid metabolism in proximal and distal soybean cell populations. *Mol. Plant-Microbe Interact.*, 1991, 4, 60 – 68.

Grayer, R. J.; Kokubun, T. Plant fungal interactions: the search for phytoalexins and other antifungal compounds from higher plants. *Phytochem.,* 2001, 56, 253-263.

Hain, R.; Bieseler, B.; Kindle, H.; Schröder, G.; Stöcker, R. Expression of a stilbene synthase gene in *Nicotiana tabacum* results in synthesis of the phytoalexin resveratrol. *Plant Mol. Biol.*, 1990, 15, 325-335.

Hain, R.; Reif, H.J.; Krause, E.; Langebartels, R.; Kindl, H.; Vornam, B.; Wiese, W.; Schmelzer, E.; Schreier, P.H.; Stöcker, R.H.; Stenzel, K. Disease resistance results from foreign phytoalexin expression in a novel plant. *Nature*, 1993, 361, 153–156.

Hart, J. H. Role of phytostilbenes in decay and disease resistance. *Annu. Rev. Phytopathol.*, 1981, 19, 437-458.

He, X.Z.; Dixon, R. Genetic manipulation of isoflavone 7-O-methyltransferase enhances biosynthesis of 4′-O-methylated isoflavonoid phytoalexins and disease resistance in alfalfa. *Plant Cell,* 2000, 12, 1689-1702.

Heather I. M.; Paiva, N.L.; Dixon, R. A.; Hirsch A. M. Chalcone synthase transcripts are detected in alfalfa root hairs following inoculation with wild-type *Rhizobium meliloti. The American Phytopathological Society,* 1997, 10 (1), 50-58.

Helmut K. B.; Barz, W. Accumulation of isoflavones and pterocarpan phytoalexins in cell suspension cultures of different cultivars of chickpea (*Cicer arietinum*). *Plant Cell Reports*, 1987, 6, 55-59.

Hipskind, J.D.; Paiva. N.L. Constitutive accumulation of a resveratrol-glucoside in transgenic alfalfa increases resistance to *Phoma medicaginis. Mol. Plant-Microbe Interact.*, 2000, 13, 551-562.

Holton, T. A., and Cornish, E. C. Genetics and biochemistry of anthocyanin biosynthesis. *Plant Cell,* 1995, 7, 1071-1083.

Ibraheem, F.; Gaffoor, I.; Chopra, S. Flavonoid phytoalexin dependent resistance to anthracnose leaf blight requires a functional *yellow seed1* in *Sorghum bicolor. Genetics*, 2010, in press.

Keen, N. T. *Evaluation of the role of phytoalexins.* In: Plant Disease Control, (Eds.) R.C. Staple, G. H. Toenniessen. Oxford Press, New York, Weily, 1981, pp. 155-177.

Kobayashi, S.; Ding, C.K.; Nakamura, Y.; Nakajima, I.; Matsumoto, R. Kiwifruits (*Actinidia deliciosa*) transformed with a *Vitis* stilbene synthase gene produce piceid (resveratrol-glucoside). *Plant Cell Rep.*, 2000, 19, 904–910.

Kodama, O.; Miyakawa , J.; Akatsuka, T.; Kiyosawa, S. Sakuranetin, a flavanone phytoalexin from ultraviolet-irradiated rice leaves. *Phytochem.*, 1992, 31, 3807-3809.

Kodan, A., Kuroda, H.; Sakai, F. Simultaneous expression of stilbene synthase genes in Japanese red pine (*Pinus densiflora*) seedlings. *J. Wood Sci.*, 2001, 47, 58–62.

Lamb, C. J.; Lawton, M.A.; Dron, M.; Dixon, R.A. Signals and transduction mechanisms for activation of plant defenses against microbial attack. *Cell*, 1989, 56, 215-224.

Langcake, P.; Pryce, R.J. The production of resveratrol by *Vitis vinifera* and other members of the Vitaceae as a response to infection or injury. *Physiol. Plant Pathol.*, 1976, 9, 77–86.

Langcake, P. Disease resistance of *Vitis* spp. and the production of the stress metabolites resveratrol, e-viniferin, a-viniferin and pterostilbene. *Physiol. Plant Pathol.*, 1981, 18, 213– 226.

Lin, W.C.; Lu, C.F.; Wu, J.W.; Cheng, M.L.; Lin, Y.M.; Yang, N.S. Transgenic tomato plants expressing the *Arabidopsis AtNPR1* gene display enhanced resistance to a spectrum of fungal and bacterial diseases. *Transgenic Res.*, 2004, 13, 567–581.

Lo, S.C.; Weiergang, I.; Bonham, C.; Hipskind, J.; Wood K. Phytoalexin accumulation in sorghum: Identification of a methyl ether of luteolinidin. *Physiol. Mol. Plant Pathol.*, 1996, 49, 21-31.

Lo, S.C.; Nicholson, R.L. Reduction of light-induced anthocyanin accumulation in inoculated sorghum mesocotyls - Implications for a compensatory role in the defense response. *Plant Physiol.*, 1998,116, 979-989.

Lorenzen, P.; Nelke, B.; Hänssler, G.; Mühlbach, H.P.; Thomzik, J.E. Transfer of a grapevine stilbene synthase gene to rice (*Oryza sativa* L.). *Plant Cell Rep.*, 1997, 16, 668–673.

Lozovaya,V.V.; Lygin, A.V.; Li, S.; Hartman, G.L.; Widholm, J.M. Biochemical response of soybean roots to *Fusarium solani f. sp. glycines* infection, *Crop Sci.* 2004, 44, 819–826.

Lozovaya, V.V.; Lygin, A.V.; Zernova, O.V.; Widholm, J.M. Genetic engineering of plant root disease resistance by modification of the phenylpropanoid pathway. *Plant Biosystems*, 2005, 139(1), 20 – 23.

Lyon, G.D., Occurrence of rishitin and phytuberin in potato tubers inoculated with Erwniia carotovora var. atroseptica. *Physiol. Plant Pathol.*, 1973, 2, 411 416.

Lyon, F. M.; Wood, R. K. S. Production of phaseollin, coumestrol and related compounds in bean leaves inoculated with *Pseudomonas* spp. *Physiol. Plant Pathol.*, 1975, 6, 117.

Malathi, P.; Viswanathan, R.; Padmanaban, P.; Mohanraj, D.; Ganeshkumar, V.; Salin, K.P. Differential accumulation of 3-deoxy anthocyanidin phytoalexins in sugarcane varieties varying in red rot resistance in response to *Colletotrichum falcatum* infection. *Sugar Tech.*, 2008, 10(2), 154-157.

Mayama, S.; Matsura, Y.; Iida, H.; Tani, T. The role of avenalumin in the resistance of oat to crown rust, *Puccinia coronata* f. sp. *avenae*. *Physiol. Plant Pathol.*, 1982, 20, 189.

Misaghi, I.J. *Alterations in phenol metabolism caused by disease.* Physiology and Biochemistry of plant-pathogen interactions, Plenum Publishing Corporation, New York. 1982, p.111.

Muller, K.O.; Borger, H. Experimentelle untersuchungen uber die Phytophthora – Resistenz der kartoffel. Arb. Biol. Reischsanet. *Land Forstwirtsch Berlin - Dahlem*, 1940, 23, 189-190.

Nagai, N.; Kitauchi, F.; Shimosaka, M.; Okazaki, M. Cloning and sequencing of a full-length cDNA coding for phenylalanine ammonia-lyase from tobacco cell culture. *Plant Physiol.*, 1994, 104, 1091–1092.

Nicholson, R.L.; Hammerschmidt, R. Phenolic compounds and their role in disease resistance. *Ann. Rev. Phytopath.*, 1992, 30, 369-389.

O'neill, T. M.; Mansfield, J. W. Mechanisms of resistance to *Botrytis* in narcissus bulbs. *Physiol. Plant Pathol.*, 1982, 20, 243.

Oommen, A.; Dixon, R.A.; Paiva, N.L. The elicitorinducible alfalfa isoflavone reductase promoter confers different patterns of developmental expression in homologous and heterologous transgenic plants. *Plant Cell*, 1994, 6, 1789–1803.

Overkamp, S.; Hein, F.; Barz W. Cloning and characterization of eight cytochrome P450 cDNAs from chickpea (*Cicer arietinum* L.) cell suspension cultures. *Plant Sci.* 2000, 12, 101–108.

Paiva, N.L.; Oommen, A.; Harrison, M.J.; Dixon, R.A. Regulation of isoflavonoid metabolism in alfalfa. *Plant Cell Tissue Organ Cult.*, 1994, 38, 213–220.

Pellegrini, L.; Rohfritsch, O.; Fritig, B.; Legrand, M. Phenylalanine ammonia-lyase in tobacco - Molecular cloning and gene expression during the hypersensitive reaction to tobacco mosaic virus and the response to a fungal elicitor. *Plant Physiol.*, 1994, 106, 877–886.

Perrin, D. R. The structure of phaseollin. *Tetrahedron Lett.*, 1964, 1, 29-35.

Perrin, D.R.; Whittle, C.P.; Batterham, T.J. The structure of phaseollin. *Tetrahedron Lett.*, 1972, 1873-76.

Pichersky, E.; Gershenzon, J. The formation and function of plant volatiles: perfumes for pollinator attraction and defense. *Curr. Opin. Plant Biol.*, 2002, 5, 237–243.

Pichersky, E.; Noel, J.P.; Dudareva, N. Biosynthesis of plant volatiles: nature's diversity and ingenuity. *Science*, 2006, 311, 808–811.

Preisig-Mu¨ ller R.; Schwekendiek, A.; Brehm, I.; Reif, H.J.; Kindl, H. Characterization of a pine multigene family containing elicitor-responsive stilbene synthase genes. *Plant Mol. Biol.*, 1999, 39, 221–229.

Robbins, M.P. ; Hartnoll, J.; Morris P. Phenylpropanoid defense responses in transgenic Lotus corniculatus- Glutathione elicitation of isoflavan phytoalexins in transformed root cultures. *Plant Cell Reports*, 1991, 10, 59-62

Rosemann, D.; Heller, W.; Sandermann, H. Biochemical plant responses to ozone: II. Induction of stilbene biosynthesis in Scots pine (*Pinus sylvestris L.*) seedlings. *Plant Physiol.*, 1991, 97, 1280-1286.

Rupprich, N.; Kindl, H. Stilbene synthases and stilbene carboxylate synthases, I. Enzymatic synthesis of 3,5,4-trihydroxystilbene from p-coumaroyl-CoA and malonyl-CoA. *Hoppe-Seyler's Zeitschrift Physiologische Chemie*, 1978, 359, 165-172.

Schro¨der, J.; Lanz, T.; Schro¨der, G. *Genes for the biosynthesis of stilbene-type phytoalexins.* In:Lamb, C. and Beachy, R. (Eds.), *Plant Gene Transfer*. New York: A.R. Liss, Inc, 1989, pp. 129.

Schwekendiek, A.; Pfeffer, G.; Kindl, H. Pine stilbene synthase cDNA, a tool for probing environmental stress. *FEBS Lett.*, 1992, 301, 41–44.

Senthil, S.; Graham, M.Y.; Yu, O.; Graham, T.L. RNA interference of soybean isoflavone synthase genes leads to silencing in tissues distal to the transformation site and to enhanced susceptibility to *Phytophthora sojae. Plant Physiol.* 2005, 137, 1345–1353.

Senthil, S.; Hu, X.; Lu, G.; Odell, J. T.; Yu, O. The promoters of two isoflavone synthase genes respond differentially to nodulation and defense signals in transgenic soybean roots. *Plant Mol. Biol.*, 2004, 54, 623–639.

Skipp, R.A.; Bailey, J. A. The effect of phaseollin on the growth of *Colletotrichum lindemuthianum* in bioassay designed to measure fungitoxicity. *Physiol. Plant Pathol.*, 1976, 9, 253.

Smith, D.A. Some effects of the phytoalexin, kievitone, on the vegetative growth of *Aphanomyces euteiches, Rhizoctonia solani* and *Fusarium solani* f. sp. *phaseoli. Physiol. Plant Pathol.*, 1976, 9, 45.

Snyder, B.A.; Nicholson, R.L. Synthesis of phytoalexins in sorghum as a site specific response to fungal ingress. *Sci.*, 1990, 248, 1637-1639.

Sotheeswaran, S.; Pasupathy, V. Distribution of resveratrol oligomers in plants. *Phytochem.*, 1993, 32, 1083–1092.

Sparvoli, F.; Martin, C.; Scienza, A.; Gavazzi, G.; Tonelli, C. Cloning and molecular analysis of structural genes involved in flavonoid and stilbene biosynthesis in grape (*Vitis vinifera* L.). *Plant Mol. Biol.*, 1994, 24: 743–755.

Szankowski, I.; Briviba, K.; Fleschhut, J.; Schonherr, J.; Jacobsen, H.J.; Kiesecker, H. Transformation of apple (*Malus domestica* Borkh.) with the stilbene synthase gene from grapevine (*Vitis vinifera* L.) and a PGIP gene from kiwi (*Actinidia deliciosa*). *Plant Cell Rep.*, 2003, 22, 141–149.

Teasdale, J.; Daniels, D.; Davis, W. C.; Eddy, R. Jr.; Hadwiger, L. A. Physiological and cytological similarities between disease resistance and cellular incompatibility responses. *Physiol. Plant Pathol.*, 1974, 54, 690.

Thomzik, J.E. Transformation in oilseed rape *Brassica napus* L. In: Bajaj, Y.P.S. (Ed.), *Biotechnology in Agriculture and Forestry* Springer, Berlin/Heidelberg/New York. 1993, pp. 171–182.

Thomzik, J.E. Synthesis of a grapevine phytoalexin in transgenic tomatoes (*Lycopersicum esculentum* Mill.) conditions resistance against *Phytophthora infestans. Physiol. Mol. Plant Pathol.*, 1997, 51, 265–278.

Van Den Heuvel, J.; Grootveld D. Formation of phytoalexins within and outside lesions of *Botrytis cinerea* in French bean leaves. *Neth. J. Pl. Path.*, 1980, 86, 27-35.

Van Etten, H.D. Antifungal activity of pterocarpans and other selected isoflavonoids. *Phytochem.*, 1976, 15, 655-659.

Varns, J.L.; Kuc, J.; Williams, E.B. Terpenoid accumulation as a biochemical response of the potato tuber to *Phytophthora infestans. Phytopathol.*, 1971, 61:174.

Wanner, L.A.; Li, G.; Ware, D.; Somssich, I.E.; Davis, K.R. The phenylalanine ammonia-lyase gene family in *Arabidopsis thaliana. Plant Mol. Biol.*, 1995, 27, 327–338.

Wasmann, C.C.; Van Etten, H.D. Transformation-mediated chromosome loss and disruption of a gene for pisatin demethylase decrease the virulence of *Nectria haematococca* on pea. *Mol. Plant–Microbe Interact.*, 1996, 9, 793–803.

Wharton, P.S.; Nicholson, R.L. Temporal synthesis and radiolabelling of the sorghum 3-deoxyanthocyanidin phytoalexins and the anthocyanin, cyanidin 3-dimalonyl glucoside. *New Phytologist*, 2000, 145, 457-469.

Wiese, W.; Vornam, B.; Krause, E.; Kindl, H. Structural organization and differential expression of three stilbene synthase genes located on a 13 kb grapevine DNA fragment. *Plant Mol. Biol.,* 1994, 26, 667–677.

Yu O. Metabolic engineering of the plant phenylpropanoid pathway. *Encyclopedia of Plant and Crop Sci.*, 2006, 1-4.

Yun, D.J.; Hashimoto, T.; Yamada, Y. Metabolic engineering of medicinal plants - Transgenic Atropa belladonna with an improved alkaloid composition. *Proc Natl Acad Sci USA,* 1992, 89, 11799-11803.

Zhong, T.W.; Li, D. Yun, C.S.; Hong, D.S.; Quing, Y.E.S.; Cai, L.L. Rice transformation with a phytoalexin gene and bioassay of the transgenic plants. *Acta Botanica Sinica.* 1998, 40(9), 803-808.

Zhu, Y.J.; Agbayani, R.; Jackson, M.C.; Tang, C.S.; Moore, P.H. Expression of the grapevine stilbene synthase gene VST 1 in papaya provides increased resistance against diseases caused by *Phytophthora palmivora. Planta,* 2004, 220, 241–250.

In: Genetically Engineered Crops
Editors: Lindsay M. Grover

ISBN: 978-1-61122-526-6

Chapter 9

PLANT PROTECTION THROUGH RNAI: ALTERNATIVE FOR PESTICIDES AND CHEMICAL APPROACHES TO PEST CONTROL

S.V. Ramu[1], Rohini Sreevathsa[1], S.K.Sarangi[2] and M. Udayakumar[2]

[1]Dept of crop physiology ,UAS, GKVK, Bangalore-65, India
[1]Dept of Biotechnology, Bangalore University. Bangalore, India

ABSTRACT

RNA interference (RNAi) has become a powerful genetic approach to systemically dissect gene function on a genome wide scale. Double stranded RNA (dsRNA) mediated gene silencing has been proven as a potential approach to control several insect pests including fruit fly, nematode, cotton bollworm, and corn rootworm. RNA interference (RNAi) induced by hairpin RNAs are new line of defence against coleopteran and lepidopteron pests. It was shown that insect larvae fed with plant material expressing dsRNA for the insect genes triggered the RNAi pathway in insect larvae and blocked the expression of target genes. The resulting reduction in larval growth and subsequent death has opened up new options for insect control involving the insect genes perse. Hence, protein targets for insect control strategies can be minimized or completely ignored. Prospecting for novel resistance mechanisms depend on the plant generating an effective form of double-stranded RNA in the absence of an endogenous target gene without detrimental effect to itself. Although the amazing popularity of RNAi as a biotechnology tool is certainly justified, the underlying biology is still being worked out and the relative advantages and disadvantages of the approach are being efficiently addressed using emerging Bioinformatic tools and many genome sequences. In this review we would like to highlight the attempts of pest control using gene silencing or RNAi technology, few strategies, target genes for RNAi in insects, target site of action, and probable likely targets for RNAi in insects and plants for potential durable pest resistance mechanisms.

Key words;- RNAi, Nematode, Insect , Bacillus thuriengenesis

INTRODUCTION

With the growing population, there is a continuous need to increase food production. One strategic way to increase food production would be to increase the yield in crops by minimizing the pest associated losses. There has been enormous competition between plants and pests since millions of years. Insect pest management took a new deviation with the advent of transgenic crops developed expressing cry genes from Bacillus thuringiensis. Numerous Bt strains producing various endotoxins are in commercial use as conventional sprayable formulations and transgenic crops in a number of of developed and developing countries. However, development of insect resistance to transgenic crops producing Bt toxins poses a major threat to their sustainable use in agriculture. Therefore there is a need for alternate pest control strategies. The late ninties marked the beginning of novel technology, RNAi. Ever since, the technology is being extended as a provision of valuable crop traits like resistance to viruses, bacteria and nematodes. The ability of the RNAi technology for the development of insect resistant transgenic plants would be a very useful alternate pest control strategy. Our emphasis in this review is to elaborate on the utility of RNAi as strategy for pest control.

RNAI AN EMERGING TOOL IN BIOTECHNOLOGY

RNAi (RNA interference) is a predominant tool in functional genomics in the recent years. It refers as post-transcriptional gene silencing in plants (Vander et al.1990) quelling in fungi (Cogoni et al. 1996) and RNA interference in animals (Fire et al. 1998). The introduction of homologous double stranded RNA (dsRNA) to specifically target a gene's product, resulting in null or hypomorphic phenotypes. This dsRNA is degraded into approximately 21-nucleotide RNAs, known as 'small interfering RNAs' (siRNAs). The effectiveness of RNAi for functional genomics has been demonstrated in the nematode *Caenorhabditis elegans,* (Louisa Mathew et,al 2004) but has yet to be fully utilized in plants. The mechanism of siRNA biogenesis and function are thought to be highly conserved in almost all the eukaryotes including plants and animals (Shanfa et al. 2004). It was discovered that presence of dsRNA, formed from the annealing of sense and antisense strands forms homologous dsRNA acts as precursor for producing the interfering activity (Fire et al., 1998). The RNA silencing pathways are all associated with small (~20 to 30nt) RNAs that function as specificity factors for inactivating homologous sequences by variety of mechanisms (Danesh moazed 2009). The generation of endogenous RNAi from dsRNA resulting from convergent transcription (Sense-antisense RNA Base –pairing) or through transcription through inverted repeat sequences (Hairpin RNA formation) in which siRNA are produced from double stranded RNA by an RNase III termed DCL in plants (Dicer in animals) and then incorporated in to a RNA induced silencing complex (RISC) in which siRNA play a guiding role in sequence specific cleavage of target mRNAs. The siRNA signal is found to spread along the mRNA target, which results in the production of secondary siRNA and the induction of transitive RNA silencing. The artificial production and mechanism of RNAi has been shown schematically in figure 1. The most interesting aspects of RNAi are, dsRNA, rather than single-stranded antisense RNA as the interfering agent is highly sequence specific and remarkably potent (only a few dsRNA molecules per cell are required for effective interference). The interfering activity (presumably the dsRNA) can cause interference in cells and tissues far away from the site of introduction.

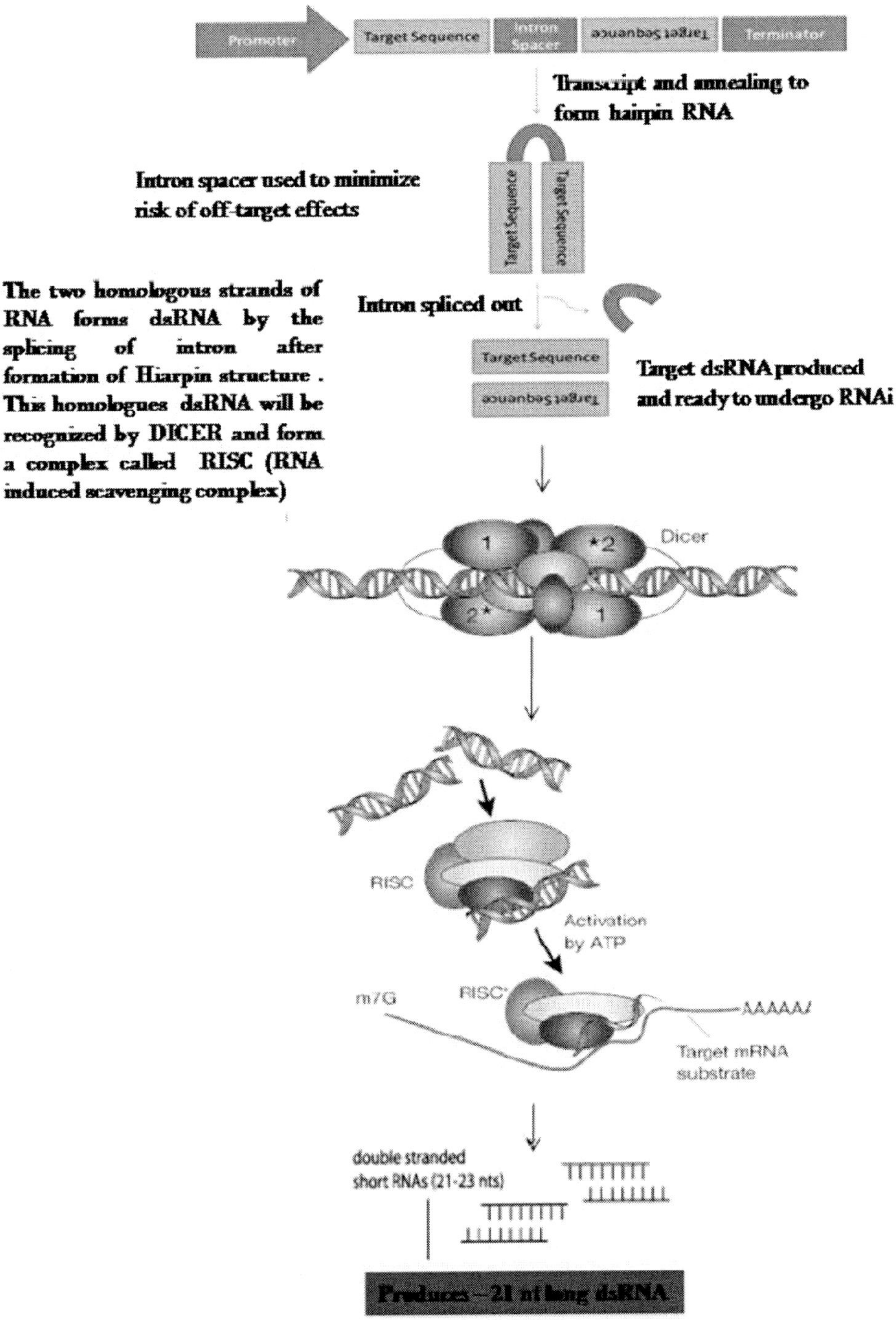

Figure 1. A schematic representation of the production of dsRNA from a RNAi vector. Specific promoter, inverted repeats and an intron spacer are used to produce the best RNAi effects.

RNAi has been exploited in plants for applications ranging from functional genomics to provision of valuable crop traits, such as resistance against viruses, bacteria and nematodes. Despite having been considered for many years, application of RNAi technology to give resistance to herbivorous insects has been realized.

Conventionally, antisense-mediated gene silencing has been widely used in the analysis of gene function in plants. Although antisense-mediated gene silencing is an RNAi-related phenomenon, hpRNA-induced RNAi has been shown to be much more efficient. In a hpRNA-producing vector, the target gene is cloned as an inverted repeat spaced with an unrelated sequence and is driven by a strong promoter, such as the 35S CaMV promoter for dicots or the maize ubiquitin 1 promoter for monocots Müller 2006). When an intron is used as the spacer, which is essential for stability of the inverted repeat in *Escherichia coli*, the efficiency becomes very high with almost 100% of transgenic plants show gene silencing. RNAi can be used against a vast range of targets; 30 and 50 untranslated regions (UTRs) as short as 100 nt could be efficient targets of RNAi. RNAi pathway is an innate immune pathway of invertebrates such as nematodes, trypanosomes, hydra, planaria and insects (Bosher et al. 2000). It has been reported that RNAi functions as a gatekeeper in mosquitoes, modulating arbovirus replication to allow virus transmission but preventing virus concentrations that could lead to fitness costs and pathogenic effects. Consequently, RNAi is potentially a major factor determining the vector competence of mosquitoes for arboviruses (Khoo et al. 2010). Although RNAi is not a knockout but a knockdown technology, its high efficiency and ease of application makes it applicable to genome-wide analysis of gene function. In plants, RNAi is often achieved by a transgene that produces hairpin RNA (hpRNA) with a dsRNA region (figure 1). Artificial production of RNAi in plants using genetic engineering have led to development of strategies to silence the plant genes or foreign target genes by producing small dsRNA. Several Bioinformatic software's are available these days to design and predict the gene sequence for RNAi against target mRNA.

The delivery methodology in the majority of insect studies has been microinjection of nanogram amounts of dsRNA, synthesized in vitro, into the insect haemoceol (Dzitoyeva, S. et al. 2001). This method of delivery contrasts with the situation in C. elegans, where RNAi effects can be produced by feeding bacteria expressing dsRNA (Fire et al.1998 Timmons, et al. 2001), or even by soaking nematodes in dsRNA solution (Tabara, H. et al 1998). Microinjection of dsRNA in insects was considered to be necessary to produce RNAi effects because the complete genome sequence for D. melanogaster (and, subsequently, for other insects) has shown that they lack genes encoding RNA dependent RNA polymerase (RdRP). RdRP is the enzyme necessary for the siRNA amplification step that leads to persistent and systemic RNAi effects (Sijen, T. et al. 2001). The absence of RdRP in insects predicts that any effects of RNAi will be limited to cells that have taken up dsRNA and will require continuous input of dsRNA to persist. Injection of dsRNA into the body cavity, where it can circulate through the haemolymph, allows short-term effects on gene expression in most cells can be assessed. In animals and especially in insects, RNAi is usually accomplished by injection of double-stranded RNA (dsRNA), but in nematodes it may also be triggered by a diet containing dsRNA.

RNAI IN INSECTS

RNAi is regularly applied in the field of entomology to study the RNAi mechanism and the function, regulation and expression of gene cascades, mostly in Drosophila melanogaster, Tribolium castaneum and Bombyx mori (Figure 2) (Huvene et al.2010). The laboratory experiments have been conducted in which dsRNA is injected directly in the organism, which is not applicable to control insect pests in the field. For efficient insect control, the organism should be able to autonomously take up the dsRNA, through feeding and digestion in the gut. The insect gut is divided into three regions; foregut, midgut and hindgut. Of these the first two are the initial oral regions of the insect and are chitin-lined, so that their surfaces do not present areas of exposed cells (although receptors and transporters are present to allow processes such as taste recognition in the mouth cavity and water transfer in the hindgut to occur). The midgut itself is responsible for nutrient absorption, whereas excretion and water balance take place primarily in the Malpighian tubules attached to the hind end (Daniel and Gatehouse 2008) The insect midgut consists of a single layer of columnar cells with microvilli, endocrine cells, and stem cells at the base, grouped in the so-called nidi. The midgut is designed to absorb nutrients from the gut lumen with its large absorption area created by the microvilli, with many channels and endocytosis apparati (Hakim et al., 2010).

In recent years new insights into the dsRNA uptake mechanisms in insects has been studied with emphasis on uptake through midgut and in body tissues. RNAi can be divided in to cell-autonomous and noncell- autonomous RNAi. i) In case of cell-autonomous RNAi, the silencing process is limited to the cell in which the dsRNA is introduced/expressed and encompasses the RNAi process within individual cells, dsRNA is cleaved by a RNase III, often called Dicer, into 21–25 nt-long short interfering RNA duplexes (siRNA). These siRNAs are incorporated in the RNA induced silencing complex (RISC); after discarding the passenger strand, the RISC will bind to a homolog mRNA, cutting it and thereby hindering translation. ii) In case of non-cell-autonomous RNAi, the interfering effect takes place in tissues/cells different from the location of application or production of the dsRNA. There are two different kinds of non-cell-autonomous RNAi: environmental RNAi and systemic RNAi (Whangbo and Hunter 2008, Hunter 2007). In environmental RNAi dsRNA is taken up by a cell from the environment. Therefore, this process can also be observed in unicellular organisms. Systemic RNAi can only take place in multicellular organisms because it includes processes in which a silencing signal is transported from one cell to another or from one tissue type to another. In multicellular organisms environmental RNAi can be followed by systemic RNAi.

For the efficient application of RNAi in insect control, researchers should focus on non-cell-autonomous RNAi. The insect will have to internalize the dsRNA of a target gene through feeding. In order to silence the target gene, this dsRNA must be taken up from the gut lumen into the gut cells, representing environmental RNAi. If the target gene is expressed in a tissue outside of the gut, the silencing signal will also have to spread via cells and tissues, which is systemic RNAi. The two good examples for the dsRNA uptake in insects are the transmembrane channel-mediated uptake and endocytosis – mediated uptake. More detailed information on these two mechanisms are been reviewed by Huvenne et al. (2010).

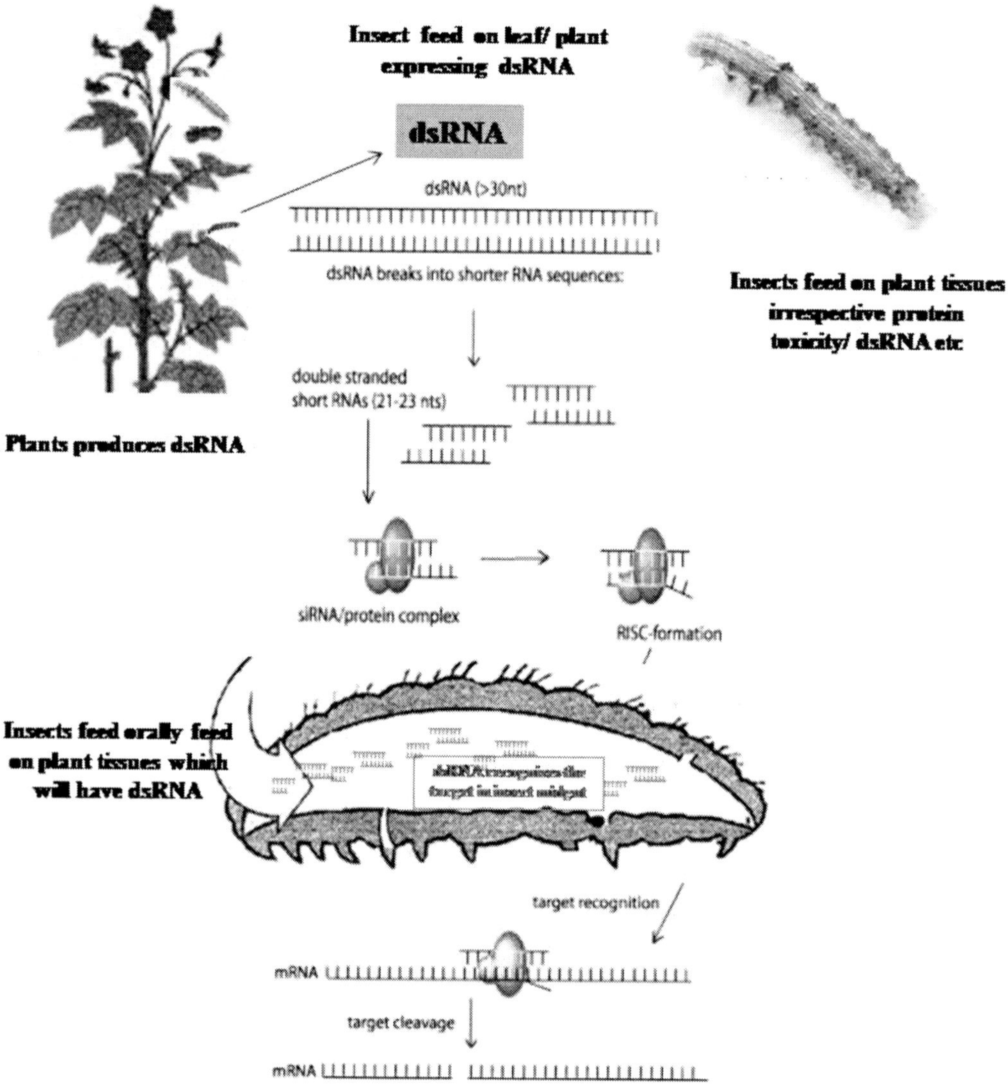

Figure 2. Mechanism of RNAi in insects. The dsRNA produced from the plants enters into the insect midgut. The midgut epithelial cells recognizes the dsRNA, cleaves the target mRNA and degrades it thereby killing the insect.

RNA- Mediated Crop Protection Against Insects

Despite the start of RNAi a decade ago, the application the technology to give resistance to herbivorous insects has only been realized recently. Among transgenic approaches to insect pest management, *Bacillus thuringiensis* (*Bt*) toxins has shown magnificent success. The Cry family of *Bacillus thuringiensis* insecticidal proteins is widely utilized in spray formulations and transgenic approaches to control insect pests (Karl HJ Gordon et.al 2006). In 2007 two reports published in nature biotechnology described transgenic plants producing dsRNAs directed against insect genes. These plants showed resistance against *cotton bollworm*

(Helicoverpa armigera: Lepidoptera) and *Western corn rootwarm(WCR: Diabrotica virgifera virgifera Leconte:* Coleoptera) which are the major economically important agricultural pests (figure3). The success behind this approach is i) Identification of specific target gene for RNAi and Development of a suitable construct for expression *in planta.*

ii) Expression of dsRNA *in planta* and delivery of sufficient amounts of intact dsRNA for uptake by the insect (Daniel et al 2008).

Mao et al. 2007 identified a *cytochrome P450* gene (CYP6AE14) from cotton bollworm (*Helicoverpa armigera*), which permits this herbivore to tolerate otherwise inhibitory concentrations of the cotton metabolite, gossypol. CYP6AE14 is highly expressed in the midgut and its expression correlates with larval growth when gossypol is included in the diet. When larvae are fed with plant material expressing double-stranded RNA (dsRNA) specific to CYP6AE14, levels of this transcript in the midgut decrease and larval growth is retarded. Both effects are more dramatic in the presence of gossypol. Insects feeding plant material expressing dsRNA may be a general strategy to trigger RNA interference and could find applications in entomological research and field control of insect pests.

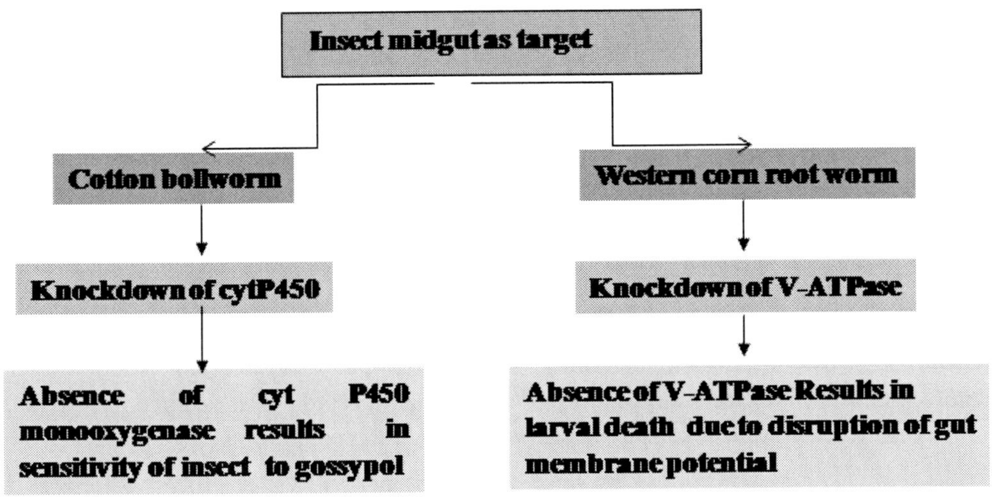

Figure 3. An overview of RNAi approaches used to develop insect resistant plants. Large dsRNA and siRNA cleavage products are expressed throughout the plant and are orally delivered to insect herbivores feeding on transgenic plant material. For gene silencing to initiate in targeted insect pests, large dsRNAs and siRNAs must persist in the insect midgut in sufficient quantities for uptake into cells.

Baum *et al.* first used feeding assays, in which larvae were fed an artificial diet supplemented with specific dsRNAs, to screen a large number of essential insect genes in cDNA libraries. A total of 290 potential targets were identified and corresponding dsRNAs were synthesized invitro, their effects on larval performance were determined by delivery in artificial diet feeding trails. These assays identified 14 genes whose knockdown by low levels of dsRNA resulted in insect stunting and mortality. Transgenic corn expressing one of these dsRNAs, targeted to a subunit of the midgut enzyme vacuolar ATPase (V-ATPase), showed protection against WCR infestation comparable to that provided by a *Bt* transgene. When dsRNAs designed to target WCR genes were tested on larvae from two other insect pests, mortality declined with decreasing sequence identity between the WCR genes and their orthologs in the other species, indicating that gene silencing was potentially very selective.

In turn, insects have several, often large families of enzymes (P450s, glutathione *S*-transferases and carboxylesterases) able to detoxify plant chemical cocktails. Some of these have been implicated in the evolution of pesticide resistance, for example through tissue-specific transcriptional upregulation. Such genes are attractive targets for applying the ingestible dsRNA approach to managing resistance against plant allelochemicals or insecticides. Indeed, Mao *et al. 2007,* have corroborated the conclusion of Baum *et al.2007* that the strategy works for multiple midgut-expressed genes in rootworms by demonstrating silencing of a glutathione *S*-transferase gene in bollworm. These findings strongly suggest that plant-delivered silencing of insect-detoxifying genes could be a powerful strategy for controlling insect pests.

A dsRNA-based *in vivo* delivery system was tested on model organism, *Reticulitermes flavipes*, the most common subterranean termite in North America. In an effort to test target genes with pest control potential, a nutrition-related cellulase gene (*Cell-1*) and two soldier-inhibitory hexamerin genes (*Hexa-1 & 2*) were checked. Suppression of these target genes impacted vital biological processes of termites including but not limited to feeding, juvenile hormone modulation, development, and differentiation (Bakhetia, M. *et al* 2005.*).* Eventually, these biological impacts led to individual death and / or compromised fitness. In addition, the information regarding dsRNA stability, an oral delivery system and validation of gene silencing effects can be found in the same review.

Further we discuss few reported RNAi cases in different organisms targeting different organs in insects. However to date, RNAi has only been successfully demonstrated in a small number of Lepidoptera. Aminopeptidase N from *Manduca sexta* was the first molecule to be tentatively identified as a *cry* toxin-binding protein. This amino peptidase was silenced and decreased sensitivity towards the binding of *cry1C* protein was observed (Rajagopal et al., 2002).

The functional contribution of cloned subunits to insect nicotinic acetylcholine (ACh) receptors has been difficult to determine using heterologous expression. Therefore, the subunit composition of naturally expressed functional receptors was studied using RNA interference. The nicotinic alpha subunit, *Manduca* ACh Receptor Alpha 1 (MARA1) can be detected in neuronal cultures isolated from the ventral nerve cord of fifth instar larvae of *Manduca sexta* by in situ hybridization. It's presence correlates with large ACh induced, Ca^{2+} responses. The expression of MARA1 is downregulated by treatment with dsRNA which significantly reduced both the number of responding cells and the amplitude of remaining Ca^{2+} responses (Vermehren et al., 2001).

Hemolin from the Giant silkmoth, *Hyalophora cecropia*, identified as a bacteria-inducible molecule and a member of the immunoglobulin superfamily, is present in oocytes and embryos. Bettencourt et al., 2002 used RNA interference (RNAi) to investigate *H. cecropia* gene function *in vivo* and demonstrated that Hemolin is crucial for the normal development of embryos. When RNAi-females were mated, no larvae emerged from their eggs and when dissected, the eggs revealed malformed embryos.the silencing of *Hemolin* gene expression leads to embryonic lethality.

In *Bombyx mori* (Quan et al., 2002), injection of double-stranded RNA (dsRNA) corresponding to the silkworm *white* gene (*Bmwh3*) into preblastoderm eggs of the wild-type silkworm induced phenotypes similar to those observed with mutants of the *white egg 3* locus. The induced phenotypes were characterized by the presence of white eggs and translucent larval skin. Similarly in Grasshopper, injection of dsRNA corresponding to the eye colour gene *vermilion* into first instar nymphs triggered suppression of ommochrome formation in the eye lasting through two instars equivalent to 10–14 days in absolute time, which concludes that systemic dsRNA application elicits specific and long term gene silencing in juvenile grasshopper instars (Ying Dong et.al 2005).

Fabrick et.al 2005, demonstrated that RNAi can be utilized in the lepidopteran, *Plodia interpunctella.* A cDNA for tryptophan oxygenase was cloned and it was seen that silencing of tryptophan oxygenase through RNAi during embryonic development resulted in loss of eye-color pigmentation in lepidopteran insects. Members of the tryptophan oxygenase family are highly conserved, which was predicted by comparison with genomic sequences from *Drosophila melanogaster, A. gambiae,* and *T. castaneum.*

CONCLUSION

The use of RNAi technology in the protection of crops is now being demonstrated. However, like any other upcoming technology, the RNAi strategy for pest control also has advantages and disadvantages. The technology is advantageous because,

- The technology has helped in identification of a new set of targets which can be used in the control of pests. Because these targets have previously not been identified, there should be no acquired resistance in the pest species.
- The double stranded RNA used in RNAi against the novel targets is a biodegradable product as compared to the known chemically synthesised pesticides, such as DMSO etc. The biodegradable nature of the constructs makes them more environmentally sound.
- In contrast to conventional pesticides, no professional assistance would be required in order to treat the relevant areas, due to the more safe nature of the DNA and RNA constructs, compositions and host cells of the invention.
- Minimum disruption of human activity would be required since the double stranded RNA region containing constructs are designed such that they will have no adverse effects or only minor effects on gene expression outside of the target pest population.
- Although sufficient insight has been gained in terms of uptake mechanisms, it still needs to be proven even though the feeding experiments are very promising. There are still many challenges to overcome in this new exiting field of dsRNA exploiting the application of RNAi for the control of pest management.

However, in the course of popularizing this technology, one has to focus on certain aspects.

- The RNAi mediating dsRNA has to match the target with a high degree of nucleotide sequence identity in order to effectively down regulate expression and thus control the pest. Thus, specificity can be achieved by designing double stranded RNA molecules in which one strand has high homology to the target sequence but which strand has only low homology to the orthologous sequence in a mammalian species, such as a human. This specificity is greater than can be achieved with conventional chemical pesticides.

- RNAi does not necessarily provide an immediate effect in terms of killing the pest, rather the effects are mediated effectively but require time for the double stranded RNA to be associated with its target. The RNAi effect may result in killing the pest at a later moment and not directly upon contact, such as Noviflumuron (which is a chitin synthesis inhibiter). Thus, the use of RNAi may allow more facile control of large infestations of pests such as insects and/or arachnids because there is less chance of a shock effect being propagated amongst the pests where they may encounter a large number of dead pests in the vicinity of the insecticide and/or arachnicide (James, C 2006).
- The use of multiple targets at the same time may provide more efficacious control of pest populations and reduce the possibility of acquired resistance. The targets may be common to a number of pest species providing broad scale treatment.
- Insect neuroendocrine genes like AChE (acetyl choline esterase), prothoracicotropic hormone (PTTH) (Xu Jun 2007), Ecdystetroid UDP-Glucosyltransferase genes (Yan-Hang 2007) can be the target sites for the RNAi mediated gene silencing.
- This technology can be exploited towards resistance against sap-sucking homopteran pests such as aphids, leafhoppers and whitefly where Bt approaches have been proved difficult.

Before these exciting possibilities can be realized, some significant challenges remain.

Are all detoxifying genes equally amenable to this approach? Will this technology extend to other insect groups? Will ingestible RNAi be circumvented by sequence polymorphisms in pest populations? Will the success seen in the lab translate into effective pest control in the field?

For WCR at least, there is reason for optimism, as the assay involved measuring root damage on the crop itself. Nonetheless, the long-term effectiveness of the approach will need to be established. Likewise, the bollworm studies await rigorous assessment of how effectively cotton plants are protected. Despite these questions, there is no doubt that researchers and farmers have grounds to look forward to a new era in insect pest control.

To fully capitalize on the potential of RNAi, we need to become better at predicting and controlling its effects. RNA interference induced in insects after ingestion of plant-expressed hairpin RNA offers promise for managing devastating crop pests. Transgene–encoded ingestible dsRNA may one day stand alongside *Bt* transgenes in insect management programs.

References

Baum, JA; Thierry Bogaert; William Clinton; Gregory R Heck; Pascale Feldmann, Control of coleopteran insect pests through RNA interference. *Nature Biotech*, 2007, 25: 1322–1326.

Bettencourt R, Terenius Oand Faye I. *Hemolin* gene silencing by ds-RNA injected into Cecropia pupae is lethal to next generation embryos. *Insect Molecular Biology*, 2002, 11: 3, 267 – 271

Bosher JM, Labouesse M: RNA interference: genetic wand and genetic watchdog. *Nature Cell Biol* 2000, 2:31-36.

Cogoni C, Irelan J T, Schumacher M, Schmidhauser T J, Selker E U, and Macino G, Transgene silencing of the al-1 gene in vegetative cells of Neurospora is mediated by a cytoplasmic effector and does not depend on DNA-DNA interactions or DNA methylation. *EMBO Journal,* 1996, 15(12), 3153–3163

Cynthia CH Khoo; Joseph Piper; Irma Sanchez-Vargas; Ken E Olson and Alexander WE Franz. The RNAl interference pathway affects midgut infection- and escape barriers for Sindbis virus in *Aedes aegypti*. BMC Microbiology, 2010, 10:130

Daniel R.G. Price and John A. Gatehouse. RNAi-mediated crop protection against insects. *Trends in Biotechnology*, 2008, Vol.26 No.7

Fire A, A potent and specific genetic interference by double stranded RNA in Caenorhabditis elegans. *Nature*, 1998, *391*: 806–811 .

Gordon, KHJ and Waterhouse, PM. (2007). RNA interference induced in insects after ingestion of plant-expressed hairpin RNA offers promise for managing devastating crop pests. *Nature Biotech*, 25, 1231-1232.

H. Huvenne, G. Smagghe Mechanisms of dsRNA uptake in insects and potential of RNAi for pest control: A review, *Journal of Insect Physiology*, 2010, 56 227–235

Hakim, R.S., Baldwin, K., Smagghe, G., Regulation of midgut growth, development, and metamorphosis. *Annual Review of Entomology*, 2010. 55, 112408-085450.

Hiroaki Tabara, Alla Grishok, and Craig C. Mello RNAi in C. elegans: Soaking in the genome sequence. *Science*, 1998, 282, 430–431

James C Problem formulation and hypothesis testing for environmental risk assessments of genetically modified crops. Global review of commercialized transgenic crops. *Current Science*, 2003, 84: 303–309.

Jeffrey A. Fabrick, Michael R. Kanost, and James E. Baker, RNAi-induced silencing of embryonic tryptophan oxygenase in the Pyralid moth, *Plodia interpunctell*. J Insect Sci. *2005,* 2004; 4: 15.

Jose, A.M., Hunter, C.P., Transport of sequence-specific RNA interference information between cells. *Annual Review of Genetics*, 2007, 41, 305–330.

Jun XU, Jian Ya SU, JinLiang S, WeiHua X, Molecular characterization and developmental expression of the gene encoding the prothoracicotropic hormone in th beet armyworm, *Spodoptera exigua. Springer science in China series C-Life sciences*, 2007, vol 50, No 4, 466-472.

Lisa Timmons, Donald L. Court, and Andrew Fire Ingestion of bacterially expressed dsRNAs can produce specific and potent genetic interference in Caenorhabditis elegans. *Gene*, 2001, 263, 103–112

Manjula Bakhetia M. RNA interference and plant parasitic nematodes. *Trends in Plant Science,* 2005, 10: 362–367.

Mao YB. Silencing a cotton bollworm P450 Monooxygenase gene by plant- mediated RNAi impairs larval tolerance of gossypol. *Nature Biotech*. 2007, 25, 1307–1313.

Mathew L Comparative and Functional Genomics. 2004. 5: 240–244. *www.interscience.wiley.com*. DOI: 10.1002/cfg.396.

Müller AE Applications of RNA interference in transgenic plants Perspectives in Agriculture. Veterinary Science. *Nutrition and Natural Resources*, 2006, 1(24): 13.

Quan, G.X., Kanda, T., Tamura, T.,. Induction of the white egg 3 mutant phenotype by injection of the double-stranded RNA of the silkworm white gene. *Insect Molecular Biology*, 2002, 11, 217–222.

Rajagopal, R., Sivakumar, S., Agrawal, N., Malhotra, P., Bhatnagar, R.K., Silencing of midgut aminopeptidase N of Spodoptera litura by double-stranded RNA establishes its role as Bacillus thuringiensis toxin receptor. *Journal of Biological Chemistry*, 2002, 277, 46849–46851.

S Dzitoyeva, N; Dimitrijevic and H Manev; Intra-abdominal injection of double-stranded RNA into anesthetized adult Drosophila triggers RNA interference in the central nervous system. *Mol. Psychiatry*, 2001, 6, 665–670

Si YH, Fang MG, Huang Y, Zheng FL, Wang HZ Construction and characterization of a *Helicoverpa armigera* Nucleopolyhedrovirus Bacterial artificial chromosome with deletion of Ecdysteroid UDP-Glucosyltrasferase gene. *J. Biosci. Biotech.Biochem*, 2007, 71(10): 2435-2441.

Timmons, L and Fire, A. (1998). Specific interference by ingested dsRNA. Nature, 395, 854

Titia Sijen, Jamie Fleenor, Femke Simmer, Karen L. Thijssen, Susan Parrish, Lisa Timmons, Ronald H.A. Plasterk and Andrew Fire' On the role of RNA amplification in dsRNAtriggered gene silencing. *Cell*, 2001, 107, 465–476

Van der Krol A.R,. Mur L A, Beld M, Mol J.N.M and Stuitje A. R. Flavonoid Genes in Petunia: Addition of a Limited Number of Gene Copies May Lead to a Suppression of Gene Expression, *THE PLANT CELL*, 1990, 2, (4) 291-299

Vermehren A, Qazi S, Trimmer BA. The nicotinic α subunit MARA1 is necessary for cholinergic evoked calcium transients in *Manduca* neurons. *Neuroscience Letters, 2001*, 313: 113-116.

Waterhouse PM., Graham MW, Wang MB. *Virus resistance and gene silencing in plants can be induced by simultaneous expression of sense and antisense RNA*. Proc Nat Acad Sci USA, 1998, *95*: 13959–13964 .

Whangbo, J.S., Hunter, C.P.,. Environmental RNA interference. *Trends in Genetics*, 2008, 24, 297–305.

Ying Dong and Markus Friedrich, Nymphal RNAi: systemic RNAi mediated gene knockdown in juvenile grasshopper. *BMC Biotechnology*, 2005, 5: 25

INDEX

A

B

C

D

E

F

G

H

M

N

O

P

Q

R

S

T

U

V

W

Y

Z